Additional Mathematics
Pure and Applied

in SI units

A. Godman, B. Sc. (Hons.)
J.F. Talbert, M.A.

with answers

LONGMAN

LONGMAN GROUP LIMITED
LONDON

*Associated companies, branches and representatives
throughout the world*

© Longman Malaysia Sdn. Bhd. 1971 and 1973.

First published † 1971
Second impression (with corrections) † 1972
New edition † 1973
New impression † 1974 *(twice)*
New impression † 1975 *(thrice)*
New impression † 1976 *(twice)*
New impression † 1977 *(twice)*

ISBN 0 582 69408 6

Text set in 10/12 pt Times New Roman
Printed by New Art Printing Co (Pte) Ltd.

Preface

This book is designed to cover the work in Pure Mathematics and Mechanics for the revised Cambridge Additional Mathematics syllabus and the work for Papers 1 and 2 (Pure) and Paper 3 (Applied) of the new London Additional Mathematics syllabus. These two syllabi overlap to a considerable extent and in each some choice of topics can be made. The teacher or student will therefore be able to select from this book a course suited to his needs.

The treatment is straightforward but comprehensive and does not attempt to outrun the standard of the syllabus. It is assumed that the purpose of the work is to give an introduction to further important ideas and techniques in Pure Mathematics, applying these ideas to simple problems, and to introduce the basic principles of Applied Mathematics. In Applied Mathematics it is presumed that most students will also be studying Physics and will acquire an experimental background to complement their mathematical studies. The treatment in Pure Mathematics is not meant to be rigorous, as the time available does not usually permit this, but to be sufficiently clear cut to allow fairly rapid progress in grasping the basic techniques and to lay a good basis for further work. Within the limits of space (and time) it is hoped that there are sufficient examples for the student to acquire confidence in handling the techniques through practice.

The topics have been taken in what seemed to the authors a reasonably logical but still flexible order. With obvious exceptions many of the Chapters are independent of each other and the sequence can be rearranged to suit the tastes of the teacher or student and the requirements of the syllabus. The book can be used as a one-year post O level course or taken concurrently with O level according to local practice.

SI units (as now required by examinations) have been used throughout but not to the exclusion of such practical or common sense units as the centimetre or kilometre/hour. The notation and style of printing of these units conforms to the now accepted standard, though both the notations m/s and m s^{-1} have been used, to accustom the student to differing conventions.

We are grateful to the following Examining Boards for permission to reproduce questions from papers published by them: some of these questions have been metricized by us.

Cambridge Local Examinations Syndicate (C)
Oxford Local Examinations (O)
University of London (L)

A.G.
J.F.T.

Note on this Edition
This new Edition (1973) includes 15 Revision Papers and the new Pure Mathematics topics in the Cambridge Additional Mathematics Syllabus: Relations and Functions, and Parametric Equations.

NOTE ON SI UNITS AND NOTATION

SI (Système International) metric units have been used exclusively in this book. SI is the latest version of the metric system, agreed internationally in 1960, and is being adopted as the national measuring system in the UK. It will also be used exclusively in examinations in Additional and A Level Mathematics.

SI units used in this book are based on the three primary units:

Length 1 **metre** (1 **m**),

Mass 1 **kilogram** (1 **kg**)

and Time 1 **second** (1 **s**).

The advantage of SI is that all other standard units (e.g. of velocity, acceleration, force, momentum, etc.) are derived directly from these primary units and form a consistent system. The formulae used in Applied Mathematics, for instance, are only valid if these standard units are used for the quantities therein.

Multiples of the three primary units which are also used:

Length   1 **km** (kilometre) $= 1000$ or 10^3 m

1 **mm** (millimetre) $= 0.001$ or 10^{-3} m

1 dm (decimetre) $= 0.1$ or 10^{-1} m $\left.\vphantom{\begin{matrix}a\\b\end{matrix}}\right\}$ These are used

1 cm (centimetre) $= 0.01$ or 10^{-2} m less often.

Mass   1 **g**   (gram) $= 10^{-3}$ kg

1 **t**   (tonne) $= 10^3$ kg

Time   1 **h**   (hour) $= 60$ min

1 **min** (minute) $= 60$ s

The abbreviations given above are official and are always used. Note that (**1**) the same symbols are also used for the plural; for example 5 metres is written as 5 m and not as 5 ms (which would signify 5 milliseconds): (**2**) no full stop is placed after the symbol (unless of course it ends a sentence).

Other units derived from the primary units:

Quantity	Standard Unit	Other units used
Area	1 m^2 (square metre)	1 mm^2 (square millimetre) 1 cm^2 (square centimetre) 1 $m^2 = 10^4\ cm^2$ $= 10^6\ mm^2$
Volume	1 m^3 (cubic metre)	1 mm^3 (cubic millimetre) 1 cm^3 (cubic centimetre) 1 dm^3 (cubic decimetre) (Used less often) 1 $m^3 = 10^6\ cm^3$ $= 10^9\ mm^3$
Velocity **Speed**	1 metre per second written as 1 $m\ s^{-1}$ or 1 m/s	1 $km\ h^{-1}$ (km/h) (kilometre per hour) 1 $km\ h^{-1} = \frac{5}{18}\ m\ s^{-1}$
Acceleration	1 metre per second per second, written as 1 $m\ s^{-2}$ or 1 m/s^2	—
Force	1 N (newton)	1 kN (kilonewton) $= 10^3$ N (See p. 353 for note on superseded units of force)
Moment of **force**	1 N m (newton metre)	—
Momentum ⎫ **Impulse** ⎭	1 N s (newton second)	—
Work, energy	1 J (joule)	—
Power	1 W (watt) $= 1\ J\ s^{-1}$	1 kW (kilowatt) $= 10^3$ W 1 MW (megawatt) $= 10^6$ W

In the printing of this book, the recommendations of *BS* 1991 : *Part I* : 1967 and the pamphlet *Metrication in Secondary Education* (Royal Society 1969) have been followed.

Quantities are printed in *italics* (sloping) type : unit symbols are printed in roman (upright) type: vectors are printed in ***italic*** bold type.

Examples

an acceleration of $5g$ m s^{-2} ($5 \times$ acceleration due to gravity)

a mass $m = 5$ g (5 grams)

a force of P N (P newtons)

a length of x m (x metres)

vectors ***u*** and ***AB***

Complex units such as a rate of flow of 4 cubic metres per second are printed either as 4 m^3 s^{-1} or 4 m^3/s. (The first is preferred.)

g (the acceleration due to gravity) is taken as 9·8 m s^{-2} and often as 10 for a convenient approximation.

The usual mathematical symbols are employed. Note the symbol $\approx$, meaning 'approximately equal to'.

Numbers of more than 4 digits are printed in groups of three (counting from the decimal point): for example, 12 500, 0·000 17, 3 450 000.

Contents

Pure Mathematics

1 Indices and Logarithms

Positive Integral Indices

You know that 4^3 means $4 \times 4 \times 4$: so $a^3 = a \times a \times a$. Here a is the **base** and 3 is the **index** or **power** or **exponent**. The rules for working with indices which are positive integers (1, 2, 3, ...) will be well known from previous work. They are:

Rule 1:

$$\boxed{a^m \times a^n = a^{m+n}}$$

Example

$$5^3 \times 5^4 = 5 \times 5 \times 5 \times 5 \times 5 \times 5 \times 5 = 5^7$$

Rule 2:

$$\boxed{a^m \div a^n = a^{m-n}}$$

Example

$$7^5 \div 7^3 = \frac{7 \times 7 \times 7 \times 7 \times 7}{7 \times 7 \times 7} = 7^2$$

Rule 3:

$$\boxed{(a^m)^n = a^{mn}}$$

Example

$$(4^3)^2 = 4^3 \times 4^3 = 4^6$$

We can extend the meaning of indices to cover zero, negative and rational (fractional) indices, assuming that the above rules must always apply.

Zero Index

To find a meaning for 7^0.
By *Rule 1*, $7^3 \div 7^3 = 7^0$ but obviously $7^3 \div 7^3 = 1$. Therefore 7^0 must mean 1.

Similarly $\boxed{a^0 = 1}$ for all values of a (excluding 0).

Negative Indices

To find a meaning for 5^{-2}.
By *Rule 2*, $5^2 \div 5^4 = 5^{-2}$;

but $\quad 5^2 \div 5^4 = \dfrac{5 \times 5}{5 \times 5 \times 5 \times 5} = \dfrac{1}{5 \times 5} = \dfrac{1}{5^2}$

So $5^{-2} = \dfrac{1}{5^2}$ i.e. a negative index denotes a reciprocal.

Similarly $\boxed{a^{-n} = \dfrac{1}{a^n}}$ for all values of a (excluding 0).

Examples

$$3^{-3} = \frac{1}{3^3} = \frac{1}{27}$$

$$(-2)^{-3} = \frac{1}{(-2)^3} = \frac{1}{-8} = -\frac{1}{8}$$

$$10^{-2} = 1/10^2 = 1/100 = 0.01$$

Standard Index Form for Numbers

We can express decimal numbers (particularly very large or very small numbers) in a uniform and economical way, based on powers of 10. For example, 57 300 could be written $5.73 \times 10\,000$ or 5.73×10^4. The number is now said to be in **standard index form**: any number is in standard index form when written in the form $a \times 10^n$, where $1 \leqslant a < 10$ and n is a positive or negative integer.

Examples

$$0.000\,53 = 5.3 \div 10\,000 = 5.3 \times 10^{-4}$$
$$0.001\,05 = 1.05 \times 10^{-3}$$
$$3.5 \times 10^{-5} = 0.000\,035$$
$$3.5 \times 10^3 = 3500$$

You will appreciate the convenience of standard form if you care to write out in full, for example, the mass of an electron, 9×10^{-28} g, or the number of electrons per second in one ampere, 6×10^{18}.

Rational (Fractional) Indices

To find a meaning for $9^{\frac{1}{2}}$.

By *Rule 3*, $(9^{\frac{1}{2}})^2 = 9^1 = 9$

Therefore $9^{\frac{1}{2}} = \sqrt{9} = \pm 3$

Similarly $a^{\frac{1}{2}} = \sqrt{a}$ and by repeating the method we can show that

$$\boxed{a^{\frac{1}{n}} = \sqrt[n]{a}}$$ for all values of a (excluding 0)

Examples

$$27^{\frac{1}{3}} = \sqrt[3]{27} = 3$$

$$4^{-\frac{1}{2}} = 1/4^{\frac{1}{2}} = 1/\sqrt{4} = \pm\frac{1}{2}$$

Again, by *Rule 3*, $8^{\frac{2}{3}} = (8^{\frac{1}{3}})^2 = (\sqrt[3]{8})^2 = (2)^2 = 4$

Alternatively, $\quad 8^{\frac{2}{3}} = (8^2)^{\frac{1}{3}} = (64)^{\frac{1}{3}} = \sqrt[3]{64} = 4$

Hence for any rational number $\dfrac{m}{n}$, $\quad a^{m/n} = (\sqrt[n]{a})^m$ OR $\sqrt[n]{(a^m)}$

In numerical calculations, it is usually simpler to take the first form, i.e., find the root first.

Examples $\quad 100^{1\frac{1}{2}} = 100^{\frac{3}{2}} = (\sqrt{100})^3 = (10)^3 = 1000;$

$$32^{-\frac{2}{5}} = \frac{1}{32^{2/5}} = \frac{1}{(\sqrt[5]{32})^2} = \frac{1}{(2)^2} = \frac{1}{4}.$$

Exercise 1.1

Find the values of:

1 8^0 **2** 5^{-1} **3** 10^{-3} **4** $8^{\frac{1}{3}}$

5 $100^{\frac{1}{2}}$ **6** $8^{\frac{2}{3}}$ **7** $27^{-\frac{2}{3}}$ **8** $(-32)^{\frac{1}{5}}$

9 $(\frac{1}{9})^0$ **10** $(-27)^{-\frac{2}{3}}$ **11** $(625)^{-\frac{1}{4}}$ **12** $(x^3)^{-\frac{2}{3}}$

13 $(-343)^{-\frac{1}{3}}$ **14** $(1000)^{-\frac{5}{3}}$

Convert to standard index form:

15 $92\,000\,000$ **16** 530 **17** $0 \cdot 000\,003\,4$ **18** $0 \cdot 000\,000\,000\,23$

19 $215\,000$ **20** $0 \cdot 0101$ **21** 1020 **22** $3\,460\,000\,000$

Write in ordinary decimal form:

23 $2 \cdot 3 \times 10^5$ **24** $2 \cdot 3 \times 10^{-3}$ **25** $4 \cdot 05 \times 10^{-6}$ **26** $1 \cdot 3 \times 10^8$

Simple Exponential Equations

Example 1

Solve the equation $3^x = 81$

Here the unknown (x) is an index or exponent and 3^x is an **exponential** function with base 3. We can solve such equations simply if both sides can be expressed in terms of the same base. A more general method will be shown later (p. 10).

Here, 81 is known to be 3^4. Therefore $3^x = 3^4$ and so $x = 4$.

Example 2

Solve $8^x = 0 \cdot 25$

We must notice that 8 and $0 \cdot 25$ can both be expressed to base 2.

Then $(2^3)^x = \frac{1}{4} = 2^{-2}$; so $3x = -2$ and $x = -\frac{2}{3}$.

Example 3

Solve $2^{2x} + 4(2^x) - 32 = 0$.

We must notice that $2^{2x} = (2^x)^2$. If we set $2^x = p$, the equation will be seen to be a quadratic in p: $p^2 + 4p - 32 = 0$, i.e. $(p + 8)(p - 4) = 0$ which gives $p = -8$ or $p = 4$.

Hence $2^x = -8$ or $2^x = 4$. The first has no solution and the second gives $x = 2$.

Example 4

Solve the equation $2^{2x+3} + 1 = 9(2^x)$

Rewrite as $2^3(2^x)^2 - 9(2^x) + 1 = 0$ and put $2^x = p$; then we have the quadratic equation $8p^2 - 9p + 1 = 0$ which leads to $p = \frac{1}{8}$ or $p = 1$. So $2^x = \frac{1}{8}$ or $2^x = 1$ which gives $x = -3$ or $x = 0$.

Exercise 1.2

Solve the equations:

1	$2^x = 32$	**2**	$5^x = 625$
3	$4^x = 0.5$	**4**	$9^x = \frac{1}{729}$
5	$10^x = 0.0001$	**6**	$32^x = 0.25$
7	$2^{2x} - 5(2^x) + 4 = 0$	**8**	$3^{2x} - 12(3^x) + 27 = 0$
9	$2^{2x+1} - 33(2^x) + 16 = 0$	**10**	$3^{2x+1} + 26(3^x) - 9 = 0$
11	$2^{2x+1} - 15(2^x) = 8$	**12**	$2^{2x+2} + 8 = 33(2^x)$

The Graph of the Exponential Function

Example 5

Solve $2^x = 5$.

Clearly it is not possible to express 5 to base 2 exactly; we shall therefore obtain an approximate solution graphically. This involves drawing the graph of the function 2^x. Take values of x in the range $-2 \leqslant x \leqslant 3$ (i.e. values of x from -2 to $+3$ inclusive).

x	-2	-1	0	$\frac{1}{2}$	1	2	3
2^x	$\frac{1}{4}$	$\frac{1}{2}$	1	1.4	2	4	8

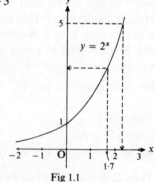

The table gives a few basic values for x and 2^x. Plot these points and join up as smoothly as possible to obtain the exponential curve (fig 1.1). From your graph

(i) solve the equation $2^x = 5$;

Fig 1.1

(ii) find the value of $2^{1\cdot7}$.

Both your results can only be approximate; they should be about **(i)** 2·32 **(ii)** 3·25

Exercise 1.3

Draw the graph of $y=3^x$ in the range $-2\leqslant x\leqslant 2$. From your graph find the approximate values of **(i)** $3^{0\cdot8}$, **(ii)** $3^{1\cdot2}$ and **(iii)** solve the equation $3^x=5$. Notice the shape of this curve compared with that of 2^x; this is the typical exponential curve shape.

Logarithms

Closely connected with the exponential function a^x is the **logarithm** function. If $y=a^x$ then we define x as the logarithm of y to the base a (written $\log_a y$).

$$\boxed{\text{If } y = a^x \text{ then } x = \log_a y}$$

For example, $3^2 = 9$; therefore $2 = \log_3 9$;

$$10^2 = 100, \text{ so } 2 = \log_{10} 100$$

$p^3 = r$, therefore $3 = \log_p r$.

Again, if $4 = \log_x 19$, then $x^4 = 19$.

Such conversions from the exponential form to the logarithmic form and backwards should be practised: e.g., if $x = \log_3 81$, then $3^x = 81$ and so $x = 4$. Hence $\log_3 81 = 4$.

Exercise 1.4

Write in logarithmic form:

1 $7^2 = 49$ **2** $2^5 = 32$ **3** $10^0 = 1$
4 $10^3 = 1000$ **5** $5^{-2} = \frac{1}{25}$

Write in index form and find the value of x:

6 $x = \log_2 8$ **7** $x = \log_3 27$ **8** $x = \log_4 64$
9 $x = \log_5 125$ **10** $x = \log_2 \frac{1}{8}$ **11** $x = \log_{10} 0\cdot001$

Find the values of:

12 $\log_{10} 10$ **13** $\log_3 3$ **14** $\log_8 64$
15 $\log_3 1$ **16** $\log_2 \frac{1}{2}$ **17** $\log_2 0\cdot25$
18 $\log_2 32$ **19** $\log_3 243$ **20** $\log_{10} 0\cdot01$
21 $\log_2 2^x$ **22** $\log_a a^x$

The Graph of the Logarithm Function

$y = a^x$ is equivalent to $x = \log_a y$. The same graph serves for both (fig 1.2). Interchanging x and y gives the graph of $y = \log_a x$ with the axes as shown in fig 1.3. Now turn this graph about the line $y = x$ and obtain the graph of $y = \log_a x$ with the axes in the usual position (fig 1.4).

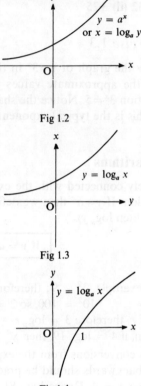

Fig 1.2

Fig 1.3

Observe that (1) the function $\log_a x$ does not exist if x is negative; (2) for $0 < x < 1$, $\log_a x$ is negative.

For example, $(3)^{-2} = \frac{1}{9}$;
so $\log_3 (\frac{1}{9}) = -2$.

Fig 1.4

Operating Rules for Logarithms

Let $P = a^m$, then $m = \log_a P$ and let $Q = a^n$, then $n = \log_a Q$.

1 $PQ = a^m \times a^n = a^{m+n}$

Then

$$\boxed{\log_a PQ = m + n = \log_a P + \log_a Q}$$ 　　　　　 *Rule 1*

For example, $\log_a 6 = \log_a 3 + \log_a 2$.
[NB: it is NOT true that $\log_a (P + Q) = \log_a P + \log_a Q$].

2 $\dfrac{P}{Q} = \dfrac{a^m}{a^n} = a^{m-n}$

Therefore

$$\boxed{\log_a \left(\frac{P}{Q}\right) = m - n = \log_a P - \log_a Q}$$ 　　　　　 *Rule 2*

For example, $\log_a 5 = \log_a \frac{10}{2} = \log_a 10 - \log_a 2$

3 $P^n = (a^m)^n = a^{mn}$

Therefore $\boxed{\log_a P^n = mn = n \log_a P}$ *Rule 3*

For example on any base, $\log 9 = \log 3^2 = 2 \log 3$;
$$\log \sqrt{6} = \log 6^{\frac{1}{2}} = \tfrac{1}{2} \log 6.$$

These are the rules you have become familiar with in previous practical calculations using logarithms. They are similar to the three rules for indices.

Two Special Logarithms

1 For any base a, $a^0 = 1$. Therefore $\log_a 1 = 0$. The log of 1 to any base is zero (the graph of $\log_a x$ cuts the x-axis at 1).

2 Again, $a^1 = a$. Therefore $\log_a a = 1$. The logarithm of the base always equals 1.

Example 6

Simplify $\log_4 9 + \log_4 21 - \log_4 7$

$$\log_4 9 + \log_4 21 - \log_4 7 = \log_4 (9 \times 21 \div 7)$$
$$= \log_4 27 = \log_4 3^3$$
$$= 3 \log_4 3$$

Example 7

If $\log_7 2 = 0{\cdot}356$ *and* $\log_7 3 = 0{\cdot}566$ *find the value of*
$$2 \log_7 (\tfrac{7}{15}) + \log_7 (\tfrac{25}{12}) - 2 \log_7 (\tfrac{7}{3}).$$

The expression $= \log_7 (\tfrac{7}{15})^2 + \log_7 (\tfrac{25}{12}) - \log_7 (\tfrac{7}{3})^2$

$$= \log_7 \left(\frac{(\tfrac{7}{15})^2 \times \tfrac{25}{12}}{(\tfrac{7}{3})^2} \right)$$

$$= \log_7 \left(\frac{7}{15} \times \frac{7}{15} \times \frac{25}{12} \times \frac{3}{7} \times \frac{3}{7} \right)$$

$$= \log_7 \frac{1}{12} = \log_7 1 - \log_7 12$$

$$= -\log_7 (2^2 \times 3)$$

$$= -2 \log_7 2 - \log_7 3$$

$$= -(2 \times 0{\cdot}356 + 0{\cdot}566)$$

$$= -1{\cdot}278.$$

Exercise 1.5

Simplify:

1 $\log_5 \sqrt{3}$	**2** $\log_2 \sqrt[3]{5}$	**3** $\log_4 27$	**4** $\log_a x^3$
5 $\log_r 1000$	**6** $\log_2 5^3$	**7** $\log_{10} 10\,000$	**8** $\log_7 49$
9 $\log_3 (\frac{1}{8})$	**10** $\log_6 121$	**11** $\log_x x^4$	**12** $\log_4 4^x$
13 $\log_5 12 + \log_5 10$		**14** $2\log_7 9 - \log_7 81$	
15 $\log_3 24 + \log_3 15 - \log_3 10$		**16** $\log_7 98 - \log_7 30 + \log_7 15$	
17 $\log_5 \sqrt{6} + \log_5 9$		**18** $\log_a (\frac{5}{7}) + 2\log_a (\frac{7}{6}) - \log_a (\frac{5}{6})$	

If $\log_5 2 = 0.431$ and $\log_5 3 = 0.682$, find the values of

19 $\log_5 6$	**20** $\log_5 1.5$	**21** $\log_5 27$	**22** $\log_5 (\frac{2}{3})$
23 $\log_5 8$	**24** $\log_5 10$	**25** $\log_5 \sqrt{3}$	**26** $\log_5 12$
27 $\log_5 13.5$	**28** $\log_5 (\frac{1}{4})$	**29** $\log_5 (\frac{3}{8}) + 2\log_5 (\frac{4}{5}) - \log_5 (\frac{2}{5})$	
30 $\log_5 100$			

Common Logarithms and Practical Calculations

As you already know, numerical calculations can often be reduced in labour by using logarithms (though the results will only be approximate). The logs used are to base 10, called **common** logarithms, and are found in the usual tables. Base 10 (which is the base of our decimal system of numbers) is used for the following reason.

Suppose it is given that $\log_{10} 3.2 = 0.505$.

Then $\qquad \log_{10} 32 \quad = \log_{10} (3.2 \times 10) = \log_{10} 3.2 + \log_{10} 10$
$\qquad\qquad\qquad\qquad = 0.505 + 1 = 1.505$

Similarly $\quad \log_{10} 3200 \quad = \log_{10} 3.2 + \log_{10} 10^3$
$\qquad\qquad\qquad\qquad = 0.505 + 3\log_{10} 10 = 3.505$

So for any number expressible in the standard form 3.2×10^n, $\log_{10} 3.2 \times 10^n = n + 0.505$. The part 0.505 (the **mantissa**) is always the same and can be put in a table against 32: the **characteristic** n, can be found by inspection from the standard form. On any other base $\log_a 32$ would be quite different from $\log_a 3.2$ and the economy of common logarithms would be lost.

Note that for $\log_{10} (3.2 \times 10^{-n})$ we write $-n + 0.505$ which is written as $\bar{n}.505$ (negative characteristic but positive mantissa).
If base 10 is intended we usually just write $\log x$.

Tables of common logarithms can be used to calculate logs to other bases: this is the method referred to on p. 5.

Example 8

Find $\log_7 12$.

No.	Log
1·079	0·0331
0·8451	$\bar{1}$·9269
	0·1062

Let $x = \log_7 12$.

Then $7^x = 12$.

Now take logarithms (base 10) of both sides.

Then $\log 7^x = \log 12$

or $x \log 7 = \log 12$

Hence $x = \dfrac{\log 12}{\log 7} = \dfrac{1·0792}{0·8451} = 1·278$.

(The final division is done using logs as shown).

Example 9

Solve the equation $(0·3)^{x+1} = (0·7)^x$

Taking logs, $\log (0·3)^{x+1} = \log (0·7)^x$

or $(x + 1) \log (0·3) = x \log (0·7)$

that is, $x \log (0·3) + \log (0·3) = x \log (0·7)$

Solving for x, $x = \dfrac{\log (0·3)}{\log (0·7) - \log (0·3)}$

$$= \frac{\bar{1}·4771}{\bar{1}·8451 - \bar{1}·4771}$$

$$= \frac{-1 + 0·4771}{(-1 + 0·8451) - (-1 + 0·4771)}$$

$$= \frac{-0·5229}{0·3680} = -1·421$$

(Note that $\bar{1}·4771$ must be expressed as $-0·5229$ before proceeding with the calculation).

Exercise 1.6

Find the value of:

1 $\log_3 5$ 2 $\log_2 7$ 3 $\log_5 20$ 4 $\log_7 35$

5 $\log_8 10$ 6 $\log_3 (\frac{1}{8})$ 7 $\log_{12} 94$ 8 $\log_3 9·4$

9 $\log_2 0·47$

Solve the equations:

10 $2^x = 5$ 11 $3^x = 5$

(Compare with your graphical answers, p. 6 – 7).

12 $1·8^x = 2·7$ 13 $5^x = 10$ 14 $3^{x+2} = 7$

15 $2^{x-1} = 5^x$ 16 $3^{x+2} = 5^{x-1}$ 17 $(2·5)^{x-3} = (0·4)^x$

18 $(0·37)^{2x-1} = 9$ 19 $(0·4)^{x+3} = (0·5)^{2x}$ 20 $3^{x+2} = 5 \times 2^x$

Exercise 1.7 *Miscellaneous*

1 Find the values of (a) $27^{-\frac{5}{3}}$ (b) $(1\cdot44)^{-\frac{1}{2}}$ (c) $(1\frac{9}{16})^{-\frac{3}{2}}$ (d) $(-32)^{0\cdot4}$
(e) $100^{-\frac{3}{2}}$.

2 Without using tables, solve the equation $8x^{-\frac{2}{3}} = \frac{2}{25}$. (C)

3 Find the value of n if (i) $3^{-n} = 9$; (ii) $3^n = \frac{1}{27}$; (iii) $3^{2n} = 27$;
(iv) $9^n = \frac{1}{3}$; (v) $27^n = \frac{1}{9}$; (iv) $(-9)^n = 3^{-4}$.

4 Use tables to calculate the values of $\log_3 21$ and $7^{2\cdot5}$.

5 Find the value of $\dfrac{\log x^3}{\log x^2}$ (both logs to base a).

6 Find x if $(2\cdot7)^{3x-1} = 5^x$.

7 (i) Find x if $(\log_{10} x)^2 - \log_{10} x = 0$; (ii) find y if $\log_3 y = 2\log_3 7$.

8 Solve the equations (i) $2^x = 0\cdot4$; (ii) $\log_{10}(x^2 + 6x + 28) = 2$. (L)

9 If $3 + \log_{10} x = 2\log_{10} y$, express x in terms of y. (C)

10 If $p = \log_7\left(\frac{14}{15}\right)$, $q = \log_7\left(\frac{21}{20}\right)$ and $r = \log_7\left(\frac{49}{50}\right)$, what are the values
of (a) $p + q - r$, (b) $p + 3q - 2r$, given that $\log_7 2 = 0\cdot356$ and
$\log_7 3 = 0\cdot566$?

11 Solve the equation $(0\cdot2)^x = (0\cdot5)^{x+7}$.

12 Without using tables, evaluate
$\frac{2}{3}\log 27 + 5\log 2 + \frac{1}{3}\log 8 + 3\log\frac{5}{2} - 2\log 3 - 3\log 10$. (C)

13 If $\log x = 2 + \log y = \log\frac{4}{9} - \log\frac{9}{125} + 2\log\frac{9}{5}$ where all logarithms are to base 10, find x and y without using tables. (C)

14 If money is invested at compound interest, P units invested at
$r\%$ compound interest per annum for n years will amount to
$P\left(1 + \dfrac{r}{100}\right)^n$.

If \$150 is invested at $2\cdot5\%$, after how many years will it be doubled?
(Substitute in the formula and obtain $150(1\cdot025)^n = 300$ and then
solve for n).

15 If £120 is invested at $3\cdot5\%$ compound interest per annum, after
how long will it amount to £200? (Use the formula from question
14).

16 Solve the equation $A = e^{kt/2}$ if $A = 35\cdot6$, $e = 2\cdot718$ and $k = 5$.

17 Prove that $\log_a b \times \log_b a = 1$. (Let $x = \log_a b$ and consider logs
to base b).

18 Show that $x = a^{\log_a x}$.

19 Prove that $\log_2 7 = \dfrac{\log_{10} 7}{\log_{10} 2}$.

20 Generalise the result of question 19 by proving that
$$\log_a b = \frac{\log_c b}{\log_c a}.$$

21 Solve $(0\cdot1)^x = (0\cdot2)^5$. (C)

2 Surds

36 is a perfect square as $\sqrt{36} = \pm 6$; so is $6\frac{1}{4}$ as $\sqrt{6\frac{1}{4}} = \pm 2\frac{1}{2}$. 125 is a perfect cube as $\sqrt[3]{125} = 5$. Most numbers however do not have exact roots. It is not possible to find $\sqrt{3}$, $\sqrt[5]{2}$, $\sqrt[3]{5}$ etc exactly and such roots are called **surds**. They are *irrational* numbers as they cannot be expressed as fractions. (A rational number is of the form a/b where a and b are integers). At the end of this chapter you can study a proof of why $\sqrt{2}$, for example, cannot be a fraction. Approximate values can of course be obtained from tables but it is often simpler to work with the surds themselves.

In this chapter we shall deal only with surds which are square roots and $\sqrt{}$ is taken to mean the positive square root.

Manipulation of Surds

1 Multiplication of surds

Clearly $\sqrt{9} \times \sqrt{16} = 3 \times 4 = 12$. Now $\sqrt{9 \times 16} = \sqrt{144} = 12$. So $\sqrt{9} \times \sqrt{16} = \sqrt{9 \times 16}$.

This rule is quite general: $\boxed{\sqrt{m} \times \sqrt{n} = \sqrt{mn}.}$

We can use this rule to simplify surds.

Examples

Simplify $\sqrt{40}$, $\sqrt{12}$, $\sqrt{12} + \sqrt{27} - \sqrt{48}$.

$$\sqrt{40} = \sqrt{4} \times \sqrt{10} = 2\sqrt{10} \quad \text{(also equal to } 2 \times \sqrt{2} \times \sqrt{5}$$
$$\text{but this is a more complicated form).}$$
$$\sqrt{12} = \sqrt{4} \times \sqrt{3} = 2\sqrt{3}$$
$$\sqrt{12} + \sqrt{27} - \sqrt{48} = 2\sqrt{3} + 3\sqrt{3} - 4\sqrt{3} = \sqrt{3}$$

NB $\sqrt{a + b} \neq \sqrt{a} + \sqrt{b}$ and $\sqrt{a - b} \neq \sqrt{a} - \sqrt{b}$. Avoid these common errors.

2 Division of surds

$$\frac{\sqrt{144}}{\sqrt{16}} = \frac{12}{4} = 3; \quad \sqrt{\frac{144}{16}} = \sqrt{9} = 3. \quad \text{So } \frac{\sqrt{144}}{\sqrt{16}} = \sqrt{\frac{144}{16}}$$

Once again this rule is a general one: $\boxed{\dfrac{\sqrt{m}}{\sqrt{n}} = \sqrt{\dfrac{m}{n}}.}$

Example

$$\sqrt{\frac{8}{9}} = \frac{\sqrt{8}}{\sqrt{9}} = \frac{2\sqrt{2}}{3}.$$

3 Conjugate surds

$\sqrt{a} + \sqrt{b}$ and $\sqrt{a} - \sqrt{b}$ are specially related surds, called *conjugate* surds: their product is always a rational number, as

$$(\sqrt{a} + \sqrt{b})(\sqrt{a} - \sqrt{b}) = (\sqrt{a})^2 - (\sqrt{b})^2 = a - b.$$

Examples

$$(5 - \sqrt{3})(5 + \sqrt{3}) = 25 - 3 = 22$$

$$(2\sqrt{2} + 3)(2\sqrt{2} - 3) = (2\sqrt{2})^2 - 9 = 8 - 9 = -1.$$

4 Rationalizing the denominator

Example 1

Reduce $\dfrac{3}{\sqrt{5}}$ *to a simpler form for calculation, if* $\sqrt{5} \approx 2\cdot236$.

$$\frac{3}{\sqrt{5}} = \frac{3}{\sqrt{5}} \times \frac{\sqrt{5}}{\sqrt{5}} = \frac{3\sqrt{5}}{5}$$

Hence

$$\frac{3}{\sqrt{5}} \approx \frac{3 \times 2\cdot236}{5} = \frac{6\cdot708}{5} = 1\cdot322.$$

Example 2

Simplify $\dfrac{1 + \sqrt{2}}{2\sqrt{2} - 3}$.

Multiply the numerator and denominator of this fraction by the conjugate surd of the denominator $(2\sqrt{2} + 3)$: this process is called rationalizing the denominator.

Then

$$\frac{1 + \sqrt{2}}{2\sqrt{2} - 3} = \frac{1 + \sqrt{2}}{2\sqrt{2} - 3} \times \frac{2\sqrt{2} + 3}{2\sqrt{2} + 3}$$

$$= \frac{2\sqrt{2} + 2\sqrt{2}\sqrt{2} + 3 + 3\sqrt{2}}{(2\sqrt{2})^2 - (3)^2}$$

$$= \frac{5\sqrt{2} + 4 + 3}{8 - 9} = \frac{7 + 5\sqrt{2}}{-1}$$

$$= -(7 + 5\sqrt{2}).$$

Example 3

Evaluate $\dfrac{1}{(1 - \sqrt{3})^2} - \dfrac{1}{(1 + \sqrt{3})^2}.$

$$\frac{1}{(1 - \sqrt{3})^2} = \frac{1}{1 - 2\sqrt{3} + 3}$$

$$= \frac{1}{4 - 2\sqrt{3}}$$

$$= \frac{4 + 2\sqrt{3}}{(4 - 2\sqrt{3})(4 + 2\sqrt{3})}$$

$$= \frac{4 + 2\sqrt{3}}{16 - 12}$$

$$\frac{1}{(1 + \sqrt{3})^2} = \frac{1}{4 + 2\sqrt{3}}$$

$$= \frac{4 - 2\sqrt{3}}{16 - 12}$$

Hence the original expression $= \dfrac{4 + 2\sqrt{3}}{4} - \dfrac{4 - 2\sqrt{3}}{4} = \sqrt{3}.$

Example 4

Express $\dfrac{2\sqrt{3} + 3\sqrt{2}}{2\sqrt{3} - 3\sqrt{2}}$ *in the form* $a + b\sqrt{c}.$

$$\frac{2\sqrt{3} + 3\sqrt{2}}{2\sqrt{3} - 3\sqrt{2}} = \frac{(2\sqrt{3} + 3\sqrt{2})(2\sqrt{3} + 3\sqrt{2})}{(2\sqrt{3} - 3\sqrt{2})(2\sqrt{3} + 3\sqrt{2})}$$

$$= \frac{12 + 12\sqrt{6} + 18}{(2\sqrt{3})^2 - (3\sqrt{2})^2}$$

$$= \frac{30 + 12\sqrt{6}}{12 - 18}$$

$$= -(5 + 2\sqrt{6})$$

(i.e., $a = -5, b = -2, c = 6$).

Example 5

Find the square roots of $49 - 12\sqrt{5}$.

When a compound surd such as $a + b\sqrt{5}$ is squared, the result is $a^2 + 2ab\sqrt{5} + 5b^2$, that is, $(a^2 + 5b^2) + 2ab\sqrt{5}$, which is another compound surd of the same type involving $\sqrt{5}$.

Hence the square root of $49 - 12\sqrt{5}$ must be of the form $\pm(a + b\sqrt{5})$.

$$a^2 + 5b^2 + 2ab\sqrt{5} \text{ is identical to } 49 - 12\sqrt{5}.$$

Hence $a^2 + 5b^2 = 49$ (i)

and $2ab = -12$ (ii)

From (ii) $a = -\dfrac{6}{b}$. Substitute this in equation (i).

Then $\dfrac{36}{b^2} + 5b^2 = 49$

which gives $5b^4 - 49b^2 + 36 = 0$.

Thus $(5b^2 - 4)(b^2 - 9) = 0$

and therefore $b^2 = \frac{4}{5}$ or $b^2 = 9$.

Hence $b = \pm\dfrac{2}{\sqrt{5}}$ or $b = \pm 3$.

The corresponding values of a are $\mp 3\sqrt{5}$ or ∓ 2.

These pairs of values of a and b are set out in this table:

a	$-3\sqrt{5}$	$+3\sqrt{5}$	-2	$+2$
b	$+\dfrac{2}{\sqrt{5}}$	$-\dfrac{2}{\sqrt{5}}$	$+3$	-3

The first pair gives the root as $-3\sqrt{5} + \dfrac{2}{\sqrt{5}}\sqrt{5} = 2 - 3\sqrt{5}$

The second pair gives $+3\sqrt{5} - \dfrac{2}{\sqrt{5}}\sqrt{5} = -(2 - 3\sqrt{5})$

The third pair gives $-(2-3\sqrt{5})$ and the fourth pair gives $2-3\sqrt{5}$. The required answer is therefore $\pm(2-3\sqrt{5})$. There is no need to work out all the possibilities in detail, as one pair of values of a and b will give the numerical value of the root.

Example 6

Find the square roots of $14 + 4\sqrt{6}$.

Let one root be $a + b\sqrt{6}$.

Then $(a+b\sqrt{6})^2 = a^2 + 6b^2 + 2ab\sqrt{6}$ which is identical to $14 + 4\sqrt{6}$.

Hence $a^2 + 6b^2 = 14$ (i)   and $2ab = 4$ (ii).

From (ii) $a = \dfrac{2}{b}$ and in (i) $\dfrac{4}{b^2} + 6b^2 = 14$

which leads to $\qquad 3b^4 - 7b^2 + 2 = 0$.

This gives $\qquad (3b^2 - 1)(b^2 - 2) = 0$

which makes $\qquad b = \pm\dfrac{1}{\sqrt{3}}$ or $b = \pm\sqrt{2}$.

The corresponding values of a are $\pm 2\sqrt{3}$ or $\pm\sqrt{2}$.

The roots are therefore $\pm(\sqrt{2} + 2\sqrt{3})$.

Exercise 2.1

Simplify:

1 $\sqrt{72}$ **2** $\sqrt{50}$ **3** $\sqrt{18}$ **4** $\sqrt{20}$

5 $\sqrt{1000}$ **6** $\dfrac{\sqrt{32}}{\sqrt{8}}$ **7** $\sqrt{\dfrac{1}{8}}$ **8** $\dfrac{\sqrt{200}}{\sqrt{50}}$

9 $\sqrt{864}$ **10** $\sqrt{45}$ **11** $\sqrt{50} + \sqrt{32} - \sqrt{8}$

12 $\sqrt{27} - \sqrt{12} + 2\sqrt{75}$ **13** $\sqrt{2} \times \sqrt{8}$ **14** $\sqrt{5} \times \sqrt{15}$

15 $\sqrt{21} \times \sqrt{27}$ **16** $4\sqrt{3} \times 2\sqrt{3}$ **17** $8\sqrt{2} \div \sqrt{8}$ **18** $(2 + \sqrt{3})^2$

19 $(1 - \sqrt{3})^2$ **20** $(2\sqrt{2} - 1)^2$ **21** $\sqrt{1\frac{1}{2}} \times 2\sqrt{3}$

22 $(3\sqrt{2} + 4)^2$ **23** $(\sqrt{2})^3$ **24** $(2\sqrt{2} - 3)(3\sqrt{2} + 1)$

25 $(\sqrt{2} - 3)^2 (2\sqrt{2} + 5)$

If $\sqrt{2} = 1\cdot414$ and $\sqrt{3} = 1\cdot732$, evaluate to 3 significant figures:

26 $\dfrac{1}{\sqrt{2}}$ **27** $\dfrac{2}{\sqrt{3}}$ **28** $\dfrac{5}{\sqrt{2}}$ **29** $\sqrt{8}$

30 $\dfrac{1}{\sqrt{8}}$ **31** $\sqrt{27}$ **32** $\dfrac{1}{\sqrt{12}}$ **33** $\dfrac{1}{\sqrt{3} - \sqrt{2}}$

Find the square roots of:

34 $19 + 6\sqrt{2}$ **35** $67 - 12\sqrt{7}$ **36** $17 - 4\sqrt{15}$

37 $35 - 12\sqrt{6}$ **38** $89 + 24\sqrt{5}$

By rationalizing the denominators simplify:

39 $\dfrac{1}{\sqrt{2} - 1}$ **40** $\dfrac{2}{\sqrt{5} - 1}$ **41** $\dfrac{1}{2\sqrt{3} - 1}$ **42** $\dfrac{\sqrt{2} + 1}{\sqrt{2} - 1}$

43 $\dfrac{\sqrt{3} - \sqrt{2}}{\sqrt{3} + \sqrt{2}}$ **44** $\dfrac{2\sqrt{3} + 1}{3\sqrt{2} - 1}$ **45** $\dfrac{3\sqrt{5} - \sqrt{2}}{2\sqrt{5} + 3\sqrt{2}}$

46 Express $\dfrac{3\sqrt{2} - 2\sqrt{3}}{3\sqrt{2} + 2\sqrt{3}}$ in the form $a - b\sqrt{c}$.   (C)

47 Express $\dfrac{1 + \sqrt{2}}{\sqrt{5} + \sqrt{3}} + \dfrac{1 - \sqrt{2}}{\sqrt{5} - \sqrt{3}}$ in the form $a\sqrt{5} + b\sqrt{6}$.   (C)

48 Given that $h = \log 2$ and $k = \log 7$, express $\log \sqrt[3]{392}$ in terms of h and k. Express as a surd the number x such that $\log x = \dfrac{(4h - k)}{3}$.

49 In triangle ABC, AB $=$ BC $= 1$ unit and $\angle$ B $= 90°$. Show that $\angle$ A $= \angle$ C $= 45°$. Express $\cos 45°$ and $\sin 45°$ in surd form.

50 Triangle ABC is equilateral with side 2 units. Bisect angle A by AD where D lies on BC. By considering triangle ABD express $\sin 60°$, $\tan 60°$, $\cos 30°$ and $\tan 30°$ as surds.

Equations involving Surds

Example 7

Solve the equation $\sqrt{2x + 1} - \sqrt{x} = 1$.

Squaring both sides, $(2x + 1) - 2\sqrt{2x + 1}\sqrt{x} + x = 1$

or $3x - 2\sqrt{2x + 1}\sqrt{x} = 0$

Isolating the surd, $3x = 2\sqrt{2x + 1}\sqrt{x}$

Squaring again, $9x^2 = 4(2x + 1)(x) = 8x^2 + 4x$

Hence $x^2 - 4x = 0$ or $x(x - 4) = 0$

which gives $x = 0$ or $x = 4$.

It is essential to test both answers in the original equation: the process of squaring may introduce values which do not satisfy the *original* equation. In the above example, both answers satisfy the equation. Here is an example where only one solution will satisfy the original equation.

Example 8

Solve the equation $\sqrt{3x + 4} - \sqrt{x + 5} = 1$.

Squaring both sides, $(3x + 4) - 2\sqrt{3x + 4}\sqrt{x + 5} + (x + 5) = 1$

$$4x + 8 = 2\sqrt{3x + 4}\sqrt{x + 5}$$

or $2x + 4 = \sqrt{3x + 4}\sqrt{x + 5}$

Squaring again, $4x^2 + 16x + 16 = 3x^2 + 19x + 20$

or $\qquad\qquad\qquad x^2 - 3x - 4 = 0$

which gives $\qquad\qquad (x - 4)(x + 1) = 0$

i.e., $\qquad\qquad\qquad x = 4 \text{ or } x = -1$

The solution $x = 4$ satisfies the original equation as $\sqrt{16} - \sqrt{9} = 1$; the solution $x = -1$ does not satisfy the original equation (test this). Hence $x = 4$ is the only solution.

Exercise 2.2

Solve the equations and check each answer:

1 $\sqrt{x} - \sqrt{x - 2} = 1$ $\qquad\qquad$ **2** $\sqrt{2x + 4} = 1 + \sqrt{x + 9}$

3 $\sqrt{x + 9} - \sqrt{x + 4} = 1$ $\qquad\quad$ **4** $\sqrt{3x + 1} - \sqrt{x + 4} = 1$

5 $\sqrt{3x + 4} - \sqrt{x + 2} = \sqrt{x - 3}$ $\quad$ **6** $\sqrt{5x + 1} + \sqrt{x + 1} = \sqrt{10x + 6}$

7 $\sqrt{6x + 1} - \sqrt{x} = \sqrt{2x + 1}$ $\qquad$ **8** $\sqrt{5x + 1} - \sqrt{3x} = \sqrt{x - 2}$

9 $\sqrt{x^2 + 5x + 2} = 1 + \sqrt{x^2 + 5}$

Proof: $\sqrt{2}$ *is not a rational number, i.e. a fraction.*

Suppose $\sqrt{2}$ were in fact equal to $\frac{a}{b}$, where a and b are integers without a common factor, i.e. $\frac{a}{b}$ has been cancelled to its lowest terms. We can prove that this cannot be so, as it leads to a contradiction.

If $\sqrt{2} = \frac{a}{b}$, then by squaring, $2 = \frac{a^2}{b^2}$ i.e. $2b^2 = a^2$.

Then $2b^2$ must be an even number (any number multiplied by 2 must be even): and hence a^2 must also be even.

But if a^2 is even, then a must also be even: only the square of an even number can itself be even. So $a = 2c$, say.

Then $2b^2 = a^2 = (2c)^2 = 4c^2$ and hence $b^2 = 2c^2$.

Now we see that $2c^2$ is even and so b^2 must also be even: and in that case b must be even.

So we have now proved that **both** a and b are even numbers. This contradicts the original statement that $\frac{a}{b}$ was in its lowest terms. If we return to the beginning and start afresh we must reach the same position over and over again. The only possible conclusion is that $\sqrt{2}$ cannot be a fraction and hence must be an irrational number.

Similar proofs can be devised for other irrational numbers: make up one yourself for, say, $\sqrt{3}$ or $\sqrt{10}$, etc.

3 Simultaneous Equations, Relations, Functions, Remainder Theorem

Simultaneous Linear Equations

To solve a pair of simultaneous equations in two variables, such as

$$3x - 4y = 5$$
$$2x + 3y = 7,$$

you will be familiar with the methods of **elimination** or **substitution**. These are linear equations. When drawn graphically, they represent two straight lines, intersecting at the point representing their solution. Two such equations will therefore always have a unique solution, except when they represent parallel lines, e.g.,

$$3x + 4y = 5$$
$$6x + 8y = 7.$$

Simultaneous linear equations in 3 variables

It is not possible to solve uniquely the *pair* of equations in 3 variables:

$$2x + 3y - z = 56 \quad \text{(i)}$$
$$x + 2y + 3z = 7 \quad \text{(ii)}$$

If one variable (say x) was eliminated we should be left with a linear equation in y and z which would have an infinite number of solutions. However if we are given three independent equations a solution can be obtained in certain cases.

Example 1

Solve
$$x + 2y + z = 8 \quad \text{(i)}$$
$$2x - y + 3z = 9 \quad \text{(ii)}$$
$$3x + 4y - z = 8 \quad \text{(iii)}$$

A general method is: from any pair eliminate one of the variables; repeat for another pair. Then solve the two equations obtained.

From (i) and (ii) eliminate y:
$$x + 2y + z = 8 \quad \text{(i)}$$
$$4x - 2y + 6z = 18 \quad \text{(ii)} \times 2$$

Hence
$$5x \qquad + 7z = 26 \quad \text{(iv)}$$

Now take (ii) and (iii) $\big\}$
and eliminate y $\big\}$
$$8x - 4y + 12z = 36 \quad \text{(ii)} \times 4$$
$$3x + 4y - z = 8 \quad \text{(iii)}$$

Hence
$$11x \qquad + 11z = 44$$

or
$$x \qquad + z = 4 \quad \text{(v)}$$

Now solve (iv) and (v): we find $z = 3$, $x = 1$.
To obtain y, substitute in any one of the original equations: $y = 2$.
Hence the solution is $x = 1$, $y = 2$, $z = 4$.

A geometrical representation of the three equations would require 3 dimensions: to the usual coordinate axes for x and y, we add a third axis, the z-axis, which is at right angles to both. Each equation will then represent a **plane** in space. In the above example the three planes intersect at a single point. However these planes may all intersect in a common line or in parallel lines or two or more of the planes may themselves be parallel: a unique solution will not be obtained in these cases. Consider the following example.

Example 2

Solve $x + 2y + z = 8$ (i)

$2x - y + 3z = 6$ (ii)

$4x + 3y + 5z = 22$ (iii)

Eliminate y from (i) and (ii):

$$5x + 7z = 20 \quad \text{(iv)}$$

Eliminate y from (ii) and (iii):

$$10x + 14z = 40 \quad \text{(v)}$$

But (iv) and (v) cannot be solved uniquely as they are the same equation: hence the original equations have no unique solution.

Exercise 3.1

Solve the simultaneous equations:

1 $x + y + z = 2$
$2x - y + 3z = 9$
$3x + y - 2z = -2$

2 $2x + y - z = -2$
$x - y - z = -6$
$x + 2y + 3z = 7$

3 $2x + 3y + z = 9$
$x - y + 4z = 11$
$3x + y - 5z = -23$

4 $x + y + z = 4$
$2x - y + 6z = 8$
$x + 2y - 3z = 0$

5 $x + y + z = 0$
$5x - 3y + 5z = 12$
$10x + y - 5z = 6$

6 $2x + 3y + z = -1$
$x - y - z = -8$
$3x + y + 4z = 11$

7 Solve the equations $2x + 6y + 5z + 1 = 0$
$30(x - y - z) + 11 = 0$
$6(x + y) - 10z - 9 = 0.$ (C)

8 The expression $ax^2 + bx + c$ has values of 4, 3, 18 when $x = 0$, 1, -2 respectively. Find a, b and c.

9 The curve $y = px^2 + qx + r$ passes through the points $(1, 4), (2, 12)$ and $(-3, 32)$. (This means that $x = 1$, $y = 4$ etc., satisfy the given equation.) Find p, q and r.

Simultaneous Linear and Quadratic Equations

Example 3

Solve the equations $x + y = 5$ (i)
$$x^2 + y^2 = 13 \text{ (ii)}.$$

Note that (i) is a linear equation but (ii) is a quadratic equation (non-linear of the 2nd degree). Graphically (ii) will represent a curve (in this case a circle). Now a line can cut a circle in at most 2 points. We should therefore expect 2 sets of solutions (fig 3.1).

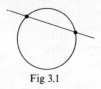

Fig 3.1

A general method of solution is to make one of the variables the subject of the linear equation; then substitute this value in the non-linear equation.

From (i) we obtain $x = 5 - y$ (iii)
Then in (ii) $(5 - y)^2 + y^2 = 13$
which gives $25 - 10y + y^2 + y^2 = 13$
or $2y^2 - 10y + 12 = 0$
or $y^2 - 5y + 6 = 0$
Hence $(y - 3)(y - 2) = 0$, giving $y = 3$ or $y = 2$
Obtain x from (iii): $x = 2$ or 3.
The two sets of solutions are thus:
$$x = 2, \ y = 3 \text{ or } x = 3, \ y = 2.$$
The next example is less straightforward.

Example 4

Solve $3x + 2y = 10$ (i)
$$x^2 - xy + y^2 = 21 \text{ (ii)}.$$

From (i) we obtain $y = \dfrac{10 - 3x}{2}$ (iii) which we substitute in (ii).

$$\therefore \qquad x^2 - x\left(\frac{10 - 3x}{2}\right) + \left(\frac{10 - 3x}{2}\right)^2 = 21$$

$$\therefore \quad x^2 - \frac{x(10 - 3x)}{2} + \frac{100 - 60x + 9x^2}{4} = 21$$

Clearing fractions, $4x^2 - 2x(10 - 3x) + 100 - 60x + 9x^2 = 84$
and so, finally, $19x^2 - 80x + 16 = 0$ (iv)

It can reasonably be expected that the solutions will be rational numbers: hence we should expect (iv) to factorise. This may take a little patience and trial and error.

We obtain $(19x - 4)(x - 4) = 0$ and hence $x = \frac{4}{19}$ or 4.

From (iii) we now obtain the values of y and the solution sets are

$$x = \tfrac{4}{19},\ y = \tfrac{89}{19} \text{ or } x = 4,\ y = -1.$$

Notes: For (iii) we made y the subject, but equally we could have chosen x: either gives a fraction for substitution in (ii). In practice, choose the subject so as not to have a fraction, if possible.

As it would be rather laborious to check the first solution set in (ii), it is essential to work carefully through the algebraic manipulation. If it seems impossible to factorize the resulting quadratic (iv) suspect that you may have made a previous mistake and check through your working again.

Exercise 3.2

Solve these simultaneous equations:

1 $x + y = 5,\ xy = 6$
2 $x - 2y = 1,\ x^2 + y^2 = 29$ (Substitute for x)
3 $2x + y = 5,\ x^2 - xy = 12$
4 $3x - y = 7,\ x^2 - xy + y^2 = 7$
5 $5x + y = 9,\ 3xy + y^2 = -5$
6 $3x + 2y = 13,\ 3x^2 + y^2 = 31$
7 $2x - 3y + 11 = 0,\ 2x^2 - xy = 36$
8 $2x - 5y = 1,\ x^2 - xy + 3y^2 = 9$
9 $2x + 3y = 7,\ y^2 = 26 - 2x^2$
10 $5x + 3y + 7 = 0,\ 3y^2 = x^2 - 4y + 3$

Special cases

The above general method can always be used. In certain very special cases however, the following less laborious method can be used.

Example 5

Solve $3x - 2y = 4,\ 6x^2 - xy - 2y^2 + 36 = 0$.

$$3x - 2y = 4 \qquad\qquad \text{(i)}$$
$$6x^2 - xy - 2y^2 + 36 = 0 \qquad \text{(ii)}$$

Factorize the second degree terms in (ii):

$(3x - 2y)(2x + y) + 36 = 0$ and from (i) $3x - 2y = 4$.

Therefore $4(2x + y) + 36 = 0$
which gives $2x + y + 9 = 0$ (iii)
Now solve (i) and (iii). We obtain $x = -2$, $y = -5$, only one set of solutions.

Example 6

Solve $x + 3y = 2$, $x^2 - 2xy + y^2 = 36$.

Note that the second equation is really $(x - y)^2 = 36$ and thus
$x - y = \pm 6$ (the double value is essential)
Now solve the pairs of linear simultaneous equations:

$$\begin{array}{ll} x + 3y = 2 \\ x - y = 6 \end{array} \quad \text{and} \quad \begin{array}{ll} x + 3y = 2 \\ x - y = -6 \end{array}$$

which gives the solution sets: $x = 5$, $y = -1$ and $x = -4$, $y = 2$.
Look out for such special cases, as some work can be saved.

Exercise 3.3

Solve:
1 $x + y = 3$, $x^2 - y^2 = 15$
2 $3x - 2y = 10$, $3x^2 + xy - 2y^2 = 50$
3 $2x - y = 8$, $x^2 + 2xy + y^2 = 1$
4 $x - y = 4$, $x^2 - xy = 12$
5 $x + 2y = 1$, $3x^2 + 5xy - 2y^2 = 10$
6 $3x + 2y = 17$, $x^2 - 4xy + 4y^2 = 9$

We now look at two very important ideas in Mathematics, **relations** and **functions**. The student will have already met the word function and will have some idea of its meaning but this section seeks to give precision to the concept.

Relations

Consider the **set** of numbers $\{3, 5, 7, 9\}$. A set is any clearly defined collection of objects. They can be listed as shown and written inside $\{\ \}$ brackets or they can be specified in words. For example, a set is described as "the set of all positive multiples of $3 < 20$", which would produce the set $\{3, 6, 9, 12, 15, 18\}$. A set can of course contain an infinite number of objects in which case it cannot be listed in full. For example, the set of all natural numbers (whole numbers) will then be written as $\{1, 2, 3, 4, ...\}$, the dots showing that the set continues without end.

In the set $\{3, 5, 7, 9\}$, examine the relation "is greater than". $5 > 3$,

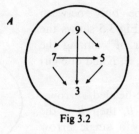

Fig 3.2

$9 > 3$, $7 > 5$ and so on. We can show this relation neatly in a diagram, called an **arrow** diagram (fig 3.2). The boundary line (any closed curve) surrounds all the members of the set and *only* the members of this set. An arrow is drawn between the related numbers so that the direction of the arrow agrees with the *sense* of the relation, i.e., from the greater number to the smaller.

$$5 \xrightarrow{\text{is greater than}} 3 \text{ and so on.}$$

Example 7

Illustrate the relation "divides exactly" on the set A where A = $\{2, 5, 6, 8, 10\}$.

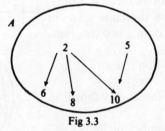

Fig 3.3

At a first glance fig 3.3 is the result. 2 divides 6 exactly so an arrow goes from 2 to 6 and so on. But this is not complete, because $2 \div 2$ is also exact and in fact each member of the set divides itself. We show this by adding a small loop to each number (fig 3.4).

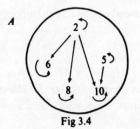

Fig 3.4

Relations need not be only between properties of numbers. Fig 3.5 shows the relation "is the brother of" among the set of 5 boys *A, B, C, D, E, F* and one girl *G*.

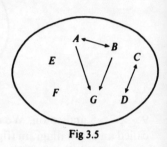

Fig 3.5

Here we use double-headed arrows. If *A* is the brother of *B*, then *B* is the brother of *A*, but only a single arrow connects *B* to *G* (as *G* is the *sister* of *B*). *E* and *F* have no arrows as neither has a brother.

Exercise 3.4

Draw arrow diagrams, using loops and double-headed arrows where necessary, to illustrate the relations:

1 "is double" on the set {2, 3, 4, 6, 8}.
2 "is half of" on the set {2, 3, 4, 6, 8}.
3 "is perpendicular to" on the set of lines in fig 3.6.
4 "is less than" on the set {1, 5, 6, 9}
5 "is 3 more than" on the set {2, 4, 5, 6, 7}.
6 "is 3 less than" on the same set as in Question 5. How do the two diagrams compare?
7 "is the square of" on the set {1, 2, 3, 4, 5, 6, 9, 16}.
8 "is equal to" on the angles shown in fig 3.7.
9 "has the same number of letters as" on the set of words {CARD, LETTER, GONE, MAKE, SEE}.

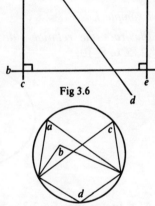

Fig 3.6

Fig 3.7

10 "is North of" on the set of towns {Moscow, Colombo, Singapore, Cairo, New York, Mexico, Rio de Janeiro}.

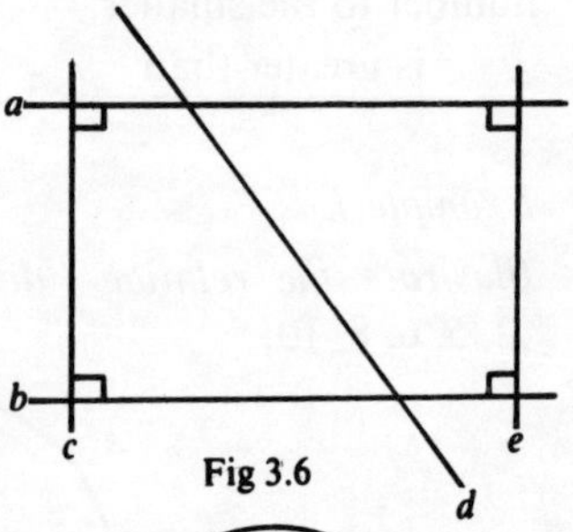

Relations between two sets

A relation need not exist only between members of one set. Two sets can be involved.

Example 8

Boys a, b, c, d are taught Maths by Mr T: boys b, c, d, e History by Mr V: boys a, c, e, f Geography by Mr X. Illustrate the relation "is taught by" between the sets Boys = {a, b, c, d, e, f} and Teachers = {T, V, X}.

Boys   **Teachers**

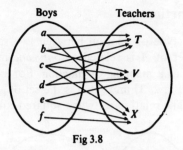

Fig 3.8

Fig 3.8 shows the arrow diagram produced. An arrow proceeds from each boy to his teachers.

Example 9

An arrow diagram is drawn to illustrate the marks obtained by some boys in a test (fig. 3.9). **Boys**   **Marks**

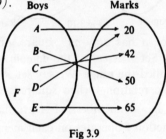

Fig 3.9

The relation is "has ...... marks".
Note: Boy F has no arrow. Perhaps he missed the test.

Exercise 3.5

Illustrate the following relations between sets by arrow diagrams.

1 "is the capital of" between the set of towns {Washington, London, Nairobi, Delhi, Paris} and the set of countries {England, France, India, Kenya, USA}.

2 "is the son of" between the sets of fathers and boys, where Mr A has 3 sons a, b, c; Mr B has no sons; Mr C has 2 sons d, e and Mr D has one son f.

3 "is greater than" from set A = {3, 4, 5, 6, 7} and set B = {1, 2, 4, 7}. (**Note:** This could have been illustrated as before by taking one set A and B combined = {1, 2, 3, 4, 5, 6, 7} but it is often convenient to have two sets.)

4 "is exactly divisible by" from the set A = {2, 3, 4, 5, 6, 7} to the set B = {2, 3}.

5 "is the square of" from the set {1, 4, 9, 16} to the set {−2, −1, 0, 1, 2, 3}.

Mapping or Function

A very important type of relation between two sets A and B arises when *each* member of A is related to *only one* member of B. This relation is known as a **mapping** or **function**. For example in fig 3.10, set $A = \{-2, -1, 1, 2, 3\}$ and set B is the set of all positive integers $\{1, 2, 3, \ldots\ldots\}$. The relation is "when squared equals".

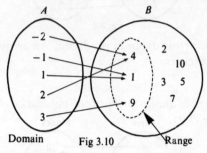

Domain   Fig 3.10   Range

As each member of A has only *one* square, only one arrow leaves each number in A. Set A is called the **domain** and the part of set B (a **subset** of B) is called the **range**. We say set A is **mapped into** set B. If we symbolise the relation "when squared equals" by f, we can write $A \xrightarrow{\ f\ } B$, i.e. all members of A are mapped into B by the function f. The function f operates on each member of A to produce a unique result in B. So $-2 \xrightarrow{\ f\ } 4$ and 4 is called the **image** of -2. This is conveniently written as $f(-2)=4$, stating that f operating on -2 produces an image 4. Similarly $2 \xrightarrow{\ f\ } 4$ as $f(2)=4$ and so on. Note carefully the distinction between an ordinary relation and a function or mapping (figs 3.11 and 3.12).

RELATION   FUNCTION or MAPPING

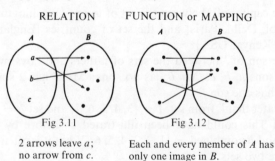

Fig 3.11   Fig 3.12

2 arrows leave a;   Each and every member of A has
no arrow from c.   only one image in B.

This point must be clearly understood to grasp the idea of a function.
Note: A function is always a relation but a relation is not necessarily a function.

Summary

A relation f is a function (or mapping) from a set *A* to a set *B* when:
(1) The entire set *A* is the domain of f.
(2) f produces only one image in *B* for each member of the domain.

The relation f may be specified either by description or by a formula or by a set of tables. For example, the function sine would produce $30° \xrightarrow{\text{sin}} 0·5$ which is usually written $\sin 30° = 0·5$. The domain will be a set of angles, and the images would normally be found from the sine tables.

Here is an example of a function specified by a formula.

Example 10

Set $A = \{1, 5, 7, 8\}$ *and set B is the set of all positive integers. The function* f *is such that if x is any member of A (usually written x∈A),* $x \xrightarrow{\quad f \quad} 2x + 3$. *Draw an arrow diagram (fig 3.13).*

$f(1) = 5$, $f(5) = 13$ and so on.
$\mathbf{f(x) = 2x + 3}$

The formula for this function $x \xrightarrow{\quad f \quad} 2x + 3$, can also be written $\mathbf{f : x \longrightarrow 2x + 3}$ which is read: f is the function such that x is mapped onto $2x + 3$ and we shall normally use this form.

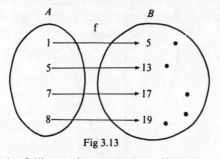

Fig 3.13

Another method of illustrating an arrow diagram, especially when the domain and range are numbers, is to use two parallel number lines.

Example 11

Show the function g: $x \longrightarrow x - 1$ *(or* $x \xrightarrow{\quad g \quad} x - 1$*) where the domain is the set of all integers (positive or negative or zero).*

Two parallel lines are marked out with the +ve and −ve integers, one for the domain and the other for the range (fig 3.14). Each number

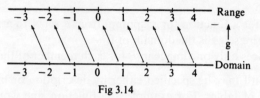

Fig 3.14

of the domain is connected to its image by an arrow. Since the domain is infinite only a token number of arrows can be drawn. The image of 1 is g(1) = 1 − 1 = 0, the image of 2 is g(2) = 1 and so on. Note the pattern formed by the arrow lines.

Repeat this example for the functions: (i) g: $x \to 2x + 1$;

(ii) h: $x \to \dfrac{x + 3}{2}$; (iii) p: $x \to x^2$, taking as domain in each case the set of all integers.

Graphs of functions

You will be familiar with the commonest method of illustrating functions – by drawing a graph with two coordinate axes, called a Cartesian graph. In this case the two number lines are placed at right angles. The domain is now called the x-axis and the range the y-axis. Fig 3.15 shows part of the graph for the function f: $x \to x^2$. The image of 2 is f(2)=4 and a point is marked with coordinates (2, 4). The first number in this pair (the x-coordinate) is taken from the domain and the second number (the y-coordinate) is the image of the first. If

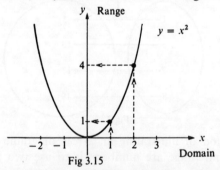

Fig 3.15

sufficient points are plotted the relation between domain and range is made visible by joining up these points to obtain a curve or a straight line and we have a graph to illustrate the function.

So we write $x \xrightarrow{\ f\ } x^2 = y$ and also $y = f(x) = x^2$ and this is the equation of the curve. Many curves can be recognised from their algebraic equations and can be quickly sketched. Otherwise a table of values is made and points plotted.

Exercise 3.6

1 In Exercise 3.5 which of the relations is a function?

2 Which of the following illustrates a function? Why not the others? (fig 3.16)

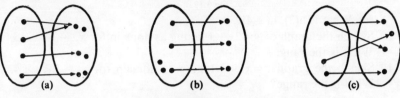

Fig 3.16

3 Illustrate with an arrow diagram the simple function $f : x \rightarrow x$. This is called the **identity** function.

4 $A = \{2, 3, 4, 5, 6\}$ and B is the set of all integers. Show the function $x \xrightarrow{\ f\ } 2x - 1$ where $x \in A$. What is the range?

5 A is the set of all natural numbers $\{1, 2, 3, \ldots\}$ and B is the set of all +ve or −ve numbers (including 0 and decimal fractions). B is the set of **real** numbers. Draw an arrow diagram for a few members of A to show the relation "is the square of". Is this a function or not? Why?

6 Draw arrow diagrams with two parallel lines to show these functions. Take as domain all ±ve integers and draw about ten arrows for each one.

(a) $x \rightarrow x + 3$;　　(b) $x \rightarrow 2x$;　　(c) $x \rightarrow \dfrac{x}{3}$;

(d) $x \rightarrow x^2 + 2$;　　(e) $x \rightarrow x^3$;　　(f) $x \rightarrow 2 - x$;

(g) $x \rightarrow (x-1)(x-2)$;　(h) $x \rightarrow 5 - x$.

7 Does the relation "is the cube root of" where the domain is all real numbers produce a function or not?

8 By trial find a formula for the functions shown below:

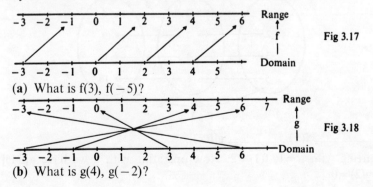

(a) What is $f(3)$, $f(-5)$?

(b) What is $g(4)$, $g(-2)$?

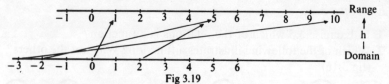

Fig 3.19

(c) What is h(3), h(−4)?

9 Sketch the graph of $y = 2x + 3$ taking as domain for x, $-3 \leqslant x \leqslant 5$. What is the range?

10 Sketch the graph $y = x^2 + 3$ taking as domain for x, $-2 \leqslant x \leqslant 4$. What is the range?

11 If $f : x \rightarrow \dfrac{2x - 3}{5}$ what is the image of 6? Find the values of f(6), f(−1), f(0), f(12).

12 Show in an arrow diagram with two parallel lines, the function $x \rightarrow 2^x$. Take as domain the set $\{-2, -1, 0, 1, 2, 3, 4\}$.

13 Explain why a table of logs or sines or tangents is a functional table. What is the domain in each case?

14 For the function $f(x) = x^3 - 2x^2 + 1$ calculate f(−1), f(0), f(2).

15 If $f(x) = 2x^3 - x + 3$, find f(−2), f(3).

Composite Functions

Functions can be combined to give a **composite** function (sometimes called a function of a function).

Example 12

Two functions are given: f: $x \rightarrow x + 1$ and g: $x \rightarrow 3x$.

Illustrate the composite function derived by operating with f first on x and then with g on the result.

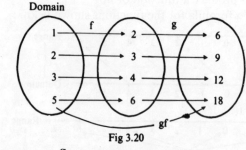

Fig 3.20

$$x \xrightarrow{\quad f \quad} f(x) \xrightarrow{\quad g \quad} g[f(x)]$$

Starting with $x = 1$, $f(1) = 2$. Now operating with g we have g[f(1)] $= g(2) = 6$.

The composite function is produced from $g[f(x)]$ which is written **gf(x)**.

Note that this is written as gf but f is the first function to operate, followed by g. (It does NOT mean $g \times f$).

Instead of calculating the results one by one, we can produce a formula for the composite function gf.

$$x \xrightarrow[\text{(add 1)}]{f} x + 1 \xrightarrow[\text{($\times$ by 3)}]{g} 3(x + 1) = 3x + 3.$$

So gf: $x \to 3x + 3$. For example $gf(1) = 3 + 3 = 6$ as found already. Check the other results using this formula.

Is this the same function as fg? Here g is done first followed by f.

$$x \xrightarrow[\text{($\times$ by 3)}]{g} 3x \xrightarrow[\text{(add 1)}]{f} 3x + 1. \text{ So fg}: x \to 3x + 1 \text{ which is not the}$$

same function as gf. The order of a composite function is important.

Example 13

If $f : x \to 3x$ *and* $g : x \to \sin x°$ *state the functions* fg *and* gf.

fg: here g is done first.

$$x \xrightarrow{\quad g \quad} \sin x° \xrightarrow{\quad f \quad} 3 \sin x°$$

gf: here f is done first.

$$x \xrightarrow{\quad f \quad} 3x \xrightarrow{\quad g \quad} \sin (3x)°$$

Exercise 3.7

1 If $f : x \to x - 1$ and $g : x \to 2x$ give formulae for the composite functions fg and gf in the same manner, what are the values of fg (2), gf(2), fg(−1) and gf(−1)? Illustrate the composite functions using arrow diagrams with three parallel lines taking as domain in each case the set $\{-1, 0, 2, 5\}$.

2 If $f : x \to x^2$ and $g : x \to x + 1$, find fg and gf. What are the values of fg(−2), fg(2), gf(−2) and gf(2)?

3 If $f : x \to x + 1$ what is the function ff? Find the values of ff(−1), ff(0) and ff(3).

4 If $a : x \to x + 1$ and $b : x \to 2x$, express in terms of a and b the composite functions $x \to 2x + 1$ and $x \to 2(x + 1)$.

5 If $f : x \to 2x$ and $g : x \to \cos x°$, express in terms of f and g the functions $x \to 2 \cos x°$ and $x \to \cos (2x)°$.

6 If $f : x \to 2x$ and $g : x \to \sin x°$, find the values of f(10), g(10), fg(10), gf(10), fg(15) and gf(15).

7 If $f : x \to 2x$, $g : x \to x + 1$ and $h : x \to \dfrac{x}{2}$, build up formulae for the composite functions fgh, hfg, gfh and hgf.

8 f is the function which takes the nearest whole number $\leqslant$ the positive square root of x, where $x \in$ set of all positive whole numbers. Thus $f(10) = 3$, $f(4) = 2$. Find $f(12)$, $ff(10)$, $ff(25)$ and $ff(110)$.

9 $f : x \to x + 1$ and $g : x \to x - 1$ are two functions. Find fg and gf. What do you notice? Such functions are **inverse** functions, which we shall look at in the next section.

Make up two more functions which behave in the same way.

10 If $f : x \to \dfrac{1}{x}$ and the domain of x is all integers (excluding zero), find $f(2)$, $f(\tfrac{1}{2})$, $ff(2)$, $ff(\tfrac{1}{2})$, $ff(3)$, $ff(x)$. What do you notice? Why is zero excluded from the domain?

11 $f(x) = a + bx$ where a and b are constants. If $f(1) = 7$, $f(2) = 10$, find $f(3)$.

12 The range of the function $f : x \to + \sqrt{x}$ is the set of real numbers 1 to 4 inclusive. What is the domain?

13 If $f : x \to x^2 + 2$ and $g : x \to x + 3$, find formulae for the functions fg and gf.

14 If $f : x \to x^2 + 2$, find another function g so that $fg : x \to x^2 - 2x + 3$.

Inverse Functions

If f is a function, it will map a member x of set A onto a unique member y of set B (fig 3.21). Is there a function g which will make the return journey? That is, g must map y of B onto x of A. For g to exist, the range of f in B must be the entire set B, otherwise certain members of B would have no image in A under g. Also no member of B must have more than one arrow arriving from A, otherwise there would be more than one path back to A. Hence for g to exist, the relation

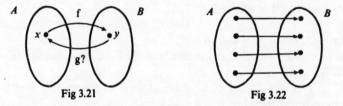

Fig 3.21   Fig 3.22

between A and B under f must be of the kind shown in fig 3.22. This type of relation is called a **one-to-one correspondence**. When this occurs, we can find the function g. g is called the **inverse** function to f and we write $\mathbf{g} = \mathbf{f}^{-1}$.

Example 13

If $f : x \rightarrow x + 2$, *find* f^{-1}.

The image of x under f is obtained by adding 2 to x. Hence the inverse must be obtained by subtracting 2 from the result.

So
$$x \xrightarrow{f} x + 2 = y$$
$$y - 2 \xleftarrow{f^{-1}} y : f^{-1} \text{ (Read from right to left).}$$

Hence f^{-1} is the function whose formula is $f^{-1} : x \rightarrow x - 2$, using x as the 'starting' value, which it is customary to do. It will make the process clearer however if a second letter is used for the intermediate stage.

We can check this by finding $f^{-1}f$.

$$f^{-1}f : x \xrightarrow{f} x + 2 \xrightarrow{f^{-1}} (x + 2) - 2 = x \text{ which is back to the}$$
original starting value. Hence f^{-1} is the inverse of f.

Many simple inverse functions can be written down at sight.
For example:

f	f^{-1}
$x \rightarrow x - 2$ (subtract 2)	$x \rightarrow x + 2$ (add 2)
$x \rightarrow 3x$ ($\times$ by 3)	$x \rightarrow \dfrac{x}{3}$ (divide by 3)
$x \rightarrow 3x - 2$ ($\times$ x by 3 then subtract 2)	$x \rightarrow \dfrac{x + 2}{3}$ (add 2 to x and then divide by 3)

Inverse functions can always be found by the following algebraic method.

Example 14

Find the function inverse to $f : x \rightarrow 3x - 2$.

Then
$$f : x \longrightarrow 3x - 2 = y$$

So
$$3x - 2 = y \text{ which gives } x = \frac{y + 2}{3}$$

Hence
$$\frac{y + 2}{3} = x \longleftarrow y : f^{-1}$$

Thus using x as the starting value, $f^{-1} : x \longrightarrow \dfrac{x + 2}{3}$

Example 15

If $f : x \longrightarrow 3 - x$, *find* f^{-1}.
$$f : x \longrightarrow 3 - x = y$$

So $3 - x = y$ or $x = 3 - y$

Hence $3 - y = x \longleftarrow y : f^{-1}$

The inverse function is therefore $f^{-1} : x \to 3 - x$. Note that this is the same function as f. Such a function is called **self-inverse**.

Exercise 3.8

1 If $f : x \to x + 3$, find $f(2)$. Now write the formula for f^{-1} and find $f^{-1}(5)$. Is your result correct?

Find the inverse functions for each of the following:

2 $x \to 2x$ **3** $x \to x - 2$ **4** $x \to 5x$ **5** $x \to \frac{1}{2}x$

6 $x \to x + 4$ **7** $x \to 2x + 1$ **8** $x \to 3x - 1$ **9** $x \to 5 - x$

10 $x \to \dfrac{1}{x}$ **11** $x \to \dfrac{x + 1}{2}$ **12** $x \to \dfrac{3x - 1}{4}$

13 $x \to 5 - 2x$ **14** $x \to \dfrac{12}{x}$

15 Which of the above functions were self-inverse?

16 If $f : x \to 3x$ and $g : x \to x + 2$, find fg. Then find the inverse of fg, written $(fg)^{-1}$. Now find f^{-1}, g^{-1} and $g^{-1}f^{-1}$. Compare this last result with $(fg)^{-1}$ What do you notice?

17 Repeat Question **16** with the functions $f : x \to x - 2$ and $g : x \to \dfrac{x}{3}$.

18 If $f : x \to 3x$ and $g : x \to \sin x°$, find formulae for f^{-1} g and gf^{-1}.

19 If $f : x \to x + 3$ and $g : x \to 2x - 1$, find the values of $f^{-1}(2)$, $g^{-1}(3)$, $g^{-1}(-1)$, $f^{-1}g(2)$ and $g^{-1}f(4)$.

20 Why has the function $f : x \to x^2$, where the domain of x is all integers, no inverse?

The Remainder Theorem

For this section we consider functions in the form of **polynomials** such as $x \to 3x^3 - 2x^2 + 5x - 1$ or $x \to 2x^2 - 7$ etc. which are collections of terms involving powers of x and constants. Since we shall be more concerned with the image rather than the actual function the notation $f(x) = 3x^3 - 2x^2 + 5x - 1$ will be used.

The values of x which produce an image of 0, i.e. which make $f(x) = 0$, are called the **zeros** of the function. If $f(x) = 3x^2 - 5x + 2$, the zeros are 1 and $\frac{2}{3}$ as $f(1) = 0$ and $f(\frac{2}{3}) = 0$. Check these statements.

Consider the function $f(x) = x^2 - 5x + 6$ and divide it by $(x - 4)$.

$$\begin{array}{r} x - 1 \\ x - 4 \overline{)\, x^2 - 5x + 6} \\ x^2 - 4x \\ \hline - x + 6 \\ - x + 4 \\ \hline 2 \end{array}$$

The process is similar to long division in arithmetic and will be familiar to you. When $x^2 - 5x + 6$ is divided by $x - 4$ the quotient is $x - 1$ and the remainder is 2.

Now calculate f(4): what do you find? $f(4) = 2$
Repeat, dividing $f(x)$ by $x - 3$: compare your remainder with the value of f(3).
Repeat again, dividing by $x + 2$: compare your remainder with the value of $f(-2)$.

These results are not coincidences. It is true that when we divide a polynomial $f(x)$ by $(x - a)$ the remainder is the value of $f(a)$. This is the **Remainder Theorem**.
Similarly if we divide $f(x)$ by $(x + a)$ the remainder will be the value of $f(-a)$.

A simple proof can be given as follows. Suppose $f(x)$ is divided by $(x - a)$.
Let the quotient be $Q(x)$ (a function of x) and let the remainder be R (which will be a number).

Then $f(x) = (x - a) \times Q(x) + R$
(Compare $15 \div 4$. Then $15 = 4 \times 3 + 3$ or $15 = 4 \times Q + R$ where the quotient $Q = 3$ and the remainder $R = 3$).
Now this is a statement of fact: it is not an equation, which is only true for certain values of x, it is an **identity**, which is true for all values of x.
Put $x = a$.
Then $f(a) = (a - a) \times Q(a) + R = 0 + R$.
Hence $R = f(a)$.
If $f(x)$ is divided by $px + q$ then $f(x) = (px + q) \times Q(x) + R$.

Now put $x = -\dfrac{q}{p}$.

Then $f\left(-\dfrac{q}{p}\right) = 0 \times Q + R$.

Hence $R = f\left(-\dfrac{q}{p}\right)$.

Examples
$f(x) = x^3 - 3x^2 + x - 5$. *What are the remainders when* $f(x)$ *is divided by* (**a**) $x - 1$, (**b**) $x + 2$, (**c**) $2x - 1$, (**d**) $3x + 2$?
(**a**) Divisor $x - 1$. Then $R = f(1) = 1 - 3 + 1 - 5 = -6$

(b) Divisor $x + 2$. Then $R = f(-2) = -8 - 12 - 2 - 5 = -27$
(c) Divisor $2x - 1$. Then $R = f(\frac{1}{2}) = \frac{1}{8} - \frac{3}{4} + \frac{1}{2} - 5 = -\frac{41}{8}$
(d) Divisor $3x + 2$. Then $R = f(-\frac{2}{3}) = -\frac{8}{27} - \frac{4}{3} - \frac{2}{3} - 5 = -\frac{197}{27}$
Check results **(a)** and **(b)** by long division.

Exercise 3.9

Find the remainders when

1 $x^2 - 5x + 1$	is divided by	**(a)** $x - 2$,	**(b)** $x + 1$
2 $2x^2 + x + 1$	,,	**(a)** $x - 1$,	**(b)** $x + 3$
3 $3x^2 - 5$	,,	**(a)** $2x - 1$,	**(b)** $x + 3$
4 $x^2 - 3x + 7$	,,	**(a)** $2x + 1$,	**(b)** $x - 5$
5 $x^3 - x^2 + x - 1$	,,	**(a)** $x - 3$,	**(b)** $2x - 1$
6 $x^3 + 3x - 6$	,,	**(a)** $3x - 1$,	**(b)** $2x + 3$
7 $2x^3 - x^2 - 5x + 1$	,,	**(a)** $x + 3$,	**(b)** $2x - 5$.

8 If $x^2 - 5x + 6$ is divided by $x - 2$ and $x - 3$, what are the remainders? Factorize the expression. Compare your factors with these two divisors. What do you notice?
9 $x^3 - 2x^2 - x + 2$ is divided by $x + 1$, $x - 1$ and $x - 2$. Find the remainders. What can you now deduce?

Use of the remainder theorem
As the last exercise foreshadowed, the remainder theorem can be used to factorize polynomials. When used for this purpose, it is sometimes called the **factor theorem**. If $f(x)$ is divided by $(x - a)$ and $f(a) = 0$, then there is no remainder on division. This means that $(x - a)$ is a factor of $f(x)$. By this method, with some intelligent guesswork, we can factorize polynomials. You can already factorize quadratic polynomials such as $3x^2 - x - 2$ by sight. The remainder, or factor, theorem is used to factorize polynomials of higher degree.

Example 16
Factorize $x^3 - 2x^2 - 5x + 6$.

Let $f(x) = x^3 - 2x^2 - 5x + 6$. By inspection, as $f(x)$ is of the 3rd degree, we would expect it to have 3 linear factors at most so that
$$f(x) = (x - a)(x - b)(x - c)$$
where a, b, c are positive or negative integers. Also by multiplying the last terms of each factor, $a \times b \times c$ numerically equals 6. Hence a, b and c are factors of 6: the possible factors of $f(x)$ will then be $(x \pm 1)$, $(x \pm 2)$, $(x \pm 3)$ or $(x \pm 6)$ and we must find the first one from these by trial.
Try $(x + 1)$:

$$f(-1) = -1 - 2 + 5 + 6 \neq 0 \quad (x + 1) \text{ is not a factor.}$$

Try $(x - 1)$:
$$f(1) = 1 - 2 - 5 + 6 = 0 \qquad (x - 1) \text{ is a factor.}$$

We can now either (A) continue trying to find the other factors from among the list above, or (B) divide $f(x)$ by $(x - 1)$ and factorize the quotient.

(A)

Try $(x - 2)$: $f(2) \neq 0$
Try $(x + 2)$: $f(-2) = 0$
So $(x - 1)$ and $(x + 2)$ are factors.

$$\therefore f(x) = (x - 1)(x + 2)(\qquad)$$

and the third factor must be $(x - 3)$ if you compare the linked products with the corresponding terms of $f(x)$.

(B)

$$
\begin{array}{r}
x^2 - x - 6 \\
x - 1 \overline{)\, x^3 - 2x^2 - 5x + 6} \\
\underline{x^3 - x^2} \\
-x^2 - 5x \\
\underline{-x^2 + x} \\
-6x + 6 \\
\underline{-6x + 6}
\end{array}
$$

Hence
$$f(x) = (x - 1)(x^2 - x - 6)$$
$$= (x - 1)(x - 3)(x + 2)$$

Method (B) is quite general and can always be used, once the first linear factor has been found. Note that the quotient may have no further linear factors. Method (A) is of limited use. For example, if the polynomial contained repeated factors, such as $(x - 1)^2$, or if there were less than the expected number of linear factors, much time might be wasted using this method.

Example 17

It is known that $(x - 1)$ and $(x - 2)$ are factors of $x^3 + ax^2 - 7x + b$, where a and b are constants. Find a and b and then the third factor.

$(x - 1)$ is a factor. Therefore, putting $x = 1$ in the polynomial,
$$1 + a - 7 + b = 0 \text{ or } a + b = 6.$$

Similarly, as $(x - 2)$ is a factor, we obtain
$$8 + 4a - 14 + b = 0, \text{ or } 4a + b = 6.$$
Solving these two linear equations, we obtain $a = 0$ and $b = 6$.

Hence the polynomial is $x^3 - 7x + 6 = (x - 1)(x - 2)(\qquad)$ and the third factor is seen to be $(x + 3)$.

Example 18

Solve the equation $2x^3 - 3x^2 - 5x + 6 = 0$.

Let $f(x) = 2x^3 - 3x^2 - 5x + 6$ and factorize $f(x)$.
Try $(x - 1)$: $f(1) = 2 - 3 - 5 + 6 = 0$

$\therefore$ $(x - 1)$ is a factor.

Using method (B), divide $f(x)$ by $(x - 1)$, obtaining

$$2x^2 - x - 6 = (2x + 3)(x - 2)$$

Hence $f(x) = (2x + 3)(x - 1)(x - 2)$. Now the zeros of $f(x)$ are $x = -\frac{3}{2}$ or $x = 1$ or $x = 2$.

These are the values of x which make $f(x) = 0$ and are therefore the roots of the given equation.

Exercise 3.10

Using the remainder theorem factorize:

1 $x^3 - 2x^2 - 5x + 6$ **2** $x^3 - 4x^2 + 5x - 2$
3 $x^3 + 2x^2 - x - 2$ **4** $x^3 - 1$
5 $x^3 + 6x^2 + 11x + 6$ **6** $2x^3 + x^2 - 5x + 2$
7 $2x^3 + 7x^2 + 8x + 3$ **8** $6x^3 - 5x^2 - 3x + 2$
9 $4x^3 + x^2 - 27x + 18$ **10** $x^4 - 5x^3 + 5x^2 + 5x - 6$

11 What is the value of k if $(x - 1)$ is a factor of $x^3 + x^2 + kx + 1$?

12 The remainder on dividing $x^3 + px^2 - x - 1$ by $(x + 1)$ is 1. Find the value of p.

13 $(x - 1)$ and $(x + 3)$ are factors of $x^3 + ax^2 + bx + 12$. Find a and b and the remaining factor.

14 $(x - 2)$ is a factor of $2x^3 + ax^2 + bx - 2$, and when this expression is divided by $(x + 3)$ the remainder is -50. Find a and b and the remaining factors.

15 When $ax^3 + bx^2 + cx + 6$ is divided by $(x - 1)$ the remainder is -8. It also has $(x - 2)$ and $(x + 3)$ as factors. Find a, b and c and the remaining factor.

Solve the equations:

16 $x^3 - 5x^2 - x + 5 = 0$ **17** $x^3 - 9x^2 + 26x - 24 = 0$
18 $2x^3 + x^2 - 13x + 6 = 0$ **19** $x^4 + 5x^3 + 5x^2 - 5x - 6 = 0$

Exercise 3.11   Miscellaneous

1 If $(x + 1)$ and $(x - 2)$ are factors of $x^3 + ax^2 - 5x + b$, find the values of a and b and hence find the remaining factor.   (C)

2 Solve the simultaneous equations $x + 4y = 5$

$$2x^2 + 21xy + 27y^2 = 0. \quad \text{(C)}$$

3 $(x^2 - 1)$ is a factor of $x^3 + ax^2 - x + b$. When the expression is divided by $(x - 2)$ the remainder is 15. Find the values of a and b.

(C)

4 For what values of a and b will the polynomial $ax^3 - x^2 + bx + 3$

have a remainder of 4 when divided by $(x - 1)$ and a remainder of 1 when divided by $(x + 4)$?

5 If $f(x) = ax^2 + bx + c$ and $f(1) = 6$, $f(-1) = 8$ and $f(-2) = 18$, find the values of a, b and c.

6 Solve the equations $x + y = 2$, $x^2 + xy + y^2 = 12$.

7 Solve the equations: $3x - y + z = 6$
$$2x + y - 3z = -2$$
$$x + 2y + 3z = 5.$$

8 Factorize $x^3 + x^2 - 6x + 4$.

9 If $(x - 2)$ is a factor of $x^3 - 4x^2 + px + 16$, find p and then the other factors.

10 When $2x^3 - 3x^2 + ax + b$ is divided by $(x - 1)$ the remainder is -2. It is also known that $(x + 1)$ is a factor of the polynomial. What is the remainder when it is divided by $(2x - 1)$?

11 Solve the equations $x + 3y = -5$, $x^2 + xy - 6y^2 + 25 = 0$.

12 The points $(1, 4)$, $(-2, 7)$ and $(3, 18)$ lie on the curve
$$y = ax^2 + bx + c.$$
Find a, b and c.

13 When $f(x)$ is divided by $(x - a)$ and $(x - b)$ the remainders are $f(a)$ and $f(b)$ respectively. What is the remainder when $f(x)$ is divided by $(x - a)(x - b)$?

(Hint: the divisor $(x - a)(x - b)$ is of the 2nd degree, so the remainder must be of the first degree, i.e., a linear expression of the type $cx + d$. We can then write the identity
$f(x) = (x - a)(x - b) \times Q(x) + cx + d$. Now put $x = a$ and $x = b$ and solve for c and d.)

14 If $(x - 1)$, $(x + 2)$ are factors of $6x^4 + ax^3 - 13x^2 + bx + 14$, find the values of a and b and the other factor when a and b have these values.   (C)

15 Solve the simultaneous equations $\dfrac{x}{3} - \dfrac{y}{10} = \dfrac{5}{6}$, $x(y - 2) = 2y + 3$.

(C)

16 Show the relation "is a prime factor of" between the sets $\{2, 3, 5, 7\}$ and $\{6, 8, 10, 14, 30\}$ in a diagram.

17 Find the inverses of the functions (i) $x \rightarrow x + 1$; (ii) $x \rightarrow 2x - 3$;
(iii) $x \rightarrow 3 - x$; (iv) $x \rightarrow \dfrac{1}{x + 1}$.

18 If $f : x \rightarrow x + 2$ and $g : x \rightarrow 3 - x$ find the functions f^{-1}, g^{-1}, fg, gf, $f^{-1}g^{-1}$, $g^{-1}f^{-1}$.

19 Illustrate with arrows using two parallel number lines the functions:
(i) $x \rightarrow x - 1$; (ii) $x \rightarrow x^2 - 1$.

20 If $f : x \to \dfrac{x-1}{3}$ find $f(2)$, $f(1)$, $f^{-1}(2)$, $f^{-1}(1)$.

21 The function f is defined as follows: $f : x \to 2 - x$ for domain $0 \leqslant x \leqslant 2$, $f : x \to x - 2$ for domain $2 \leqslant x \leqslant 4$. Draw a graph of the function f for domain $0 \leqslant x \leqslant 4$.

22 A function f is defined as $f : x \to$ (the greatest integer $\leqslant x$). Find the values of $f(8\cdot5)$, $f(\sqrt{10})$, $f(\pi)$, f (sin $45°$), $f(\cos 0°)$.

23 The function f is $f : x \to ax^2 + bx$. If $f(1) = -1$ and $f(-2) = 14$, find the values of a and b.

24 If $f : x \to \sin x°$ and $g : x \to 4x$, sketch in the same diagram with the same scales the functions gf and fg for domain $0 \leqslant x \leqslant 90$.

25 If $f : x \to 2x - 3$ and $g : x \to 3x + p$, find the value of p which makes $fg = gf$.

26 Find the range for each of the following functions if the domain of x in each case is $0 \leqslant x \leqslant 3$:

(i) $f : x \to x + 3$;   (ii) $g : x \to \dfrac{x + 1}{4}$;   (iii) $h : x \to \dfrac{1}{1 + x}$;

(iv) $k : x \to \sin (30x)°$.

27 For the function $f : x \to 2x - 3$ find a value of x so that $f(x) = x$. Explain why there can only be one such value of x.

28 If $f : x \to 2x$ and $g : x \to x + 1$, find the value of $f^{-1}g \; g^{-1}f(2)$.

29 If $f : x \to x^3 - 2x^2 - x + 3$, solve the equation $f(x) = 1$.

30 If $f : x \to \log_{10} x$ and the domain of x is $0\cdot01 \leqslant x \leqslant 10$, find the range of the function.

31 If $f : x \to x - 1$, find another function g so that $fg : x \to 3x - 1$. What will gf be then?

32 If $f : x \to x - 1$, find another function g so that gf is the identity function.

33 Draw a diagram to illustrate the relation "is a multiple of" between the sets $\{2, 5, 6, 9, 10\}$ and $\{1, 2, 3, 5\}$.

4 Sequences: Arithmetic and Geometric Progressions

Sequences

(a) 3, 7, 11, 15, ...;　　**(b)** 15, 9, 3, −3, ...;　　**(c)** 2, 6, 18, 54, ...;
(d) 1, 4, 9, 16, 25, ...

Each of the above sets of numbers is a **sequence** and each number
is a **term** of the sequence. The dots ...... indicate that the sequence
continues indefinitely. Each term (after the first) is determined by a
rule or formula and the sequence can be continued using this rule.
Can you discover the rule for each of the above sequences? Continue
each one for three more terms.

In **(a)** each term is 4 more than the preceding term: so the sequence
continues 19, 23, 27 ......
In **(b)** each term is 6 less than the preceding one: the next terms are
−9, −15, −21.
In **(c)** each term is multiplied by 3: the next terms are 162, 486, 1458.
In **(d)** each term is the square of a natural number, taken in order: so
the next terms are 36, 49, 64.

It is possible to express each term of a sequence in terms of a variable,
usually the number of the term in the sequence.
For example, in **(d)** the terms are $(1)^2, (2)^2, (3)^2, (4)^2, $ ...... and so the
nth term is n^2. The 7th term from the start will thus be $7^2 = 49$ and
so on.

For **(a)** we make a table showing the pattern of the sequence:

No. of term	1st	2nd	3rd	4th		9th		nth
Term	3	3 + 4 3 + (1)4	3 + 4 + 4 3 + (2)4	3 + (3)4	··· ...	3 + (8)4	··· ...	3 + (n − 1)4

Hence the nth term of the sequence is given by $3 + (n − 1) \times 4$
$= 3 + 4n − 4 = 4n − 1$.
Thus the 1st term $= 4 \times 1 − 1 = 3$ (as we knew), the 7th term ($n = 7$)
$= 4 \times 7 − 1 = 27$, the 12th term $= 4 \times 12 − 1 = 47$ and so on.

Write out the first 12 terms of the sequence and check using the formula.

For (c) we have this table:

No. of term	1	2	3	4	......	n
Term	2	2×3	$2 \times 3 \times 3$ 2×3^2	$2 \times 3^2 \times 3$ 2×3^3	......	$2 \times 3^{n-1}$

and the sequence can be described: nth term $= 2 \times 3^{n-1}$

For example, the 5th term $= 2 \times 3^{5-1} = 2 \times 3^4 = 162$. The 1st term $= 2 \times 3^{1-1} = 2 \times 1 = 2$.

Exercise 4.1

1 The nth term of a sequence is given by $2n - 1$. Write out the first eight terms.

The nth term of a sequence is given. Write down the first four terms of each one:

2 $3n + 5$ **3** n^3 **4** $\dfrac{n+1}{2}$ **5** 2^n

6 3^{n-1} **7** $3 + 4n$ **8** $5 - 2n$ **9** $(n + 1)^2$

10 $2n^2 + 1$ **11** $2n + 1$ **12** $(-1)^n$ **13** n^{-1}

14 $(-1)^n \times n$ **15** $3 \times 2^{n-1}$ **16** $\log n$ **17** $\dfrac{n}{n+1}$

Deduce from a table (as above) the formula for the nth term of each of the following sequences:

18 $2, 4, 6, 8, \ldots$ **19** $5, 8, 11, 14, \ldots$ **20** $9, 7, 5, 3 \ldots$

21 $1, 2, 4, 8, \ldots$ **22** $7, 11, 15, 19, \ldots$ **23** $1, 8, 27, 64 \ldots$

24 $3, -6, 12, -24, \ldots$

Now we look at two important sequences, called **progressions**.

Arithmetic Progression (AP)

In an AP the difference between any term (except the first) and its predecessor (the term immediately in front) is constant. This is the **constant difference** (d) of the AP.

Examples

$5, 8, 11, 14, \ldots$ $d = 3$

$-5, -1, 3, 7, \ldots$ $d = 4$

$12, 9, 6, 3, 0, -3, \ldots$ $d = -3$

$5, 3\frac{1}{2}, 2, \frac{1}{2}, \ldots$ $d = -1\frac{1}{2}$

In general, if we denote the first term by a, the sequence will be:

No. of term	1	2	3	4.............................n
denoted by	T_1	T_2	T_3	T_4 ....T_n
Term	a	$a + d$	$a + d + d$ $a + 2d$	$a + 2d + d$ $a + 3d$.........$a + (n-1)d$

and so for any AP $\qquad$ $\boxed{T_n = a + (n-1)d}$

Example 1

What is the 35th term of 5, 9, 13, ...?
Here $\qquad\qquad\qquad a = 5, d = 4, n = 35.$
Hence $\qquad\qquad T_{35} = 5 + (34)4$
$\qquad\qquad\qquad\qquad\quad = 141.$

Example 2

Find a formula for the nth term of 12, 5, -2, ...
$$a = 12, \ d = -7.$$
$$T_n = 12 + (n - 1) \times (-7)$$
$$= 12 - 7n + 7$$
$$= 19 - 7n.$$
Check: if $n = 1$, $T_1 = 12$; if $n = 2$, $T_2 = 5$, and so on.

Example 3

The 7*th term of an AP is* 27 *and the* 12*th term is* 47. *What is the sequence?*
$$\left.\begin{array}{l} T_7 = a + 6d = 27 \\ T_{12} = a + 11d = 47 \end{array}\right\} \text{ Solve for } a \text{ and } d.$$
$$a = 3, d = 4.$$
Hence the AP is 3, 7, 11, ...

Adding the terms of an AP
Suppose we are given the AP 3, 5, 7, ... If we add the terms together (starting from the first), $3 + 5 + 7 + \ldots$ we get a **series** of sums depending on how many terms we take. We find two useful formulae for calculating the sum of such series.
(1) What is the sum of 9 terms of the series $3 + 7 + 11 + \ldots...$? Here the series is derived from the AP 3, 7, 11, ... in which $a = 3, d = 4$ and $n = 9$ (9 terms to be added together). The 9th term, $T_9 = 3 + 8 \times 4$

= 35. The terms can be shown (fig 4.1) as the lengths of columns and the required sum will be the total length of the columns.

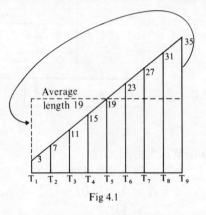

Fig 4.1

The average length is $19 = \frac{1}{2}(3 + 35)$. The columns could be equalized at 19 units length by transferring lengths above the average (on the right) to those below the average. Hence the total length would be $9 \times 19 = 171$ and this is the sum of the series.

This method can be generalized. If we have n terms of an AP, first term a, last term l (T_n), then the average term is $\frac{1}{2}(a + l)$ as above. Writing S_n for the sum of n terms, we have the formula

$$\boxed{S_n = \tfrac{1}{2}(a + l) \times n}$$ where l is the nth term (T_n).

Example 4

What is the sum of 12 terms of the AP 15, 12, 9, ...?

Here $a = 15$, $d = -3$ and $n = 12$.

$$\text{So} \quad T_{12} = l = 15 + (11)(-3) = -18$$

Then $S_{12} = \frac{1}{2}(15 - 18) \times 12 = -18$.

Check this result by actually adding the first 12 terms together.

(2) A direct formula for S_n in terms of a, d and n can now be obtained

$$T_n = l = a + (n - 1)d.$$

Hence
$$S_n = \tfrac{1}{2}(a + l) \times n$$

$$= \tfrac{1}{2}[a + a + (n - 1)d] \times n$$

and thus
$$\boxed{S_n = \frac{n}{2}[2a + (n - 1)d].}$$

When using this formula note that the series must start at the 1st term. See Example 7 below.

Example 5

What is the sum of the first n odd numbers?
The odd numbers form the sequence 1, 3, 5, 7, ... which is an AP with $a = 1, d = 2$.

Hence
$$S_n = \frac{n}{2}[2 + (n - 1)2]$$
$$= \frac{n}{2}(2 + 2n - 2) = n^2.$$

So, for example, the sum of the first 5 odd numbers $= 5^2 = 25$.

Example 6

How many terms of the AP   3, 7, 11, ... must be added together to produce a total of 300?
Here $a = 3, d = 4, S_n = 300$ and we must find n.
$$300 = \frac{n}{2}[6 + (n - 1) \times 4]$$
$$= \frac{n}{2}(6 + 4n - 4)$$
$$= 2n^2 + n$$
$\therefore 2n^2 + n - 300 = 0$ or $(2n + 25)(n - 12) = 0$ which gives $n = -12\frac{1}{2}$ or $n = 12$. Since n must necessarily be a positive integer, $n = 12$ and we must add 12 terms.

Example 7

In an AP the sum of the first eight terms is 164 and the sum of the next 6 terms is 333. Find the AP.

The sum of the first 8 terms is $S_8 = \frac{8}{2}(2a + 7d) = 164$.

The formula cannot be used for the sum of the next 6 terms as they do not start from the first term. But we know that the sum of the first 14 terms must equal $164 + 333 = 497$.

Hence
$$S_{14} = \frac{14}{2}(2a + 13d) = 497.$$

We now have a pair of simultaneous equations in a and d to solve.

$$4(2a + 7d) = 164 \text{ or } 2a + 7d = 41;$$

and
$$7(2a + 13d) = 497 \text{ or } 2a + 13d = 71$$

Solving we obtain $a = 3, d = 5$ and the AP is 3, 8, 13, ...

Arithmetic Mean

The **arithmetic mean** (AM) of two numbers a and b is their average, $\dfrac{a+b}{2}$. The AM of 7 and 14 is $10\frac{1}{2}$.

If three numbers are consecutive terms of an AP, then the middle number is the AM of the other two. If the terms are $a - d$, a, $a + d$, then the AM of the outer pair is $\frac{1}{2}(a - d + a + d) = a$. (Note that to describe three consecutive terms of an AP it is often simpler to use $a - d$, a, $a + d$ rather than a, $a + d$, $a + 2d$.)

Exercise 4.2

Find **(a)** T_{10}, **(b)** S_{10}, **(c)** T_{15}, **(d)** S_{15}, **(e)** T_n, **(f)** S_n for the following seven APs. Simplify **(e)** and **(f)** in each case as much as possible.

1 2, 4, 6, 8, ...   **2** −1, 1, 3, ...   **3** 3, 7, 11, ...   **4** 4, 12, 20, ...
5 18, 15, 12, ...   **6** −9, −4, 1, ...   **7** 2, $3\frac{1}{2}$, 5, ...
8 Which term of the AP 2, 5, 8, ... is 44?
9 How many terms of the series $2 + 5 + 8 + ...$ must be taken to obtain a sum of 155?
10 The fourth term of an AP is 13 and the tenth term is 31. Find the AP and the sum of ten terms.
11 Find the sum of all the multiples of 3 between 100 and 300 (inclusive).
12 The sum of the first ten terms of an AP is 15 and the sum of the next ten terms is 230. Find the AP.
13 If the numbers 4, a, b, $14\frac{1}{2}$ are in AP find a and b.
14 What is the value of k if the three numbers $k + 3$, $2k + 6$, 8 are three consecutive terms of an AP?

Geometric Progression (GP)

In a GP the **ratio** of each term (except the first) to its predecessor is fixed: this ratio is called the **constant ratio** (r).

For example: 2, 4, 8, 16, ... $r = 2$
              −3, 6, −12, 24, ... $r = -2$
              144, 72, 36, 18, ... $r = \frac{1}{2}$

In general, if a is the first term of a GP and r the common ratio, the sequence will be:

T_1	T_2	T_3	T_4	...	T_7	...	T_{20}	...
a	$a \times r$	$a \times r \times r$	$ar^2 \times r$	...				
	ar	ar^2	ar^3	...	ar^6	...	ar^{19}	...

And we can deduce from the pattern shown that

$$T_n = ar^{n-1}$$

Example 8

What is **(a)** the twentieth term, **(b)** the nth term of the GP 2, 6, 18, ...?
Here $a = 2, r = 3$.

(a)
$$T_{20} = 2 \times 3^{19}.$$
As this is a very large number it is best left in this index form.
Using logs we can obtain a rough approximation as
$$2 \cdot 3 \times 10^9 \text{ to two sig. figs.}$$

(b)
$$T_n = 2 \times 3^{n-1}$$

Example 9

Find the 16th term of $\frac{8}{9}$, $-\frac{4}{3}$, 2, ... and T_n.
Here $a = \frac{8}{9}$ and $r = -\frac{3}{2}$.

$$T_{16} = \frac{8}{9} \times \left(-\frac{3}{2}\right)^{15}$$

$$= -\frac{2^3}{3^2} \times \frac{3^{15}}{2^{15}}$$

$$= -\frac{3^{13}}{2^{12}} (\approx -390)$$

$$T_n = \frac{2^3}{3^2} \times \left(-\frac{3}{2}\right)^{n-1}$$

$$= \frac{2^3}{3^2} \times \left(-1 \times \frac{3}{2}\right)^{n-1}$$

$$= (-1)^{n-1} \times \frac{3^{n-3}}{2^{n-4}}$$

Example 10

The second term of a GP is $\frac{8}{9}$ and the sixth term is $4\frac{1}{2}$. Find the GP.

$$\left.\begin{array}{l} T_2 = ar = \dfrac{8}{9} \\[2mm] T_6 = ar^5 = \dfrac{9}{2} \end{array}\right\}$$ Eliminate a from these simultaneous equations by division: note this method of handling such equations.

Then
$$\frac{ar^5}{ar} = \frac{\frac{9}{2}}{\frac{8}{9}} = \frac{81}{16}$$

and hence $\quad r^4 = \dfrac{81}{16}$ and $r = \pm\dfrac{3}{2}$ $\qquad$ (Note that there are two possible values of r).

From the first equation, $a = \pm\frac{16}{27}$ and the two possible GPs are $\pm\frac{16}{27}, \frac{8}{9}, \pm\frac{4}{3}, 2, \ldots\ldots$

The sum of a GP

We can find a formula for the sum of n terms of a GP.
$$S_n = a + ar + ar^2 + \ldots\ldots\ldots + ar^{n-2} + ar^{n-1}$$
Now multiplying throughout by r,
$$rS_n = \quad ar + ar^2 + ar^3 + \ldots\ldots + ar^{n-2} + ar^{n-1} + ar^n$$
Subtracting,
$$S_n - rS_n = a \qquad\qquad\qquad\qquad\qquad\quad - ar^n$$
$$\text{all other terms disappearing.}$$
Hence $\quad S_n(1 - r) = a(1 - r^n)$ which can also be written
$$S_n(r - 1) = a(r^n - 1)$$

Therefore $\qquad \boxed{S_n = \dfrac{a(r^n - 1)}{r - 1} = \dfrac{a(1 - r^n)}{1 - r}}.$

The first form should be used if $r > 1$, the second form when $r < 1$. If $r = 1$, each term $= a$ and the sum of the series is na. Since r^n is likely to be large, results are usually left in index form.

Example 11

Find the sum of ten terms of the series $8 + 4 + 2 + \ldots$
Here $a = 8, r = \frac{1}{2}, n = 10$

$$S_{10} = \frac{8(1 - (\frac{1}{2})^{10})}{1 - \frac{1}{2}} = \frac{8(1 - \dfrac{1}{2^{10}})}{\frac{1}{2}}$$

$$= 16\left(1 - \frac{1}{2^{10}}\right)$$

$$= 16\left(1 - \frac{1}{1024}\right)$$

$$= 16 \times \frac{1023}{1024}$$

$$= \frac{1023}{64}.$$

Example 12

How many terms of the series $2 + 3 + \frac{9}{2} + \ldots$ *must be taken for the sum to exceed* 30?

The series is a GP and $a = 2, r = \frac{3}{2}$.

$$S_n = \frac{2\left[\left(\frac{3}{2}\right)^n - 1\right]}{\frac{3}{2} - 1} \quad \text{and this is to be greater than 30.}$$

Hence $\qquad 4\,[(1 \cdot 5)^n - 1] > 30$ or $(1 \cdot 5)^n - 1 > 7 \cdot 5$

Therefore $\qquad (1 \cdot 5)^n > 8 \cdot 5$.

Now taking logs, $\qquad n \log 1 \cdot 5 > \log 8 \cdot 5$

Thus $\qquad n > \dfrac{\log 8 \cdot 5}{\log 1 \cdot 5} = \dfrac{0 \cdot 9294}{0 \cdot 1761} = 5 \cdot 278.$

Since n must be an integer, we must take 6 terms for the sum to exceed 30.

Geometric Mean (GM)

The **Geometric Mean** (GM) of two numbers a and b is $\pm \sqrt{ab}$ (+ if a and b are positive, − if a and b are negative).

If three numbers are consecutive terms of a GP, then the middle number is the GM of the other two. Suppose the numbers are a, ar, ar^2. Then the GM of a and ar^2 is $\sqrt{a^2 r^2} = ar$, the middle number. This can be used to test if three numbers are in GP.

Exercise 4.3

Find **(a)** T_5, **(b)** T_9, **(c)** S_5, **(d)** S_9, **(e)** T_n, **(f)** S_n for the following GPs. Leave the results in index form.

1 $8, 4, 2, \ldots$ **2** $-3, 12, -48, ..$ **3** $32, -16, 8, \ldots$

4 $\frac{16}{27}, \frac{8}{9}, \frac{4}{3}, \ldots$

5 The third term of a GP is $2\frac{2}{3}$ and the sixth term is $\frac{64}{81}$. Find the GP.

6 The AM of two numbers is $7\frac{1}{2}$ and their GM is 6. Find the numbers.

7 The product of the first and seventh terms of a GP is equal to the fourth term and the sum of the first and fourth terms is 9. Find the GP.   (L)

8 $p + 3, p + 8, p + 18$ are the 3rd, 4th and 5th terms of a GP. Find the value of p, the common ratio and the 9th term of the progression.   (L)

9 The sum of the first two terms of a GP is $2\frac{1}{2}$ and the sum of the first four terms is $3\frac{11}{18}$. Find the GP, if r is positive.

Solution of problems

Many problems involve both APs and GPs and the following examples will illustrate the method of solution. In most cases direct use of the formulae will lead to simultaneous equations to be solved.

Example 13

The second, fourth and eighth terms of an AP are consecutive terms of a GP. Prove that the sixth, twelfth and twenty-fourth terms of the AP are also consecutive terms of a GP.

For the AP, $T_2 = a + d$, $T_4 = a + 3d$, $T_8 = a + 7d$

If these are consecutive terms of a GP, then T_4 is the GM of T_2 and T_8.

$$\therefore \qquad (a + 3d)^2 = (a + d)(a + 7d)$$

or $\qquad a^2 + 6ad + 9d^2 = a^2 + 8ad + 7d^2$

which leads to $\qquad d = a$

Then in the AP, $T_6 = a + 5d = 6a$, $T_{12} = a + 11d = 12a$, $T_{24} = a + 23d = 24a$ and these are consecutive terms of a GP as $(12a)^2 = (6a)(24a)$.

Example 14

An AP has a common difference of 3 and a GP has a common ratio of 2. A third sequence is formed by subtracting the terms of the AP from the corresponding terms of the GP. If the third term of the new sequence is 4 and the sixth term is 79, find the original AP and GP.

Let $a = $ first term of the AP and $A = $ first term of the GP.

The AP is a, $a + 3$, $a + 6$, ... and its sixth term is $a + 15$.

The GP is A, $2A$, $4A$, ... and its sixth term is $32A$.

The new sequence is $A - a$, $2A - (a + 3)$, $4A - (a + 6)$, ...

Its third term is $4A - (a + 6)$ and its sixth term is $32A - (a + 15)$.

Therefore $4A - a - 6 = 4$ and $32A - a - 15 = 79$.

Solving, $a = 2$ and $A = 3$.

The AP was 2, 5, 8, ... and the GP 3, 6, 12, ...

Exercise 4.4   *Miscellaneous*

1 If $3\frac{5}{9}$ and $40\frac{1}{2}$ are the first and last terms of a GP and there are seven terms altogether, find the second term.   (C)

2 An AP has 37 terms of which 9 is the fourth and $58\frac{1}{2}$ is the last. Find the sum of the series.

3 If x, y and z are consecutive terms of a GP, show that $\log x$, $\log y$

and log z are consecutive terms of an AP.   (C)

4 A sequence of numbers is formed by adding together corresponding terms of an AP and a GP whose common ratio is 2. The first term of the sequence is 57, the second term is 94 and the third term is 171. Find the fourth term. Find also an expression for the nth term of the sequence.   (C)

5 An AP has 14 terms. The sum of the odd terms (1st, 3rd, 5th, etc.) is 140 and the sum of the even terms (2nd, 4th, 6th, etc.) is 161. Find the common difference and the 14th term.   (C)

6 Find, as accurately as your tables will allow, the sum of the first 9 terms of the GP whose first term is 120 and whose common ratio is 1·035.   (C)

7 The sum of the first and third terms of a GP is $6\frac{1}{2}$ and the sum of the second and fourth terms is $9\frac{3}{4}$. Find the first term and the common ratio. How many terms must be taken for the sum to be greater than 8000?   (C)

8 Prove that the product of the 2nd and 3rd terms of an AP exceeds the product of the 1st and 4th by twice the square of the difference between the 1st and 2nd.   (C)

9 How many terms of the GP $3 + 5 + 8\frac{1}{3} + \ldots$ must be taken for the sum to exceed 100?   (C)

10 A new sequence is formed from an AP by adding together pairs of terms, 1st + 2nd, 3rd + 4th, 5th + 6th, etc. Find the first term and the common difference of the new sequence in terms of the original AP.

11 Show that three different numbers cannot be consecutive terms of an AP *and* a GP.

12 The sum of the first five terms of an AP is 36 and the sum of the four terms from the 6th to the 9th (inclusive) is 90. Find the AP.

13 The 1st, 4th and 10th terms of an AP are also three consecutive terms of a GP. Find the common ratio of the GP.

14 The sum of n terms of an AP is $2n$ and the sum of $2n$ terms is $3n$. Find the sum of $3n$ terms.

15 For a certain series the sum of n terms is given by the formula $S_n = n(2n - 1)$. Find the first four terms of the series and the nth term. (Hint: for the second term consider the difference between S_2 and S_1 and so on).

16 A GP has a common ratio of 2. Find the value of n for which the sum of $2n$ terms is 33 times the sum of n terms.

17 The nth term of a GP is given by 3^{n-3}. Find the first term, common ratio and the sum of 8 terms.

18 In an AP the 18th term is twice the 9th term. Find the ratio of the sum of 18 terms to the sum of 9 terms for this AP.

19 An AP has common difference 2 and a GP has common ratio 2. A new series is formed by adding together corresponding terms of these progressions. Given that the first term of this new series is 8 and that the fifth term is 91, calculate the values of the first terms of the AP and the GP. Write down an expression in terms of n, for the nth term of the new series. (C)

20 The sum of n terms of a GP is $10 - \dfrac{5}{2^{n-1}}$. Find **(i)** the sum of the first 4 terms; **(ii)** the 4th term.

21 Find the sum of n terms of the sequences **(a)** $\frac{1}{2}$, $\frac{2}{3}$, $\frac{5}{6}$, ...; **(b)** $\frac{1}{2}$, $\frac{2}{3}$, $\frac{8}{9}$, ..., simplifying your answers as much as possible.

22 If $p + 1$, $2p - 10$, $1 - 4p^2$ are three consecutive terms of an AP, find the possible values of p.

23 Find the 100th term of the sequence 1, 2, 4, 5, 7, 8, ... where the terms increase alternately by 1 and 2, and the sum of 100 terms.

24 The first two terms of a GP add up to 10 and the first three terms add up to 19. Find the GP (two answers).

25 A GP has first term 1 and common ratio 2. An AP has first term 1 and common difference 1. A new sequence is formed by adding corresponding terms. Write down the first four terms, the nth term and find the sum of the first 10 terms.

26 The nth term of a sequence is given by $3^{n-1} - n$. Write down the first five terms and find the sum of the first 8 terms.

27 If the nth term of a sequence is $2^{n-1} + n + 3$, write down the first five terms and find the sum of the first ten terms.

28 Five numbers are to be in geometrical progression, the first being 8 and the last 648. Find the common ratio and the middle number.

29 An AP contains n terms. The first term is 2 and the common difference is $\frac{2}{3}$. If the sum of the last four terms is 72 more than the sum of the first four terms, find n.

5 Quadratic Equations and Functions

Roots of a Quadratic Equation

You will remember that the roots of any quadratic equation $ax^2 + bx + c = 0$ are given by the formula

$$x = \frac{-b \pm \sqrt{b^2 - 4ac}}{2a} = \frac{-b \pm \sqrt{D}}{2a} \quad \text{where } D = b^2 - 4ac.$$

Sum and product of the roots

Many problems concerning the roots of quadratic equations can be solved without actually finding them. For example, we can find the sum and the product of the roots directly from the equation. It is usual to call the roots α and β.

Then
$$\alpha = \frac{-b + \sqrt{D}}{2a} \quad \text{and } \beta = \frac{-b - \sqrt{D}}{2a}$$

$$\alpha + \beta = \frac{-b + \sqrt{D} - b - \sqrt{D}}{2a} = -\frac{b}{a} \quad \text{and}$$

$$\alpha\beta = \left(\frac{-b + \sqrt{D}}{2a}\right)\left(\frac{-b - \sqrt{D}}{2a}\right) = \frac{b^2 - D}{4a^2}$$

$$= \frac{b^2 - (b^2 - 4ac)}{4a^2} = \frac{4ac}{4a^2} = \frac{c}{a}.$$

So for any quadratic equation $ax^2 + bx + c = 0$ with roots α and β:

$$\boxed{\alpha + \beta = -\frac{b}{a}, \quad \alpha\beta = \frac{c}{a}.}$$

Example

If the roots of $3x^2 - 4x - 1 = 0$ are α and β, then $\alpha + \beta = \frac{4}{3}$ and $\alpha\beta = -\frac{1}{3}$.

The above can be found more neatly as follows. If the equation $ax^2 + bx + c = 0$ has roots α and β then it is equivalent to the equation $(x - \alpha)(x - \beta) = 0$ as this gives $x = \alpha$ or β.
i.e. $x^2 - (\alpha + \beta)x + \alpha\beta = 0.$
Now compare with the original equation:

$$x^2 + \frac{b}{a}x + \frac{c}{a} = 0 \qquad \text{(dividing through by } a\text{)}.$$

Hence $\alpha + \beta = -\dfrac{b}{a}$ and $\alpha\beta = \dfrac{c}{a}$ as before.

Further we see that any quadratic equation can be written in the form:

$$x^2 - (\text{SUM of roots})x + (\text{PRODUCT of roots}) = 0.$$
$$ \alpha + \beta \phantom{()x + (\text{PRODUCT of roots})} \alpha\beta$$

and this form is very useful.

Example
The equation whose roots are $\frac{3}{4}$ *and* $-\frac{1}{2}$ *is* $x^2 - (\frac{1}{4})x + (-\frac{3}{8}) = 0$
i.e. $8x^2 - 2x - 3 = 0$ (*Check by solving*).

Symmetric functions of the roots

Knowing the values of $\alpha + \beta$ and $\alpha\beta$ for a given equation, we can calculate the values of other symmetrical functions of the roots.

Example 1
If α, β *are the roots of* $2x^2 - x - 2 = 0$, *find the values of* (**i**) $\alpha^2 + \beta^2$;
(**ii**) $\alpha - \beta$; (**iii**) $\alpha^2 - \beta^2$; (**iv**) $\dfrac{1}{\alpha} + \dfrac{1}{\beta}$; (**v**) $\alpha^3 + \beta^3$; (**vi**) $\alpha^3 - \beta^3$.

We know that $\alpha + \beta = \frac{1}{2}$, $\alpha\beta = -1$. Each of the new functions must be expressed in terms of $\alpha + \beta$ and $\alpha\beta$. Carefully note the methods used.

(**i**) $\alpha^2 + \beta^2 = (\alpha + \beta)^2 - 2\alpha\beta = \frac{1}{4} + 2 = \frac{9}{4}$

(**ii**) $\alpha - \beta$ cannot be found directly. We use the fact that
$$(\alpha - \beta)^2 = (\alpha + \beta)^2 - 4\alpha\beta = \frac{1}{4} + 4 = \frac{17}{4}$$
Hence $\alpha - \beta = \dfrac{\sqrt{17}}{2}$ numerically.

(**iii**) $\alpha^2 - \beta^2 = (\alpha + \beta)(\alpha - \beta) = (\tfrac{1}{2})\left(\dfrac{\sqrt{17}}{2}\right) = \dfrac{\sqrt{17}}{4}$

(**iv**) $\dfrac{1}{\alpha} + \dfrac{1}{\beta} = \dfrac{\beta + \alpha}{\alpha\beta} = \dfrac{\frac{1}{2}}{-1} = -\dfrac{1}{2}$

(**v**) $\Big\}$ $\alpha^3 + \beta^3$ and $\alpha^3 - \beta^3$. These can be factorised. Note the factors
(**vi**) $\Big\}$ for future use.

$$\alpha^3 + \beta^3 = (\alpha + \beta)(\alpha^2 - \alpha\beta + \beta^2)$$

$$\alpha^3 - \beta^3 = (\alpha - \beta)(\alpha^2 + \alpha\beta + \beta^2)$$

Then $\alpha^3 + \beta^3 = (\alpha + \beta)\left[(\alpha + \beta)^2 - 3\alpha\beta\right] = (\tfrac{1}{2})\left[(\tfrac{1}{2})^2 + 3\right] = \tfrac{13}{8}$

and $\alpha^3 - \beta^3 = (\alpha - \beta)\left[(\alpha + \beta)^2 - \alpha\beta\right] = \dfrac{\sqrt{17}}{2}\left[(\tfrac{1}{2})^2 + 1\right] = \dfrac{5\sqrt{17}}{8}$

Such methods can also be used to form new equations whose roots are functions of the roots of a given equation.

Example 2

If α, β are the roots of $3x^2 + 5x - 1 = 0$ form the equations whose roots are (i) 5α, 5β; (ii) α^2, β^2; (iii) $\dfrac{1}{\alpha}$, $\dfrac{1}{\beta}$; (iv) $\alpha + \dfrac{1}{\beta}$, $\beta + \dfrac{1}{\alpha}$.

We use the form $x^2 - $ (Sum of roots)$x + $ (Product of roots) $= 0$ to derive the required equation. In each case the sum and product of the new roots is expressed in terms of $\alpha + \beta$ and $\alpha\beta$.

From the given equation $\alpha + \beta = -\tfrac{5}{3}$ and $\alpha\beta = -\tfrac{1}{3}$.

 (i) Sum of new roots $= 5\alpha + 5\beta = 5(\alpha + \beta) = -\tfrac{25}{3}$
    Product $= 5\alpha \times 5\beta = 25\alpha\beta = -\tfrac{25}{3}$
    Hence the required equation is $x^2 + \tfrac{25}{3}x - \tfrac{25}{3} = 0$ or
    $$3x^2 + 25x - 25 = 0.$$

(ii) Sum of new roots $= \alpha^2 + \beta^2 = (\alpha + \beta)^2 - 2\alpha\beta = \tfrac{25}{9} + \tfrac{2}{3} = \tfrac{31}{9}$
    Product $= \alpha^2 \times \beta^2 = (\alpha\beta)^2 = \tfrac{1}{9}$
    New equation is $x^2 - \tfrac{31}{9}x + \tfrac{1}{9} = 0$ or $9x^2 - 31x + 1 = 0$.

(iii) Sum of new roots $= \dfrac{1}{\alpha} + \dfrac{1}{\beta} = \dfrac{\beta + \alpha}{\alpha\beta} = \dfrac{-\tfrac{5}{3}}{-\tfrac{1}{3}} = 5$

    Product $= \dfrac{1}{\alpha\beta} = -3$
    New equation is $x^2 - 5x - 3 = 0$.

(iv) Sum of new roots $= \alpha + \dfrac{1}{\beta} + \beta + \dfrac{1}{\alpha} = \alpha + \beta + \dfrac{\beta + \alpha}{\alpha\beta}$
    $$= -\tfrac{5}{3} + 5 = \tfrac{10}{3}$$
    Product $= (\alpha + \dfrac{1}{\beta})(\beta + \dfrac{1}{\alpha}) = \alpha\beta + 1 + 1 + \dfrac{1}{\alpha\beta}$
    $$= -\tfrac{1}{3} + 2 - 3 = -\tfrac{4}{3}$$
    Hence new equation is $x^2 - \tfrac{10}{3}x - \tfrac{4}{3} = 0$ or $3x^2 - 10x - 4 = 0$.

In each case the new roots are symmetrical functions of α and β. α and β themselves need not be found. If unsymmetrical functions of α and β were required (e.g. 3α and 5β) it would be necessary to know the actual values of α and β. Such problems are not considered here.

If there is a relationship between the roots, this can be used to find an unknown coefficient of the equation.

Example 3
One root of the equation $27x^2 + bx + 8 = 0$ *is the square of the other.*
Find the value of b.

Let the roots be α and α^2.

Then $\alpha + \alpha^2 = -\dfrac{b}{27}$ and $\alpha \times \alpha^2 = \dfrac{8}{27}$.

From the second of these, $\alpha^3 = \dfrac{8}{27}$ and hence $\alpha = \dfrac{2}{3}$.

Then, in the first equation, $\dfrac{2}{3} + \dfrac{4}{9} = -\dfrac{b}{27}$ which gives $b = -30$.

Exercise 5.1

Write down the sum and the product of the roots of these equations:
1 $3x^2 - x - 1 = 0$ 2 $x^2 + 4x - 1 = 0$ 3 $2x^2 - x + 5 = 0$
4 $3t^2 - 7 = 0$ 5 $px^2 - qx - r = 0$ 6 $2y^2 - (a + 3)y + a^2 = 0$
If α, β are the roots of the given equation, form the equations whose
roots are (**i**) 2α, 2β; (**ii**) α^2, β^2; (**iii**) $\dfrac{1}{\alpha}$, $\dfrac{1}{\beta}$; (**iv**) α^3, β^3; (**v**) $\alpha + 1$,
$\beta + 1$:
7 $3x^2 - x - 1 = 0$ 8 $2x^2 + 3x - 4 = 0$ 9 $3x^2 - x - 9 = 0$
10 $x^2 + x - 8 = 0$ 11 $ax^2 + bx + c = 0$ (Simplify your answers as
much as possible.)
12 Given that α, β are the roots of the equation $3x^2 - x - 5 = 0$
form the equation whose roots are $2\alpha - \dfrac{1}{\beta}$, $2\beta - \dfrac{1}{\alpha}$.
13 If α, β are the roots of $x^2 + 3x - 5 = 0$, calculate the numerical
values of (**i**) $\alpha + \beta$; (**ii**) $\alpha\beta$; (**iii**) $\alpha^2 + \beta^2$; (**iv**) $\alpha^3 + \beta^3$; (**v**) $\alpha^4 + \beta^4$.
14 If α, β are the roots of the equation $ax^2 + bx + c = 0$, show that
$a(\alpha + 1)(\beta + 1) = a - b + c$. Find the quadratic equation with
roots $\alpha + 1$, $\beta + 1$.   (L)
15 If α, β are the roots of $2x^2 - 2x - 1 = 0$, form the equation whose
roots are $\dfrac{\alpha}{\beta}$, $\dfrac{\beta}{\alpha}$.
16 One root of the equation $2x^2 - x + c = 0$ is twice the other.
Find the value of c.
17 Find the value of k for which the equation $4(x - 1)(x - 2) = k$
has roots which differ by 2.   (L)

18 If the roots of the equation $x^2 + px + 7 = 0$ are denoted by α and β, and $\alpha^2 + \beta^2 = 22$, find the possible values of p.   (C)

19 If α, β are the roots of the equation $ax^2 + bx + c = 0$, express in terms of α and β the roots of the equations (a) $ax^2 - bx + c = 0$; (b) $cx^2 + bx + a = 0$. Form the equation whose roots are $\alpha + \beta$, $\alpha\beta$.   (L)

20 Find the quadratic equation which has the difference of its roots equal to 2 and the difference of the squares of its roots equal to 5.   (L)

The Quadratic Function

The graph of the function $y = ax^2 + bx + c$ is a parabola (fig 5.1).

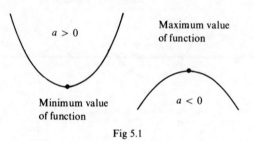

Fig 5.1

If $a > 0$, e.g. $y = 2x^2 + 3x + 7$, the function has a minimum value (turning point at the bottom of the curve).
If $a < 0$, e.g. $y = 5 - 3x - x^2$, the function has a maximum value (turning point at top of the curve).
The position of the curve relative to the x-axis depends on the nature of the roots of the associated equation $ax^2 + bx + c = 0$. The roots of the equation are the zeros of the function, i.e., the values of x where $y = 0$ or where the curve cuts the x-axis. These values are given by
$$x = \frac{-b \pm \sqrt{D}}{2a} \text{ where } D = b^2 - 4ac.$$

Types of roots of a quadratic equation

(1) If D is negative ($D < 0$ or $b^2 < 4ac$), then there is no value for $\sqrt{D}$. The function has no zeros and the curve does not cut the x-axis (fig 5.2i). We say the equation has **complex** roots. For example, $2x^2 - 3x + 5 = 0$ has complex roots as $D = (-3)^2 - 4.2.5 = -31$ and -31 has no (real) square roots.

(2) If D is positive $(D > 0$ or $b^2 > 4ac)$, then there are 2 values for $\sqrt{D}$, and the equation has two different real roots. The curve cuts the x-axis twice (fig 5.2ii). For example, $2x^2 - 3x - 1 = 0$ gives $x = \dfrac{3 \pm \sqrt{17}}{4}$, two different (irrational) values.

If D is also a perfect square, the two roots will be rational (i.e. fractional) numbers.

(3) If $D = 0$, then $x = \dfrac{-b}{2a}$ and both roots equal this value. The curve touches the x-axis where $x = \dfrac{-b}{2a}$ (fig 5.2iii).

Hence for an equation to have real roots, $D \geqslant 0$ ($\geqslant$ means greater than or equal to).

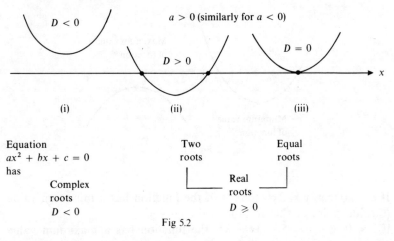

Fig 5.2

Since the value of D discriminates between the various types of roots D is called the **discriminant.**

Example 4

If the equation $x^2 - 3x + 1 = p(x - 3)$ has equal roots, what are the possible values of p?

Rewriting the equation in the standard form,
$$x^2 - x(3 + p) + (1 + 3p) = 0.$$
Here $a = 1,\ b = -(3 + p),\ c = (1 + 3p).$
For equal roots $D = 0$ or $b^2 = 4ac.$
Hence $(3 + p)^2 = 4(1)(1 + 3p)$
or $9 + 6p + p^2 = 4 + 12p$ i.e., $p^2 - 6p + 5 = 0.$
Therefore $(p - 5)(p - 1) = 0$ which gives $p = 5$ or 1.

Maximum and minimum values of a quadratic function
The turning points of the graph, which represent the maximum or minimum values of the function, can be found by the method of 'completing the square'. An alternative method using calculus will be shown later.

Example 5
What is the minimum value of $3x^2 - 2x + 1$ and for what value of x does it occur?
Writing the function as $y = 3x^2 - 2x + 1 = 3(x^2 - \frac{2}{3}x + \frac{1}{3})$ we convert the first two terms in the bracket into a perfect square.

$$x^2 - \frac{2}{3}x = (x - \frac{1}{3})^2 - \frac{1}{9} \qquad \text{(Check by expansion)}$$
Hence $$y = 3[(x - \frac{1}{3})^2 - \frac{1}{9} + \frac{1}{3}]$$
$$= 3(x - \frac{1}{3})^2 + \frac{2}{3}.$$

Now the least value of $(x - \frac{1}{3})^2$ is 0 when $x = \frac{1}{3}$. (A square cannot be negative).
So the least value of y is $\frac{2}{3}$ when $x = \frac{1}{3}$ (fig 5.3).

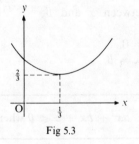

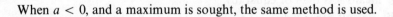

Fig 5.3

When $a < 0$, and a maximum is sought, the same method is used.

Example 6
Find the maximum value of the function $5 - x - 2x^2$.
$$y = -2x^2 - x + 5$$
$$= -2(x^2 + \frac{1}{2}x - \frac{5}{2})$$
$$= -2[(x + \frac{1}{4})^2 - \frac{1}{16} - \frac{5}{2}]$$
$$= -2(x + \frac{1}{4})^2 + \frac{41}{8}.$$
Hence, when $x = -\frac{1}{4}$, y has a maximum value of $\frac{41}{8}$ (fig 5.4).

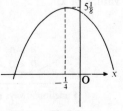

Fig 5.4

Exercise 5.2

Describe the nature of the roots of the following equations:

1 $2x^2 - x - 1 = 0$ **2** $x^2 + 5x - 1 = 0$ **3** $3x^2 - 2x + 4 = 0$
4 $4x^2 - 12x + 9 = 0$ **5** $4x^2 - x - 1 = 0$ **6** $x^2 - 4x + 4 = 0$
7 $ax^2 + bx - a = 0$

Find the maximum or minimum values (as appropriate) of the following functions and the values of x where they occur:

8 $2x^2 - x - 1$ **9** $2 - 4x - x^2$ **10** $3x^2 - x - 6$ **11** $5 + x - 2x^2$
12 If the equation $x^2 + px + 9 = 0$ has equal roots, what are the possible values of p?
13 For what values of k does the equation $kx^2 + (k + 1)x + k = 0$ have equal roots?

Range of values of a quadratic function

If the equation $ax^2 + bx + c = 0$ has roots α, β, then $D > 0$.
From the graph (fig 5.5) where a is taken as positive, we see that:

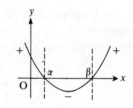

Fig 5.5

  (i) $ax^2 + bx + c < 0$
     when x lies between α and β
     i.e. $\alpha < x < \beta$.
  (ii) $ax^2 + bx + c = 0$
     when $x = \alpha$ or $x = \beta$.

Combining these two,

$$\text{If } \left.\begin{array}{l} D > 0 \\ a > 0 \end{array}\right\} \ ax^2 + bx + c \leqslant 0 \text{ when } \alpha \leqslant x \leqslant \beta$$

($\leqslant$ means less than or equal to)
(iii) For all other values of x, $x < \alpha$ or $x > \beta$, $ax^2 + bx + c > 0$.

If however, $D < 0$, then $ax^2 + bx + c$ is either always **positive** (when $a > 0$) (fig 5.6) or always **negative** (when $a < 0$) (fig 5.7) for all real values of x.

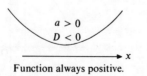

$a > 0$
$D < 0$

Function always positive.

Fig 5.6

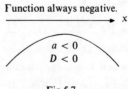

Function always negative.

$a < 0$
$D < 0$

Fig 5.7

The following examples will show the application of the above principles. It is useful to bear the graphical illustrations in mind.

Example 7
Prove that $2x^2 - 3x + 5 > 1$ *for all real values of x.*
In effect, we have to prove that $2x^2 - 3x + 4 > 0$ for all real values of x. Now $a > 0$ and $D = 9 - 32 < 0$. Hence the statement is true (graphically similar to fig 5.6).

Example 8
What is the range of values of x if $2x^2 + x \leqslant 1$?
This is equivalent to $2x^2 + x - 1 < 0$,
    i.e. $(2x - 1)(x + 1) < 0$.
The roots of $2x^2 + x - 1$ are
    $\frac{1}{2}$ and -1 (fig 5.8).
Hence if $2x^2 + x - 1 \leqslant 0$,
x has a range from -1 to $\frac{1}{2}$ (inclusive)
or symbolically $-1 \leqslant x \leqslant \frac{1}{2}$.

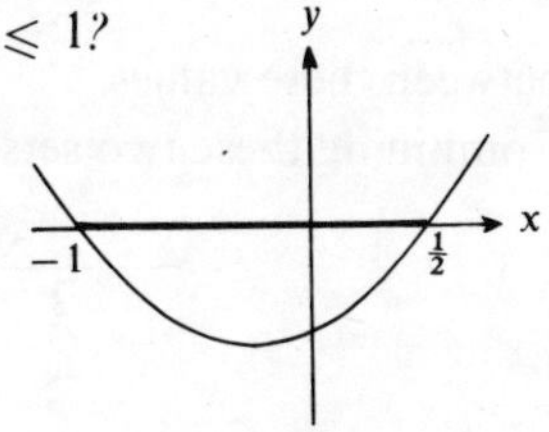

Fig 5.8

Example 9
For what values of x is $2x^2 - 3x - 4$ *less than* $3x^2 + x - 9$?
We require
    $2x^2 - 3x - 4 < 3x^2 + x - 9$
Subtracting terms from both sides,
    $0 < x^2 + 4x - 5$
Hence $(x + 5)(x - 1) > 0$.
Therefore x cannot lie between
    -5 and 1 (inclusive) (fig 5.9)
    i.e. $x < -5$ or $x > 1$.

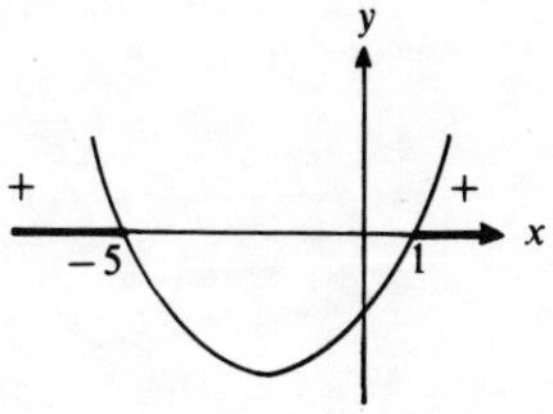

Fig 5.9

Example 10
If the equation $x^2 + (a + 1)x + a^2 - 5 = 0$ *has real roots, find the range of possible values of a.*
For real roots, $D \geqslant 0$ i.e. $b^2 \geqslant 4ac$.
$\therefore$ $(a + 1)^2 \geqslant 4(a^2 - 5)$ or $a^2 + 2a + 1 \geqslant 4a^2 - 20$
Hence $0 \geqslant 3a^2 - 2a - 21$ or $(3a + 7)(a - 3) \leqslant 0$
Hence a must lie between $-\frac{7}{3}$ and 3 (inclusive).
i.e. $-\frac{7}{3} \leqslant a \leqslant 3$
(Graphically this is similar to Example 8 above).

Example 11

For what ranges of values of x does $x^2 + x - 4$ *lie between* 1 *and* 2?

We require $1 < x^2 + x - 4 < 2$; we treat each part separately.

$$x^2 + x - 4 < 2 \text{ is equivalent to } x^2 + x - 6 < 0$$
$$\text{i.e. } (x - 2)(x + 3) < 0$$
and hence
$$-3 < x < 2.$$

$$1 < x^2 + x - 4 \text{ is equivalent to } 0 < x^2 + x - 5$$
$$\text{i.e. } x^2 + x - 5 > 0.$$

The roots of $x^2 + x - 5 = 0$ are $\dfrac{(-1 \pm \sqrt{21})}{2}$ and x must *not* lie between these values.

Combining these two sets of conditions, we have

$$-3 < x < \frac{-1 - \sqrt{21}}{2} \text{ and } \frac{-1 + \sqrt{21}}{2} < x < 2 \text{ (fig 5.10)}$$

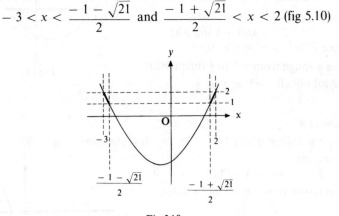

Fig 5.10

Exercise 5.3

Find the range of values of x for which

1 $x^2 + 2x \geqslant 8$ **2** $3x^2 + 5x - 2 > 0$ **3** $3x^2 + 5x < 2$
4 $2x^2 + 7x - 1 \geqslant 3$ **5** $x^2 + x > 5$ **6** $x^2 < 3x - 2$
7 For what range of values of x is $3x^2 - x + 2$ greater than $x^2 + 4x + 5$?
8 For what range of values of x does $x^2 + x - 4$ lie between -2 and 2?
9 For what range of values of x is $3x^2 + 7x + 2$ negative?
10 If $2x^2 + x + k$ is always positive for real values of x, what can be said about the value of k?

11 Show that x^2 is always greater than $2x - 3$ for all real values of x.

12 One root of the equation $2x^2 - 3ax + a = 0$ is double the other. Find the value of a.

13 Determine the range of values of p so that the roots of the equation $3x^2 - 3px + (p^2 - p - 3) = 0$ are real.   (O)

14 (i) Show that each of the expressions $x^2 + (3p + 1)x + (3p^2 + 1)$ and $x^2 + (2p + 1)x + (2p^2 + 1)$ where p is real, is not negative for any real value of x.

(ii) Show that $x^2 + (4p + 1)x + (4p^2 + 1)$ is positive for all real values of p when x is real and greater than -1.   (O)

15 (i) If α and β are the roots of the equation $ax^2 + bx + c = 0$, express $\left[\dfrac{1 + \alpha}{1 - \alpha} + \dfrac{1 + \beta}{1 - \beta} \right]$ in terms of a, b and c.

(ii) The function $ax^2 + bx + c$ is positive only when x lies between 1 and 3. Show that a is negative and find the range of values of x for which $ax^2 - bx + c$ is positive.   (L)

16 If the values of x which satisfy the quadratic equation
$$x^2 + 3 - y(x + 1) = 0$$
are real, prove that there are two values between which y cannot lie, and find these values. Hence, or otherwise, determine the values of x for which $\dfrac{x^2 + 3}{x + 1}$ has (i) a maximum value, (ii) a minimum value.   (L)

17 If s, t are the roots of the equation $2x^2 - x - 4 = 0$, find (without solving the equation) the values of (a) $s^2 + t^2$; (b) $\dfrac{1}{s} + \dfrac{1}{t}$. Construct the equation whose roots are $s^2 + \dfrac{1}{t}$, $t^2 + \dfrac{1}{s}$.

18 If one root of the equation $16x^2 + ax + 1 = 0$ is the cube of the other, find the possible values of a.

19 If one root of the equation $x^2 - ax + b = 0$ is twice the other, find b in terms of a. Form the equation whose roots are 3 and 6 and check your answer.

20 For what range of values of x is $(2x + 1)(x + 3) \geqslant 7$?

6 Trigonometrical Ratios for the General Angle: Equations

Angles

A positive angle measures a rotation in an anticlockwise direction.

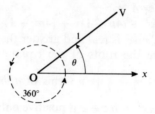

Fig 6.1

One complete revolution of the arm OV (fig 6.1) starting from an initial position Ox, is equivalent to 360°. An angle can also include more than one complete revolution. For example an angle of 920° = 2 complete revolutions (720°) + 200°. 200° in this case will be called the *basic* angle. The basic angle is obtained by removing any multiples of 360°. The basic angles will fall into one of four *quadrants* (fig 6.2).

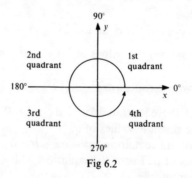

Fig 6.2

Trigonometrical Ratios for Acute Angles (First quadrant)

Fig 6.3 shows the rotating arm OV (of unit length) and the angle θ (measured in degrees) is measured from the initial line Ox. PV is perpendicular to Ox. OP is called the **x-component,** PV the **y-component,** of OV.

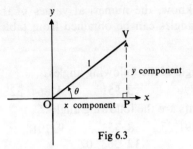

Fig 6.3

We define the trigonometrical ratios as follows:

$$\sin \theta = \frac{y}{1} = y, \ \cos \theta = \frac{x}{1} = x, \ \tan \theta = \frac{y}{x}.$$

These definitions agree with those you have already used in dealing with right-angled triangles. Fig 6.4 shows a right-angled triangle ABC.

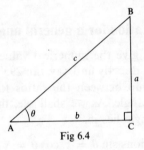

Fig 6.4

Then
$$\sin \theta = \frac{a}{c}, \ \cos \theta = \frac{b}{c}.$$

(If c is of unit length, we have the above definitions.)

Also
$$\tan \theta = \frac{a}{b} = \frac{\dfrac{a}{c}}{\dfrac{b}{c}} = \frac{\sin \theta}{\cos \theta}$$

Again, by Pythagoras' Theorem, $a^2 + b^2 = c^2$

Then
$$\frac{a^2}{c^2} + \frac{b^2}{c^2} = 1 \text{ or } \sin^2 \theta + \cos^2 \theta = 1$$

Though we have derived these relationships using an acute angle, they are identities, i.e., true for *any* angle and they should be remembered.

$$\boxed{\tan \theta = \frac{\sin \theta}{\cos \theta}} \qquad \boxed{\sin^2 \theta + \cos^2 \theta = 1}$$

As you already know, the numerical values of the trigonometrical ratios for acute angles can be obtained from tables.

Exercise 6.1

Find the basic angle of the following angles:

1 460°   **2** 790°   **3** 951°   **4** 864° 32′   **5** 763° 02′

In which quadrants are the following angles?

6 133°   **7** 291°   **8** 353°   **9** 101°   **10** 432°
11 684° 26′   **12** 368° 07′

13 From tables, write down the sines of the following angles:

    (a) 21° 24′   **(b)** 21° 22′   **(c)** 47° 40′   **(d)** 64° 58′
    (e) 3° 13′

14 From tables, write down the cosines of the angles in question 13.

15 From tables, write down the tangents of the following angles:

    (a) 32° 48′   **(b)** 56° 14′   **(c)** 18° 03′   **(d)** 78° 54′

Trigonometrical Ratios for a general angle

Since the tables only give the numerical values of the ratios for an acute angle we cannot directly find, say, tan 295° or cos 157°. We must first find a relationship between the ratios for such angles and a corresponding acute angle. As we shall see, this depends on which quadrant the basic angle lies in.

To do this, the definitions $\sin \theta = y$, $\cos \theta = x$ and $\tan \theta = \dfrac{y}{x}$, where x and y are the components of the unit rotating arm, are used throughout. Due regard is paid to the signs of x and y.

Second quadrant (fig 6.5)

Here x is negative, y is positive. θ is obtuse and the corresponding angle $(180° - \theta)$ is acute.

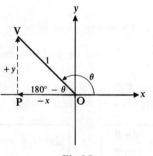

Fig 6.5

Then
$$\sin \theta = y = \sin (180° - \theta)$$
$$\cos \theta = - x = - \cos (180° - \theta)$$
$$\tan \theta = \frac{y}{-x} = - \frac{y}{x} = - \tan (180° - \theta).$$

Hence, if θ lies in the second quadrant, then

$$\sin \theta = \sin (180° - \theta)$$
$$\cos \theta = - \cos (180° - \theta)$$
$$\tan \theta = - \tan (180° - \theta).$$

Note that the
SIN is $+$

Examples
$$\sin 135° = \sin (180° - 135°) = \sin 45° = 0{\cdot}7071$$
$$\cos 168° = - \cos (180° - 168°) = - \cos 12° = - 0{\cdot}9781$$
$$\tan 98° = - \tan (180° - 98°) = - \tan 82° = - 7{\cdot}12.$$

Third quadrant (fig 6.6)
When the rotating arm OV lies in the third quadrant, both components x and y are negative. The corresponding acute angle (P$\hat{\text{O}}$V) is $(\theta - 180°)$.

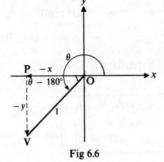

Fig 6.6

Then
$$\sin \theta = - y = - \sin (\theta - 180°)$$
$$\cos \theta = - x = - \cos (\theta - 180°)$$
$$\tan \theta = \frac{-y}{-x} = \frac{y}{x} = \tan (\theta - 180°).$$

So, if θ lies in the third quadrant,

$$\sin \theta = - \sin (\theta - 180°)$$
$$\cos \theta = - \cos (\theta - 180°)$$
$$\tan \theta = \tan (\theta - 180°).$$

Note that the
TAN is $+$

Examples

$\sin 236° = -\sin (236° - 180°) = -\sin 56° = -0.290$
$\cos 236° = -\cos (236° - 180°) = -\cos 56° = -0.5392$
$\tan 197° = \tan (197° - 180°) \quad = \tan 17° = 0.3057.$

Fourth quadrant (fig 6.7)
In this quadrant the x component is positive, the y component negative.
The corresponding acute angle $P\hat{O}V = (360° - \theta).$

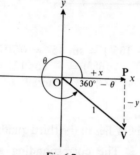

Fig 6.7

Then
$$\sin \theta = -y = -\sin (360° - \theta)$$
$$\cos \theta = \quad x = \quad \cos (360° - \theta)$$
$$\tan \theta = \frac{-y}{x} = -\frac{y}{x} = -\tan (360° - \theta).$$

So, if θ lies in the fourth quadrant,

$$\boxed{\begin{array}{l} \sin \theta = -\sin (360° - \theta) \\ \cos \theta = \quad \cos (360° - \theta) \\ \tan \theta = -\tan (360° - \theta). \end{array}}$$

Note that the
COS is +

Examples

$\sin 323° = -\sin (360° - 323°) = -\sin 37° = -0.6018$
$\cos 323° = \cos (360° - 323°) = \cos 37° = 0.7986$
$\tan 323° = -\tan (360° - 323°) = -\tan 37° = -0.7536.$

Negative angles
The rotating arm will describe a negative angle if it rotates in a clockwise direction. To convert a negative angle to a normal basic angle,
add 360° or a multiple of 360°. The value of a trigonometrical ratio
of any negative angle can then be found.

Examples

An angle of $-40°$ *is equivalent to a basic angle of* $-40° + 360° = 320°$.
Then $\sin(-40°) = \sin(320°) = -\sin(360° - 320°) = -\sin 40°$,
$\cos(-40°) = \cos 320° = \cos(360° - 320°) = \cos 40°$ and
$\tan(-40°) = \tan 320° = -\tan(360° - 320°) = -\tan 40°$.

Hence in general:

$$\begin{array}{rcl} \sin(-\theta) & = & -\sin\theta \\ \cos(-\theta) & = & \cos\theta \\ \tan(-\theta) & = & -\tan\theta \end{array}$$

and these will be true for any negative angle.

The relationships given above connecting any basic angle and a corresponding acute angle are summarised in fig 6.8. These results are very important and should be remembered.

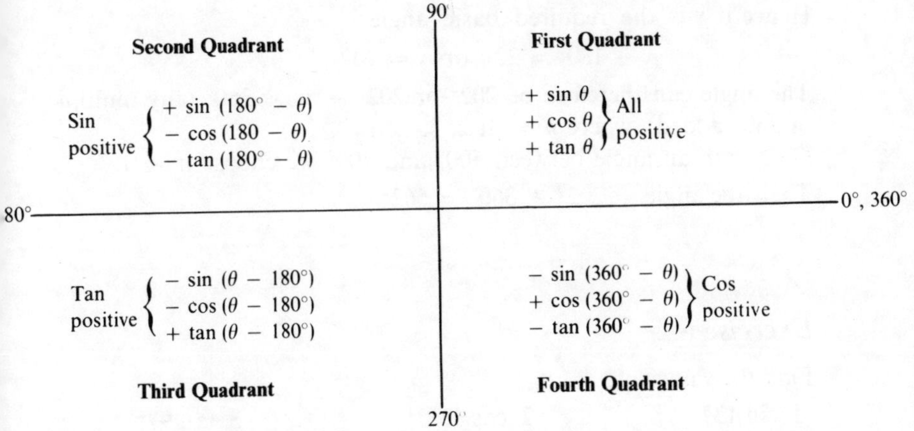

Second Quadrant

Sin positive $\begin{cases} +\sin(180° - \theta) \\ -\cos(180 - \theta) \\ -\tan(180° - \theta) \end{cases}$

First Quadrant

$\begin{array}{l} +\sin\theta \\ +\cos\theta \\ +\tan\theta \end{array} \Big\}$ All positive

Tan positive $\begin{cases} -\sin(\theta - 180°) \\ -\cos(\theta - 180°) \\ +\tan(\theta - 180°) \end{cases}$

Third Quadrant

$\begin{array}{l} -\sin(360° - \theta) \\ +\cos(360° - \theta) \\ -\tan(360° - \theta) \end{array} \Big\}$ Cos positive

Fourth Quadrant

90   270   80°   0°, 360°

Fig 6.8

Example 1

Find $\sin 494° 23'$.

$$494° 23' = 360° + 134° 23'$$
$$\sin 494° 23' = \sin 134° 23'$$
$$= \sin(180° - 134° 23')$$
$$= \sin 45° 37'$$
$$= 0·7147.$$

Example 2

If cos $x = -0.4955$, *find the values of x between* $0°$ *and* $360°$.

Since the cosine is negative, x must lie in either the 2nd or 3rd quadrants.
From tables, the corresponding acute angle is $60°$ $18'$.
Hence, for the 2nd quadrant,
$$180° - x = 60° 18' \text{ or } x = 119° 42'$$
For the 3rd quadrant,
$$x - 180° = 60° 18' \text{ or } x = 240° 18'$$
Then the required values of x are $119° 42°$ and $240° 18'$.

Example 3

The sine of an angle is -0.3746 *and its tangent is positive. Find the angle, if it lies between* $500°$ *and* $700°$.

The basic angle must lie in the third quadrant (negative sin and positive tan). The corresponding acute angle from tables is $22°$.
Hence if x is the required basic angle,
$$x - 180° = 22° \text{ or } x = 202°.$$
The angle can therefore be $202°$ or $202° + n \times 360°$ (any multiple of $360°$ added), where $n = 1, 2, 3, \ldots\ldots$
To obtain an angle between $500°$ and $700°$ we choose $n = 1$.
Then the angle $= 202° + 360° = 562°$.

Exercise 6.2

Find the values of:

1 sin $137°$	**2** cos $158°$	**3** tan $97°$
4 cos $220°$	**5** tan $320°$	**6** cos $191°$
7 tan $343° 35'$	**8** cos $234° 23'$	**9** cos $105° 52'$
10 tan $621°$	**11** cos $(-54°)$	**12** sin $482° 36'$
13 cos $631° 26'$	**14** tan $(-127° 36')$	**15** sin $(-169° 27')$

16 Find an angle in the first quadrant whose sine is
    (a) 0.7108;   **(b)** 0.9649;   **(c)** 0.4389.

17 Find an angle in the first quadrant whose tangent is:
    (a) 0.8816;   **(b)** 1.9496;   **(c)** 2.0265.

Find the value of x between $0°$ and $360°$ in the following examples:

18 sin $x = +0.9063$	**19** cos $x = +0.5948$
20 tan $x = -0.4986$	**21** sin $x = -0.5995$
22 tan $x = 1.9999$	**23** cos $x = -0.6816$

24 If θ is an angle less than 360°, sin $\theta = +0\cdot8870$ and cos θ is negative, find the value of θ.

25 The cosine of an angle is $-0\cdot9438$. The angle lies between 500° and 600°, and its sine is also negative. Find the angle.

26 The tangent of an angle is $+0\cdot68$. The angle lies between 400° and 600°, and its cosine is negative. Find the angle.

27 The cosine of an angle is $+0\cdot6788$. The angle lies between 500° and 700° and its tangent is negative. Find the angle.

28 The sine of an angle is $-0\cdot4591$. The angle lies between 550° and 750° and its cosine is negative. Find the angle.

Trigonometrical Ratios of Special Angles

The trigonometrical ratios of the angles 30°, 45°, 60° are used often in mechanics, and other branches of mathematics, so it is useful to have their values in surd form.

| 45° |

A square, ABCD, is drawn, with sides of unit length (fig 6.9). AC, a diagonal, is drawn.

$$\sin 45° = \frac{BC}{AC} = \frac{1}{\sqrt{2}}$$

$$\cos 45° = \frac{AB}{AC} = \frac{1}{\sqrt{2}}$$

$$\tan 45° = \frac{BC}{AB} = \frac{1}{1} = 1$$

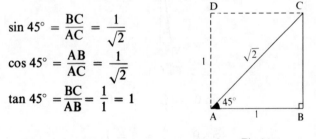

Fig 6.9

| 60° |

An equilateral triangle, ABC, of sides 2 units in length, is drawn (fig 6.10). AD is drawn from A perpendicular to BC.

$\widehat{ACD} = 60°$

$$\sin 60° = \frac{AD}{AC} = \frac{\sqrt{3}}{2}$$

$$\cos 60° = \frac{CD}{AC} = \frac{1}{2}$$

$$\tan 60° = \frac{AD}{CD} = \frac{\sqrt{3}}{1}$$

Fig 6.10

$\boxed{30°}$ $\widehat{\text{CAD}} = 30°$ (fig 6.10)

$$\sin 30° = \frac{CD}{AC} = \frac{1}{2}$$

$$\cos 30° = \frac{AD}{AC} = \frac{\sqrt{3}}{2}$$

$$\tan 30° = \frac{CD}{AD} = \frac{1}{\sqrt{3}}$$

$\boxed{0°}$ Consider the arm V_1 in fig 6.11. As the arm approaches the line Ox, the value of the angle approaches $0°$; the y-component of the arm vanishes when the angle is $0°$ and the x-component becomes the same length as the arm, i.e. equal to unity

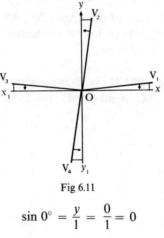

Fig 6.11

$$\sin 0° = \frac{y}{1} = \frac{0}{1} = 0$$

$$\cos 0° = \frac{x}{1} = \frac{1}{1} = 1$$

$$\tan 0° = \frac{y}{x} = \frac{0}{1} = 0$$

$\boxed{90°}$ Consider the arm V_2 in fig 6.11, which is approaching $90°$ When the angle is $90°$, the x-component of the arm becomes 0, and the y-component becomes unity.

$$\sin 90° = \frac{y}{1} = \frac{1}{1} = 1$$

$$\cos 90° = \frac{x}{1} = \frac{0}{1} = 0$$

Tan 90°, however, has no value. The tangent of an angle $= \dfrac{y}{x}$ and as the angle → (approaches) 90°, y → 1 and x → 0.

Hence $\dfrac{y}{x} = \dfrac{\text{a number} \to 1}{\text{a number} \to 0}$ which gets larger and larger. When the angle reaches 90° this ratio has no definite value. We say tan 90° is infinite and write tan 90° → $\pm \infty$.

180° Consider the arm V_3 in fig 6.11. The y-component decreases to zero as the angle approaches 180°, and the x-component approaches a value of -1.

$$\sin 180° = \frac{y}{1} = \frac{0}{1} = 0$$

$$\cos 180° = \frac{x}{1} = \frac{-1}{1} = -1$$

$$\tan 180° = \frac{y}{x} = \frac{0}{-1} = 0$$

270° Consider the arm V_4 in fig 6.11. As the vector approaches the axis Oy_1, the x-component, which is negative, approaches zero, and the y-component approaches a value of -1.

$$\sin 270° = \frac{y}{1} = \frac{-1}{1} = -1$$

$$\cos 270° = \frac{x}{1} = \frac{0}{1} = 0$$

tan 270° : As with tan 90°, tan 270° has no value.

The values for all these angles are included in fig 6.12.

	0°	30°	45°	60°	90°	120°	135°	150°	180°
sin	0	$\frac{1}{2}$	$\frac{1}{\sqrt{2}}$	$\frac{\sqrt{3}}{2}$	1	$\frac{\sqrt{3}}{2}$	$\frac{1}{\sqrt{2}}$	$\frac{1}{2}$	0
cos	1	$\frac{\sqrt{3}}{2}$	$\frac{1}{\sqrt{2}}$	$\frac{1}{2}$	0	$-\frac{1}{2}$	$\frac{-1}{\sqrt{2}}$	$\frac{-\sqrt{3}}{2}$	-1
tan	0	$\frac{1}{\sqrt{3}}$	1	$\sqrt{3}$	∞	$-\sqrt{3}$	-1	$\frac{-1}{\sqrt{3}}$	0

	210°	225°	240°	270°	300°	315°	330°	360°
sin	$-\frac{1}{2}$	$\frac{-1}{\sqrt{2}}$	$\frac{-\sqrt{3}}{2}$	-1	$\frac{-\sqrt{3}}{2}$	$\frac{-1}{\sqrt{2}}$	$-\frac{1}{2}$	0
cos	$\frac{-\sqrt{3}}{2}$	$\frac{-1}{\sqrt{2}}$	$-\frac{1}{2}$	0	$\frac{1}{2}$	$\frac{1}{\sqrt{2}}$	$\frac{\sqrt{3}}{2}$	1
tan	$\frac{1}{\sqrt{3}}$	1	$\sqrt{3}$	∞	$-\sqrt{3}$	-1	$\frac{-1}{\sqrt{3}}$	0

Fig 6.12

Graphs of the Trigonometrical Functions

The sine curve

The variation in the values of sin θ between 0° and 360° can be seen from the table in fig 6.12. Beginning with a value of 0 at 0°, sin θ increases to a value of 1 at 90°, then decreases to 0 at 180°, to -1 at 270°, and then increases to 0 at 360°. A graph showing the relationship between θ and sin θ can be drawn as follows. Make a table of values of sin θ for every 10° from 0° to 360° inclusive, using the table of sines. Note that the value of the ratio is only taken to 2 significant figures, as a greater degree of accuracy cannot be plotted. The following table shows the values from 0° to 180° only. Copy the table and complete it to 360°.

θ	0°	10°	20°	30°	40°	50°	60°	70°	80°	90°
sin θ	0	0·17	0·34	0·50	0·64	0·77	0·87	0·94	0·98	1·0

θ	100°	110°	120°	130°	140°	150°	160°	170°	180°
sin θ	0·98	0·94	0·87	0·77	0·64	0·50	0·34	0·17	0

On a sheet of graph paper, mark θ "horizontally", taking a scale of 1 inch for every 30° or 40° as convenient. On the "vertical" scale take 2 inches to represent 1 unit for values of sin θ*. Plot all the pairs of values θ, sin θ and join up to obtain a smooth curve. This is part of the sine curve (fig 6.13).

The cosine curve

Now make a similar table for cos θ, for values from 0° to 360°, at intervals of 10°. Plot the points on the same graph paper, using the same scales as for sin θ. This is part of the cosine curve, and the result is included in fig 6.13. Compare the shape of the cosine curve between 0° and 90° (by using tracing paper) with that of the sine curve between 90° and 180°. The two curves will be found to be identical in shape; the two curves are always identical in shape, but the cosine curve is 90° behind the sine curve. This difference of 90° is called the **phase difference** between the two curves. Both curves are extremely important, and they have many practical applications in electricity, radio waves, sound waves, etc.

Fig 6.14 shows the sine curve extended past the limits of 0° $-$ 360°.

---

*It is assumed that inch graph paper is being used at present. If metric paper is used, 2 cm can be taken instead of 1 inch in this and the following graphs.

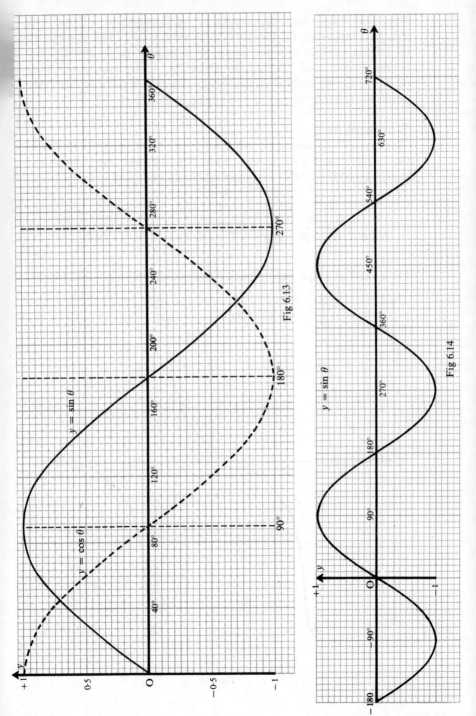

$y = \sin \theta$

$y = \cos \theta$

Fig 6.13

$y = \sin \theta$

Fig 6.14

Note that the curve is continuous in the direction of $+\theta$ and $-\theta$; it is also repetitive, i.e. the portion between 0° and 360° is repeated an infinite number of times. Draw up a table of values for cosine θ between $-180°$ and 720°, in intervals of $22\frac{1}{2}°$. Using a horizontal scale of 1 inch for 90°, and a vertical scale of 1 inch for 1 unit, draw the graph for cos θ between the two angles stated above.

The tangent curve

Make a table of values for tan θ for every 30° from 0° to 360°. Use values of 78° and 102° instead of 90°. The following table shows the values from 0° to 180°; copy the table and complete it to 360°.

θ	0°	30°	60°	78°	102°	120°	150°	180°
tan θ	0	0·58	1·73	4·70	$-4·70$	$-1·73$	$-0·58$	0

On a sheet of graph paper, mark θ horizontally taking 1 inch for 60°,

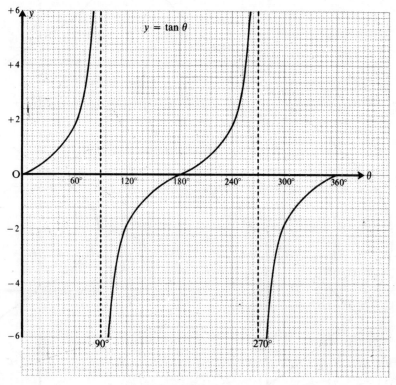

Fig 6.15

and mark a vertical scale for values of tan θ taking 1 inch for 2 units; mark the scale between $+6$ and -6 units. Plot the points and join them up to make a smooth curve, as shown in fig 6.15. Note the graph does not reach the values of $\theta = 90°, 270°$, etc; the nearer θ approaches these values, the nearer the graph approaches the vertical lines through $90°$ and $270°$, but the graph never actually touches these lines. The lines through $90°$ and $270°$ are called **asymptotes**, and the graph is said to approach them asymptotically. This is the graphical representation of the statement "as $\theta \to 90°$, so tan $\theta \to \pm\infty$".

Graphical Solution of Simple Trigonometrical Equations

The graphs that have been drawn can be used to solve simple equations involving trigonometrical ratios.

Example 4

Using a graph of sin θ, *solve the equation* sin $\theta = 0.2$, *for values of θ between $0°$ and $180°$.*

The graph of sin θ is drawn, as shown in fig 6.16; a line is then drawn horizontally through the value of 0.2 for sin θ. This line cuts the sine curve at points A and B. These points are read from the horizontal axis as $12°$ and $168°$, giving answers correct to the nearest degree.

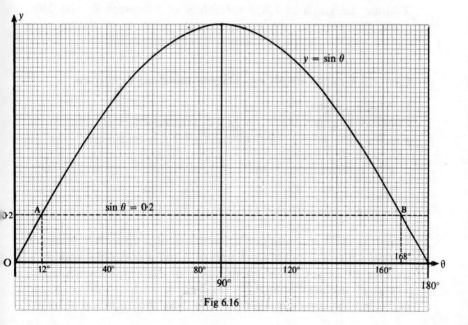

Fig 6.16

Note that if no limits are given for θ, there will be an infinite number of solutions to the equation. The line AB could be extended both ways to cut the infinite sine curve twice in each cycle, i.e., every 360°. The *general solution* of the equation $\sin \theta = 0.2$ would then be:
$12° + n \times 360°$ and $168° + n \times 360°$, where $n = 0, \pm 1, \pm 2, \pm 3, ...$
giving
$... 12° - 2 \times 360°, 12° - 360°, 12°, 12° + 360°, 12° + 2 \times 360°, ...$
i.e. $... -708°, -348°, 12°, 372°, 732°, ...$
and similarly for the other values.

Normally however solutions are only required between 0° and 360°.

Exercise 6.3

Give the answers to Questions 1 – 5 in surd form.

1 Give the value of:
  (a) $\tan 60°$; (b) $\sin 150°$; (c) $\cos 315°$; (d) $\sin 270°$; (e) $\tan 300°$.
2 Find the value of $\cos 30° \times \tan 30°$.
3 Calculate the value of $(\sin 45°)^2 + (\cos 45°)^2$.
4 Find the value of $\cos 330° \div \tan 210°$.
5 Find the value of $\cos 60° - \cos 240°$.
6 Draw the graph $\cos x$ for values between 0° and 360°; from the graph, obtain approximate solutions of the equations $\cos x = 0.45$, and $\cos x = \pm 0.65$.
7 Draw the graph of $\tan \theta$ for values of θ between 0° and 360°; use the graph to obtain approximate solutions of the equation
$$\tan \theta = -1.2.$$

General Graphs involving Trigonometrical Functions

1 Multiple angles

Consider the graph of $y = \sin 2x$. For selected values of x, the corresponding value of y is calculated, and a table is constructed.

x	0°	10°	20°	30°	40°	50°	60°	70°	80°	90°
$2x$	0°	20°	40°	60°	80°	100°	120°	140°	160°	180°
$\sin 2x$	0	0·34	0·64	0·87	0·98	0·98	0·87	0·64	0·34	0

Copy the table and continue it for values of x up to 360°. Using a vertical scale of 2 inches for 1 unit of y and a horizontal scale of 1 inch for 40° of x, plot the points, and join them by a smooth curve. The graph is shown in fig 6.17; the shape is similar to that of a sine curve, but the curve repeats itself with twice the frequency of $\sin x°$.

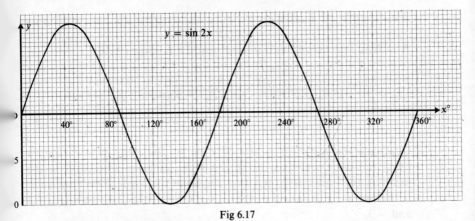

Fig 6.17

2 Added and multiplied constants

An example of an added constant is given by $y = 3 + \cos x$. A selection of corresponding values of x and y is shown in the following table.

x	$0°$	$30°$	$45°$	$60°$	$90°$	$120°$	$135°$	$150°$	$180°$
$\cos x$	1	0·87	0·71	0·5	0	−0·5	−0·71	−0·87	−1
$3 + \cos x$	4	3·87	3·71	3·5	3·0	2·5	2·29	2·13	2

Write out the table and complete it for values of x up to $360°$. Plot a graph, using the same scales as in the previous example. The shape of the curve is shown in fig 6.18; the curve is exactly the same as the cosine curve, but it is displaced 3 units above the x-axis.

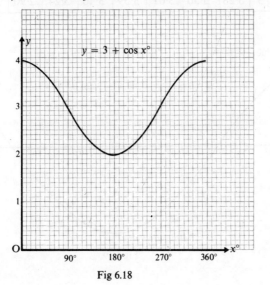

Fig 6.18

An example of a multiplied constant is given by the expression
$$y = 3 \sin x.$$
A table of corresponding values of y and x is shown below.

x	$0°$	$30°$	$60°$	$90°$
$\sin x$	0	0·50	0·87	1
$3 \sin x$	0	1·50	2·61	3

Complete the table for values of x up to $360°$, and plot the curve using the same scales as in the previous examples. The curve is the same as the sine curve, but the **amplitude** is larger. The amplitude is the extent in variation of the values of y, in this case between $+3$ and -3.

3 General graphs

An expression can be formed by adding two trigonometrical functions; for example, $y = 2 \sin x + 3 \cos x$. The table below shows the method of calculating points on the curve between values of $0°$ and $90°$ for x.

x	$0°$	$10°$	$20°$	$30°$	$40°$	$50°$	$60°$	$70°$	$80°$	$90°$
$\sin x$	0	0·17	0·34	0·50	0·64	0·77	0·87	0·94	0·98	1·00
$\cos x$	1	0·98	0·94	0·87	0·77	0·64	0·50	0·34	0·17	0
$2 \sin x$	0	0·34	0·68	1·00	1·28	1·54	1·74	1·88	1·96	2
$3 \cos x$	3	2·94	2·82	2·61	2·31	1·92	1·50	1·02	0·51	0
y	3	3·28	3·50	3·61	3·59	3·46	3·24	2·90	2·47	2

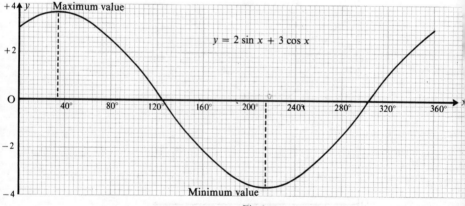

Fig 6.19

Copy the table and continue it to 360°. Plot the points, using a horizontal scale of 1 inch for 40° and a vertical scale of 1 inch for 2 units of y. Draw the graph of $y = 2 \sin x + 3 \cos x$; the result is shown in fig 6.19.

The graph shows the shape of the curve, but is not suitable for accurate work. For example, if the maximum and minimum values of y are required, the points can be determined only to the nearest 2°; the points are $x = 34°$ for a maximum and $x = 214°$ for a minimum. To obtain a more accurate result, the portions of the graph near to these points must be drawn with an enlarged scale. A table is drawn up for each required solution, using values within 2° of the indicated value of x.

x	32°	33°	34°	35°	36°
$2 \sin x$	+1·0598	+1·0892	+1·1184	+1·1472	+1·1756
$3 \cos x$	+2·5440	+2·5161	+2·4870	+2·4576	+2·4270
y	+3·6038	+3·6053	+3·6054	+3·6048	+3·6026

Mark a vertical scale of 1 inch for 0·0010 units of y, commencing at 3·6035; mark a horizontal scale of 1 inch for 1°, commencing at 32°. Plot the points and join them in a smooth curve; mark the maximum and read the corresponding values of x and y. The function has a maximum of 3·6056 units at 33° 36′, as shown in fig 6.20. Now carry out the same procedure for the minimum value of the function.

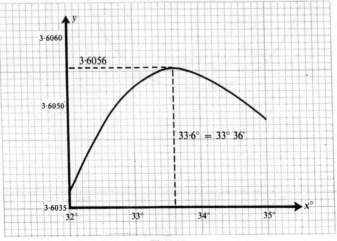

Fig 6.20

4 Reciprocal functions

An example of a reciprocal function is $y = \frac{1}{2} + \frac{1}{\sin x}$; the table below gives values for the expression between the limits of $0°$ and $180°$ for values of x.

x	$0°$	$20°$	$40°$	$60°$	$80°$	$90°$	$100°$	$120°$	$140°$	$160°$	$180°$
$\sin x$	0	0·34	0·64	0·87	0·98	1·00	0·98	0·87	0·64	0·34	0
$\frac{1}{\sin x}$	∞	2·94	1·56	1·15	1·02	1·00	1·02	1·15	1·56	2·94	∞
y	∞	3·44	2·06	1·65	1·52	1·50	1·52	1·65	2·06	3·44	∞

Mark a vertical scale of 1 inch for 1 unit of y, and a horizontal scale of 1 inch for $40°$. Plot the points and join in a smooth curve. The result is shown in fig 6.21.

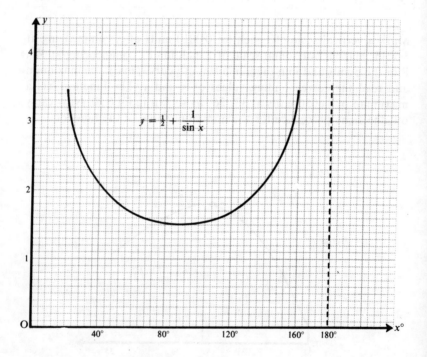

Fig 6.21

Exercise 6.4

Draw graphs of the following trigonometrical functions for values of x between $0°$ and $360°$. In each case take 1 in (or 2 cm) to represent $40°$ (or $30°$) for x. For y, take 1 in (or 2 cm) to represent 1 unit or 2 units as convenient.

1 $y = 1 + \cos 3x$ **2** $y = 3 - 2 \sin x$

3 $y = \tan x + \cos x$ **4** $y = \sin 2x - \cos 3x$

5 $y = 1 + \sin 2x - \cos x$ **6** $y = 3 \sin x + 3 \cos x$

7 $y = 2 - \dfrac{1}{\cos x}$ **8** $y = 2 \sin 3x - 3 \cos 2x$

Graphical Solution of Trigonometrical Equations

A solution of the equation $\sin \theta = 0.2$ has been given in fig 6.16. This principle can be extended to more complicated expressions. For example, $2 \sin x + 3 \cos x = 0.4$ is solved by (**1**) plotting the curve for the expression $y = 2 \sin x + 3 \cos x$, and (**2**) drawing a horizontal line for $y = 0.4$ to obtain the solution, as shown in fig 6.22. The points, A and B, where the curve and the line intersect satisfy (**a**) $y = 0.4$ and (**b**) $y = 2 \sin x + 3 \cos x$; therefore these points satisfy the condition $2 \sin x + 3 \cos x = 0.4$. Hence the approximate solutions are $117°$ and $312°$.

The equation $2 \sin x + 3 \cos x = 0$ is also solved from the graph in fig 6.22. The condition for this equation is $y = 0$, i.e. the solution is given by the intersection of the curve and the x-axis at the points C and D. The graphical solutions are $x = 125°$ or $304°$. More accurate

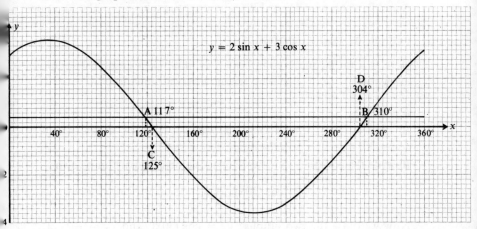

Fig 6.22

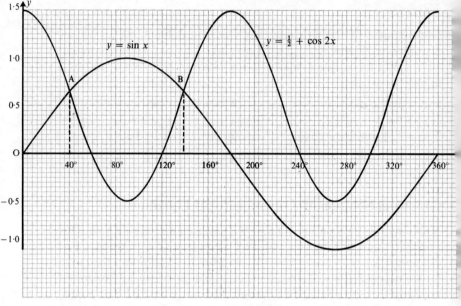

Fig 6.23

solutions can be obtained by redrawing the relevant portions of the
curve, near to the approximate solution, with enlarged scales as
described on page 69.

Solutions of equations can also be found from the intersection of
two curves. Consider the equations $y = \sin x$, $y = \frac{1}{2} + \cos 2x$. If
both curves are plotted, as shown in fig 6.23, they intersect at points
A and B, where $x = 41°$ or $139°$. These are the solutions of the equation
$\sin x = \frac{1}{2} + \cos 2x$, as at these two points y has the same value for
both of the original expressions.

Here is a rather more complicated example of a graphical solution,
involving a mixed algebraic trigonometrical equation.

Example 5

By drawing two graphs find approximate solutions of the equation
$x \cos (30x)° = 2 - x$ *in the range* $0 \leqslant x \leqslant 6$.

Rewrite the equation to leave the trigonometrical part on its own:

$$\cos (30x)° = \frac{2}{x} - 1$$

Now we draw the two graphs: $y_1 = \cos (30x)°$ and $y_2 = \frac{2}{x} - 1$.

A table of values for the given range is compiled.

x	0	1	1·5	2	3	4	5	6
$y_1 = \cos(30x)°$	1	0·87	0·71	0·50	0	−0·50	−0·87	−1
$y_2 = \dfrac{2}{x} - 1$	∞	1	0·33	0	−0·33	−0·50	−0·60	−0·67

Choosing suitable scales, these graphs are drawn with the same axes (fig 6.24). The points of intersection (where $y_1 = y_2$) give the roots of the equation $\cos(30x)° = \dfrac{2}{x} - 1$ as $x = 1·05$ (approx) and $x = 4$ in this range. A more accurate value of the first root could be obtained by drawing the two graphs between $x = 1$ and $x = 2$ to a larger scale.

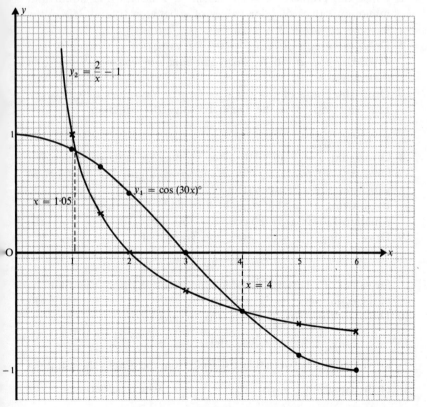

Fig 6.24

Exercise 6.5

1 Draw, on the same axes, the graphs of $y = \tan x$ and $y = 2 \sin x$ for values of x between $0°$ and $180°$. From the graphs, obtain an approximate solution to the equation $\tan x = 2 \sin x$ in that range.

2 Draw the graph of $y = \cos 3x$ for values of x between $0°$ and $360°$. Use the graph to solve the equation $\cos 3x = +0.8660$.

3 Draw the graph of $y = 4 \cos 2x$ for values of x between $0°$ and $180°$. Use the graph to solve the equation $4 \cos 2x = 1$.

4 Draw, on the same axes, the graphs of
$$y = 2 \sin x \text{ and } y = 1 + \frac{1}{\sin x}$$
for values of x between $180°$ and $360°$. Use the graph to solve the equation $2 \sin x = 1 + \dfrac{1}{\sin x}$ in that range.

5 Draw the graph of $y = 3 + 2 \sin x$. Find the maximum and minimum values of y for values of x between $0°$ and $360°$. Why has the equation $3 + 2 \sin x = 0$ no roots?

6 Draw the graph of $\cos(x - 10)°$ for values of x between $0°$ and $180°$. Use the graph to solve the equation $\cos(x - 10)° = \frac{1}{3}$ in that range.

7 Draw the graph of $y = \cos x - 7 \sin x$ for values of x between $180°$ and $360°$. Use the graph to solve the equation
$$\cos x - 7 \sin x = 2$$
in that range.

8 Given that $y = 3 \sin x° + 2 \cos x°$, make a table of values of y for $x = 0, 30, 60, \ldots\ldots, 270$. Hence draw a graph of the function y for values of x between 0 and 270.
By drawing a suitable straight line, find all the solutions of the equation $120 \sin x° + 80 \cos x° = 120 - x$.   (L)

9 Draw, using the same axes, the graphs of
$$y = \cos 30x° \text{ and } y = x - \frac{x^2}{5}$$
for values of x between 0 and 6. Find from the graphs the solutions of the equation $x^2 - 5x + 5 \cos 30x = 0$.   (L)

10 Draw in the same diagram the graphs of
$$y = \cos 30x° \text{ and } y + 1 = \frac{7}{2x + 3}$$
for values of x between 0 and 6. Find from the graphs the solutions of the equation $1 + \cos 30x° = \dfrac{7}{2x + 3}$ which lie between 0 and 6.   (L)

The Remaining Trigonometrical Ratios

The three remaining ratios are the reciprocals of the sine, cosine and tangent. They are:

$$\text{secant} = \frac{1}{\text{cosine}} \; ; \; \text{cosecant} = \frac{1}{\text{sine}} \; ; \; \text{cotangent} = \frac{1}{\text{tangent}} \, .$$

The values of these ratios in the four quadrants are determined from the reciprocals of the principal ratios.

These ratios are used in problems involving calculations where it is more convenient to use their values, read from tables, than it is to use the reciprocals of the three principal ratios. This use will be seen in Chapter 13 involving calculations in three dimensions.

Further Trigonometrical Identities

$$\sin^2\theta + \cos^2\theta = 1.$$

Now divide both sides by $\cos^2\theta$:

$$\frac{\sin^2\theta}{\cos^2\theta} + \frac{\cos^2\theta}{\cos^2\theta} = \frac{1}{\cos^2\theta} \, ;$$

but $\dfrac{\sin\theta}{\cos\theta} = \tan\theta$ and $\dfrac{1}{\cos\theta} = \sec\theta$.

Therefore
$$\boxed{\tan^2\theta + 1 = \sec^2\theta}$$

Similarly, by dividing the original identity by $\sin^2\theta$, we obtain

$$\boxed{1 + \cot^2\theta = \csc^2\theta}$$

These identities will be found useful later in solving equations.

Circular Measure

For all practical purposes, e.g. surveying, navigation, etc., angles are measured in degrees with subdivision into minutes and seconds. The choice of 360° for one complete revolution is arbitrary, and was based on the method of counting in Babylonia. (A proposed system of angular measurement in the metric system divides a right angle into 100 grades with subdivisions of 100 centigrades.) Both these systems, the sexagesimal and the metric, are based on an arbitrary unit defined either from the right angle, or from one complete revolution. The third system of angular measurement, circular measure, has no units; it is the system mainly used in mathematics.

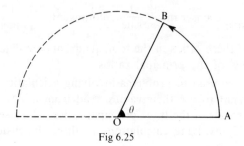

Fig 6.25

In fig 6.25, the arm rotates from the position OA to the position OB, describing an angle, θ, which is $A\hat{O}B$. In circular measure, the angle is measured by:

$$\theta = \frac{\text{arc AB}}{\text{radius OA}} = \frac{\text{arc AB}}{r} \text{ where } r \text{ is the radius.}$$

This definition of the size of an angle is without units, as the stated measurement is a ratio, i.e. a pure number.

If a unit arc is taken on a circle with a unit radius, then the value of θ in circular measure will be unity. This angle is called a **radian** (rad). The length of an arc is proportional to the angle it subtends at the centre of a circle. A semicircle subtends an angle of 180° at the centre, so the length of the arc is πr. Hence an angle of 180° measured in radians is given by:

$$\theta = \frac{\text{arc}}{\text{radius}} = \frac{\pi r}{r} = \pi \text{ rad}$$

If π rad are equivalent to 180°, then 1 rad is equivalent to $\frac{180}{\pi}$ degrees, which is 57° 17′ 45″ approximately. As π is *irrational*, (i.e. its value cannot be exactly represented by a fraction) any conversion between radians and degrees, must always give an approximate answer. The relation between circular and sexagesimal measure is given by:

$$\boxed{\pi \text{ rad } = 180 \text{ degrees}}$$

Tables of radians and degrees are available; otherwise use the relationship above. To five significant figures, $1° = 0.017\,453$ radians.

The angles for the quadrants, given in circular measure, are:

$$90° = \frac{\pi}{2} \text{ rad}; \ 180° = \pi \text{ rad}; \ 270° = \frac{3\pi}{2} \text{ rad}; \ 360° = 2\pi \text{ rad}.$$

Circular measure, used with trigonometrical ratios, is written as shown in the following examples:

$$\sin \frac{\pi}{2} = 1; \ \cos \pi = -1; \ \sin \frac{\pi}{6} = \tfrac{1}{2}; \ \tan \frac{\pi}{4} = 1; \ \cos \frac{\pi}{3} = \tfrac{1}{2};$$

$$\tan \frac{3\pi}{4} = -1; \ \sin 2\pi = 0.$$

Draw the sine and cosine curves, using a horizontal axis of 2 cm for 60°, and labelling the axis in radians, i.e. 6 cm for π rad. Use a vertical axis of 4 cm for 1 unit, and plot both curves on the same axes. Note that general angles can easily be written in circular measure, for example $(2n\pi + x)$ rad, where x is the basic angle and n is a whole number.

Arcs and sectors

(1) From the definition of circular measure, the length of an arc of a circle is derived from the equation:

$$\text{Arc} = r\theta \ (\theta \text{ in radians})$$

(2) Turn to fig 6.25; the area of the semicircle is $\frac{1}{2}\pi r^2$. The area of the sector AOB is proportional to the angle θ, as the area of any sector is proportional to its angle at the centre of the circle.

$$\therefore \ \frac{\text{Area of sector AOB}}{\text{Area of semicircle}} = \frac{\theta}{\pi} \text{ (both angles expressed in radians)}$$

$$\therefore \ \text{Area of sector AOB} = \frac{\theta}{\pi} \times \tfrac{1}{2}\pi r^2 = \tfrac{1}{2}r^2\theta$$

Example 6
The area of a sector of a circle is one-fifth of the area of the circle. If the radius of the circle is 25 cm, calculate the angle of the sector and the length of its arc.

If the area of the sector $= \frac{1}{5}$ (area of circle),
then the angle of the sector $= \frac{1}{5} (2\pi)$.

Hence the angle $= \dfrac{2\pi}{5} \text{ rad} = \dfrac{2 \times 3\cdot14}{5} = 1\cdot26 \text{ rad}.$

$\therefore$ The length of arc $= 1\cdot26 \times 25 = 31\cdot5$ cm.

Example 7

Draw the graphs of y = 3 sin x and y = x for values of x from 0 to π.

Take scales of 4 cm for 1 unit on the positive y-axis and 4 cm for $\frac{\pi}{6}$ rad

on the x-axis. Use the graph to solve the equation 3 sin x = x in this range.

Mark off the axes (Fig 6.26). Now compile a table of values for
y = 3 sin x.

x	0	$\frac{\pi}{12}$	$\frac{\pi}{6}$	$\frac{\pi}{4}$	$\frac{\pi}{3}$	$\frac{\pi}{2}$	$\frac{2\pi}{3}$	$\frac{5\pi}{6}$	π
		(15°)	(30°)	(45°)	(60°)	(90°)			
y	0	0·776	1·5	2·12	2·6	3	2·6	1·5	0

Plot these points and join up to form the curve.

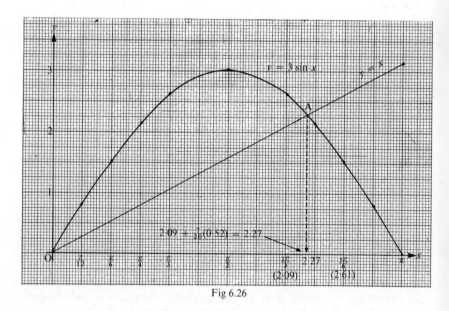

Fig 6.26

The line *y = x* passes through the origin and the point *x = π, y = π*
(π ≈ 3·14).

Mark these points and draw the line. The graphs intersect at O and A,
giving the solutions of 3 sin x = x as x = 0 and x = 2·27 (approx.). For
convenience, the table above was worked out in degree measurement

but radian measure must be used for the x-axis, the algebraic graph and the results. A statement such as 3 sin 120° = 120 would be meaningless.

As a check, 2·27 rad ≈ 130° and 3 sin 130° ≈ 2·30 which gives reasonably close agreement.

Exercise 6.6

Use tables to find the values of the following ratios:

1 cosec 70° 15′	**2** cot 232° 24′	**3** sec 104° 12′
4 cot 164° 28′	**5** sec 216° 48′	**6** cot 332° 35′

Convert the following angles to radians:

7 37° 24′	**8** 200°	**9** 38° 29′

Convert the following angles to sexagesimal measure (correct to the nearest minute):

10 1·4 rad	**11** 4·6 rad	**12** 10 rad

13 What is the angle of a sector with radius 20 cm, and length of arc 16 cm? Give your answer (i) in radians; (ii) in degrees and minutes.

14 A circle of radius 5 cm has an arc which subtends 77° 30′ at the centre of the circle. Calculate the length of the arc, to 3 significant figures.

15 An arc subtends an angle of 0·45 rad in a circle of radius 8 cm. Calculate the area of the sector.

16 The angle of a sector cut out of paper is 1·0 rad; the radius of the sector is 24 cm. The sector is folded so that the bounding radii are joined to form a right circular cone. Calculate the area of the base of this cone, correct to 2 significant figures.

17 Draw the graphs of $y = \sin x$ and $y = \dfrac{x}{2}$ for values of x from 0 to π, taking scales of 4 cm for 1 unit on the positive y-axis and 4 cm for $\dfrac{\pi}{6}$ rad on the x-axis. Use your graphs to solve approximately the equation $x = 2 \sin x$ in this range.

18 With the same axes (scales as in Question **17**) draw the graphs of $y = \cos x$ and $y = x$, taking x from 0 to $\dfrac{\pi}{2}$. Hence find an approximate solution of the equation $\cos x = x$.

19 Draw the graphs of $y = \tan x$ and $2y = 3x$ for the range $0 \leqslant x \leqslant \dfrac{\pi}{3}$, using the same scales as in Question **17**. Hence solve the equation $2 \tan x = 3x$ in this range.

20 Draw the graphs of $y = \sin x$ and $y = 1 - x^2$ from $x = 0$ to $x = \dfrac{\pi}{2}$.
Take scales of 10 cm for 1 unit on the positive y-axis and 4 cm
for $\dfrac{\pi}{6}$ rad on the x-axis. Find the solution of the equation
$1 - x^2 = \sin x$ in this range.

Trigonometrical Equations

Simple equations involving only one ratio, such as $\sin x = 0.34$, have
been discussed on p. 72 and a solution obtained by using tables.

Equations involving more than one ratio

Consider the equation $4 \sin \theta = \tan \theta$. Two ratios are involved, each
of the first degree. Replacing $\tan \theta$ by $\dfrac{\sin \theta}{\cos \theta}$ the equation can be
factorized and thus reduced to two simple equations.

$4 \sin \theta - \tan \theta = 0$ so $4 \sin \theta - \dfrac{\sin \theta}{\cos \theta} = 0$

$$\therefore \sin \theta \left(4 - \frac{1}{\cos \theta}\right) = 0$$

which gives

$\sin \theta = 0$ \qquad or \qquad $4 - \dfrac{1}{\cos \theta} = 0$

i.e. $\theta = 0°$ or $180°$ \qquad leading to $\cos \theta = \frac{1}{4}$

\qquad\qquad\qquad\qquad\qquad i.e. $\theta = 75° \ 31'$ or $284° \ 29'$

Therefore in the range $0° \leqslant \theta < 360°$ the solutions of the equation
are $0°$, $75° \ 31'$, $180°$, $284° \ 29'$.

Second degree equations

This type of equation involves only one ratio and it is similar to the
normal quadratic equation, for example,
$$\sin^2 \theta - 2 \sin \theta + 1 = 0.$$
This is similar to $x^2 - 2x + 1$, and factorizes in the same way.
$\sin^2 \theta - 2 \sin \theta + 1 = (\sin \theta - 1)(\sin \theta - 1) = (\sin \theta - 1)^2 = 0$
$\therefore \sin \theta - 1 = 0$ i.e. $\sin \theta = 1$, hence $\theta = 90°$. This is a unique solution
as no other angle between $0°$ and $360°$ has a sine of $+1$.

Example 8
Find the values of x between $0°$ and $360°$ which satisfy the equation
$15 \sin^2 x - \sin x - 6 = 0$

$$15 \sin^2 x - \sin x - 6 = (3 \sin x - 2)(5 \sin x + 3) = 0$$

∴ $3 \sin x - 2 = 0$ or $5 \sin x + 3 = 0$

i.e. $\sin x = \frac{2}{3}$ or $\sin x = -\frac{3}{5}$

$0.6667 = \sin 41° \, 49'$ from the tables

∴ $x = 41° \, 49'$ or $138° \, 11'$

$0.6 = \sin 36° \, 52'$ from the tables

∴ $x = 216° \, 52'$ or $323° \, 08'$.

Hence x can have the values of $41° \, 49'$; $138° \, 11'$; $216° \, 52'$; $323° \, 08'$.

Example 9

Find the values of x between $0°$ and $360°$ which satisfy the equation
$4 \cos^2 x - 11 \cos x + 6 = 0$.

$$4 \cos^2 x - 11 \cos x + 6 = (\cos x - 2)(4 \cos x - 3) = 0$$

∴ $\cos x = 2$ or $\frac{3}{4}$

The root, $\cos x = 2$, is discarded as it has no solution.

Hence $\cos x = \frac{3}{4}$ is the only possible root.

$0.7500 = \cos 41° \, 24'$ from the tables.

∴ $x = 41° \, 24'$ or $318° \, 36'$.

If a second degree equation involves more than one ratio, then the ratios have to be converted to one ratio only, using identities.

Consider the equation $6 \sin^2 \theta + \cos \theta - 4 = 0$. This is a second degree equation because $\sin^2 \theta$ is included. It involves more than one ratio as both $\sin \theta$ and $\cos \theta$ are included in the equation. Identities have now to be used to convert the equation to one involving one ratio only.

Now $\sin^2 \theta$ can be easily related to $\cos^2 \theta$, but $\cos \theta$ cannot be readily related to $\sin \theta$. Hence the equation is changed to one involving powers of $\cos \theta$ only. Since $\sin^2 \theta + \cos^2 \theta = 1$, $\sin^2 \theta = 1 - \cos^2 \theta$.

∴ $6 \sin^2 \theta = 6 - 6 \cos^2 \theta$

Substituting in the equation,

$$6 - 6 \cos^2 \theta + \cos \theta - 4 = 0$$
$$2 + \cos \theta - 6 \cos^2 \theta = 0$$
$$(2 - 3 \cos \theta)(1 + 2 \cos \theta) = 0$$
$$\therefore \cos \theta = \frac{2}{3} \text{ or } -\frac{1}{2}$$

If $\cos \theta = \frac{2}{3} = 0.6667$, $\theta = 48° \, 11'$ or $311° \, 49'$

If $\cos \theta = -0.5$, $\theta = 120°$ or $240°$

The solutions are $48° \, 11'$, $120°$, $240°$, $311° \, 49'$.

Example 10

Solve for values of x *between* $0°$ *and* $360°$, *the equation*
$$2 \cot^2 x + 11 = 9 \operatorname{cosec} x.$$

$$1 + \cot^2 x = \operatorname{cosec}^2 x \qquad \text{identity relating ratios}$$
$$\therefore \qquad 2 \cot^2 x = 2 \operatorname{cosec}^2 x - 2$$

Substituting in the equation,
$$2 \operatorname{cosec}^2 x - 2 + 11 = 9 \operatorname{cosec} x$$

Rearranging as a quadratic equation,
$$2 \operatorname{cosec}^2 x - 9 \operatorname{cosec} x + 9 = 0$$

Factorizing,
$$(2 \operatorname{cosec} x - 3)(\operatorname{cosec} x - 3) = 0$$

$$\operatorname{cosec} x = \tfrac{3}{2} \text{ or } 3 \ldots \ldots \text{ (Natural cosecant tables could}$$
$$\text{be used to find } x)$$

i.e. $\qquad\qquad\qquad \sin x = \tfrac{2}{3} \text{ or } \tfrac{1}{3}$

$0.6667 = \sin 41° 49' \qquad \therefore x = 41° 49' \text{ or } 138° 11'$

$0.3333 = \sin 19° 29' \qquad \therefore x = 19° 29' \text{ or } 160° 31'$

Hence the values of x are $19° 29'$; $41° 49'$; $138° 11'$; $160° 31'$.

Example 11

Solve for values of x *between* $0°$ *and* $360°$, *the following equation*:
$$2 \sin x = 3 - \frac{1}{\sin x}.$$

The equation has first to be rearranged to form a quadratic equation in $\sin x$. Multiply throughout by $\sin x$.

Then $2 \sin^2 x = 3 \sin x - 1$

Rearranging in quadratic form,
$$2 \sin^2 x - 3 \sin x + 1 = 0$$

Hence $\qquad (2 \sin x - 1)(\sin x - 1) = 0$

$\therefore \qquad\qquad\qquad\qquad \sin x = \tfrac{1}{2} \text{ or } 1$

$0.5000 = \sin 30° \qquad \therefore x = 30° \text{ or } 150$

$1.0000 = \sin 90° \qquad \therefore x = 90° \text{ (unique solution)}$

The values of x are: $30°, 90, 150°$.

Exercise 6.7

Solve the following equations for values of x between $0°$ and $360°$.

1 $5 \cos x = \cot x$ **2** $\tan x = \sin x$

3 $3 \tan x = 2 \sec x$ **4** $\sin^2 x = \tfrac{1}{4}$

5 $\tan^2 x = \tfrac{1}{3}$ **6** $\sin^2 x + 2 \sin x + 1 = 0$

7 $2 \cos^2 x + 3 \cos x + 1 = 0$

8 $2 \sin x = 1 + \dfrac{1}{\sin x}$

9 $\cot^2 x = 2 + \cot x$

10 $5 \sin^2 x - 17 \sin x + 6 = 0$

11 $\cos^2 x + \sin x + 1 = 0$

12 $\sec^2 x - 3 \tan x + 1 = 0$

13 $2 \cot^2 x - 7 \operatorname{cosec} x + 8 = 0$

14 $3 \cos^2 x = 2 \sin x \cos x$

15 $2 \cot x + \tan x - 3 = 0$

16 $2 \sin^2 x + 3 \sin x = 2$

17 $\cot x = \tan x$

18 $3 \cos^2 x = 1$

19 $\cos x + \sin x = \dfrac{1}{\cos x - \sin x}$

20 $\cot x = 3 - \tan x$

21 $3 \cos^2 x = 2 \cos x$

22 $5 \cos^2 x + 5 \sin x = 1$

23 $\sec^2 x + \tan x = 3$

24 $\sin x = 2 \cos x$

25 $3 \tan x + 4 = \dfrac{2}{\cos^2 x}$

7 The Solution of Triangles

A triangle possesses six elements – the three sides and the three angles. If any three elements (other than three angles) are given, the remaining three elements can be found. This is called solving the triangle. If three angles are given, an infinite number of similar triangles can be formed, so there is no definite solution of the triangle.

In solving triangles, two geometrical facts are useful. They are:

(1) In any triangle, the sum of the angles is 180°.

(2) In any triangle, the greatest side is opposite the greatest angle, and the least side is opposite the smallest angle.

Relationships between the sides and the angles must now be established.

The Sine Rule

In fig 7.1a, a triangle ABC has a circumcircle described around it. Consider angle A. O is the centre of the circumcircle, and BP is a diameter of the circle. PC is joined. Angle BCP is a right angle as it is the angle in a semicircle. Let the radius of the circumcircle be R.

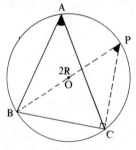

Fig 7.1a

Now $\widehat{BAC} = \widehat{BPC}$ (angles in same segment)

∴ $\sin A = \sin \widehat{BPC}$

but $\quad \sin \widehat{BPC} = \dfrac{BC}{BP}$ (from triangle BPC, right-angled at C)

$\qquad\qquad\quad = \dfrac{a}{2R}$ where a is the side opposite angle A

∴ $\qquad \sin A = \dfrac{a}{2R}$

i.e. $\qquad 2R = \dfrac{a}{\sin A}$.

If A *is an obtuse angle*
Fig 7.1b shows $\widehat{BAC}$ as obtuse. The construction is identical with that for an acute angle in fig 7.1a.

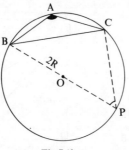

Fig 7.1b

Now $\widehat{BAC} + \widehat{BPC} = 180°$ (opp. angles of cyclic quad.)
$$\therefore \widehat{BAC} = 180° - \widehat{BPC}$$
$$\sin \widehat{BAC} = \sin (180° - \widehat{BPC})$$
$$= \sin \widehat{BPC}$$
$$\sin \widehat{BPC} = \frac{BC}{BP} \quad \text{(from triangle BPC, right-angled at C)}$$
$$= \frac{a}{2R}$$
$$\therefore \sin A = \frac{a}{2R} \quad \text{i.e. } 2R = \frac{a}{\sin A}.$$

Similarly, by considering $\widehat{B}$ or $\widehat{C}$, it can be proved that
$$\frac{b}{\sin B} = 2R \quad \text{and} \quad \frac{c}{\sin C} = 2R,$$
where $b = AC$, the side opposite angle B, and $c = AB$, the side opposite angle C.

Hence

$$\boxed{\frac{a}{\sin A} = \frac{b}{\sin B} = \frac{c}{\sin C} = 2R.}$$

This is the sine rule and should be memorised.

Note: this rule can also be used in the form: $\dfrac{\sin A}{a} = \dfrac{\sin B}{b} = \dfrac{\sin C}{c}$

The Cosine Rule

Fig 7.2 shows a triangle ABC. Consider angle A which is acute in fig 7.2a and obtuse in fig 7.2b. An altitude, CN, is drawn, of length h.

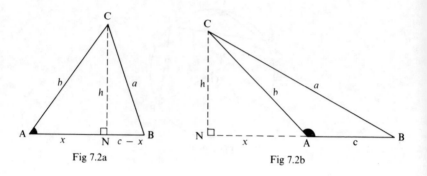

Fig 7.2a   Fig 7.2b

When $\widehat{A}$ is acute

$$a^2 = h^2 + (c - x)^2 \qquad \text{(from triangle CNB)}$$
$$= h^2 + c^2 - 2cx + x^2$$
$$b^2 = h^2 + x^2 \qquad \text{(from triangle CNA)}$$
$$\therefore \quad a^2 = b^2 + c^2 - 2cx.$$

Also $\cos A = \dfrac{x}{b}$

$$\therefore x = b \cos A$$
$$\therefore a^2 = b^2 + c^2 - 2bc \cos A.$$

When $\widehat{A}$ is obtuse

$$a^2 = h^2 + (c + x)^2 \qquad \text{(from triangle CBN)}$$
$$= h^2 + c^2 + 2cx + x^2.$$
$$b^2 = h^2 + x^2 \qquad \text{(from triangle CAN)}$$
$$\therefore a^2 = b^2 + c^2 + 2cx.$$

Also $\cos \widehat{CAN} = \dfrac{x}{b}$

$$\therefore \quad x = b \cos \widehat{CAN} = b \cos (180° - A) = -b \cos A$$
$$\therefore \quad a^2 = b^2 + c^2 - 2bc \cos A$$

Hence in either triangle $a^2 = b^2 + c^2 - 2bc \cos A$.

This is the cosine rule for angle A. By taking angles B and C, two similar formulae can be derived; the three formulae are:

$$a^2 = b^2 + c^2 - 2bc \cos A$$
$$b^2 = a^2 + c^2 - 2ac \cos B$$
$$c^2 = a^2 + b^2 - 2ab \cos C.$$

These formulae should be memorised.

The Solution of Triangles

Case I   Two sides and the included angle

Fig 7.3 shows a triangle ABC in which two sides (b and c) and the included angle (A) are given. This triangle is solved by first applying the cosine rule; in this triangle $a^2 = b^2 + c^2 - 2bc \cos A$ from which side a is calculated. Angle B is then calculated from $\dfrac{\sin B}{b} = \dfrac{\sin A}{a}$, giving $\sin B = \dfrac{b \sin A}{a}$.

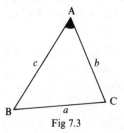

Fig 7.3

The third angle (c) can now be found by subtracting the sum of the angles A and B from 180°.

Case II   Three sides

The triangle is solved by first applying the cosine rule in the form $\cos A = \dfrac{b^2 + c^2 - a^2}{2bc}$ from which angle A is calculated. Angle B (or C) can be found using the sine rule. Note that the remaining angles could be calculated by the cosine rule, but the sine rule is (a) easier to use, with less calculation and (b) the sine rule is more convenient when using logarithms.

In any example, always sketch the triangle and mark in the given information; next, use the appropriate formula to solve the triangle, or obtain a particular piece of information.

Example 1

In a triangle XYZ, YZ = 6·7 cm, XY = 2·3 cm, XŶZ = 46° 32'.
Calculate XẐY.

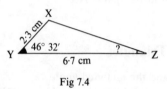

Fig 7.4

First draw the triangle, marking in the information, as shown in
fig 7.4. This is an example of two sides and the included angle being
given. Now XZ must first be found, using the cosine rule, and then
Ẑ found from the sine rule.

$$XZ^2 = YZ^2 + XY^2 - 2.YZ.XY \cos X\hat{Y}Z$$
$$= 6·7^2 + 2·3^2 - 2 \times 6·7 \times 2·3 \cos 46° 32'$$
$$= 44·89 + 5·29 - 4·6 \times 6·7 \cos 46° 32'$$
$$= 50·18 - 21·20$$
$$= 28·98$$
$$\therefore XZ = 5·383$$

No.	Log.
4·6	0·6628
6·7	0·8261
cos 46° 32'	$\bar{1}$·8375
21·20	1·3264

$$\sin X\hat{Z}Y = \frac{XY \sin X\hat{Y}Z}{XZ}$$

$$= \frac{2·3 \sin 46° 32'}{5·383}$$

∴ XẐY = 18° 04' or 161° 56'.
As XY is the smallest side of the triangle, XẐY
must be the smallest angle.
Hence XẐY = 18° 04'.

No.	Log.
2·3	0·3617
sin 46° 32'	$\bar{1}$·8608
	0·2225
5·383	0·7310
sin 18° 04'	$\bar{1}$·4915

(**Note**: log cos and log sin have been used in the working).

Example 2

In the triangle PQR, QR = 4 cm, PR = 5 cm, PQ = 7 cm. *Calculate*
the size of the largest angle in the triangle.

First draw the triangle, marking in the information, as shown in fig
7.5. The required angle is QR̂P, opposite the side of 7 cm (the longest
side).

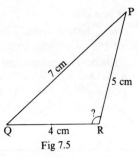

Fig 7.5

$$\cos Q\hat{R}P = \frac{QR^2 + PR^2 - PQ^2}{2\,QR.PR} \qquad \text{from the cosine rule}$$

$$= \frac{4^2 + 5^2 - 7^2}{2 \times 4 \times 5}$$

$$= \frac{16 + 25 - 49}{40}$$

$$= \frac{-8}{40} = -0.2000$$

$$0.2 = \cos 78° \ 28'$$

$$\therefore \ Q\hat{R}P = 180° - 78° \ 28' = 101° \ 32'.$$

Exercise 7.1

1 In a triangle PQR, PR = 40 cm, PQ = 50 cm, $Q\hat{P}R$ = 66°. Calculate the length of side QR.

2 If C = 10° 37', a = 156 m, b = 146 m, calculate the magnitude of angle B.

3 In triangle DEF, EF = 10 cm, DE = 9·87 cm, $D\hat{E}F$ = 29° 09'. Calculate $E\hat{D}F$.

4 In triangle ABC, a = 2x, b = 3x, B = 95°. Find c, and calculate the smallest angle.

5 In triangle ABC, side a = 6·53 cm, side b = 2·4 cm, C = 26° 14'. Calculate the remaining side and angles of the triangle.

6 The sides of a triangle are 100 cm, 90 cm, and 50 cm. Calculate the magnitude of the largest angle.

7 The sides of a triangle are 7p, 8p, 5p. Calculate the size of the smallest angle of the triangle.

8 The triangle ABC has sides with the following measurements: 0·6 km, 0·9 km, 1 km. Calculate the three angles of the triangle.

9 The sides of a triangle are 0·7x, 1·5x, 1·1x. Calculate the angles of the triangle.

Case III   Two angles and one side

Fig 7.6 illustrates the given information. The third angle is found
by subtracting the sum of the two given angles from 180°. The sine
rule is then used to determine the remaining two sides.

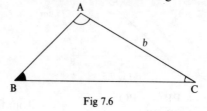

Fig 7.6

Example 3

In triangle ABC, A = 59°, B = 39°, *a* = 6·73 cm. *Find the length
of the smallest side.*

First sketch the triangle with the given information (fig 7.7).

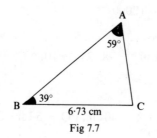

Fig 7.7

$$C = 180° - (39° + 59°) = 82°$$

Since B is the smallest angle, *b* is the smallest side.

$$\frac{b}{\sin B} = \frac{a}{\sin A}$$

$$\frac{b}{\sin 39°} = \frac{6·73 \text{ cm}}{\sin 59°}$$

$$\therefore b = \frac{6·73 \text{ cm} \times \sin 39°}{\sin 59°}$$

$$= 4·94 \text{ cm}.$$

No.	Log.
6·73	0·8280
sin 39°	$\bar{1}$·7989
	0·6269
sin 59°	$\bar{1}$·9331
4·941	0·6938

Case IV   Two sides and a non-included angle

The given information is illustrated in fig 7.8. In this diagram, the
side opposite the non-included angle is smaller than the given side
(AB). The geometrical construction to complete the triangle is shown.

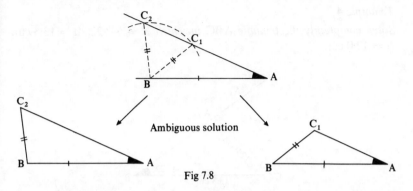

Fig 7.8

Two positions, for point C can be found, and hence two triangles, as illustrated, can be drawn from this information. This is known as the "ambiguous" case. Concentrate on the initial figure; the triangle BC_1C_2 is isosceles, hence $B\hat{C}_1C_2 = B\hat{C}_2C_1$.

Now $B\hat{C}_1 A = 180° - B\hat{C}_1C_2$

Hence $\sin B\hat{C}_1 A = \sin (180° - B\hat{C}_1C_2) = \sin B\hat{C}_1C_2$

Using the sine rule to solve the triangle will not by itself distinguish between the two solutions. The given information must be examined to decide whether the ambiguous case is involved. If it is, two values of C are found, one acute and the other obtuse. The third angle correspondingly has two values, and from these two values for the angles not originally given, the two triangles are solved using the sine rule.

The ambiguous case is not involved if:

(1) The side opposite the non-included angle is greater than the other given side.

(2) The given angle is obtuse.

These two sets of conditions are illustrated in fig 7.9. The ambiguous case is involved if the side opposite the non-included angle is smaller than the other given side and the given angle is acute.

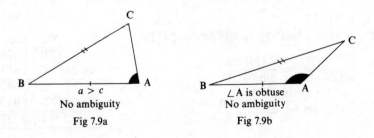

No ambiguity

Fig 7.9a   Fig 7.9b

Example 4

Solve completely the triangle ABC *in which* $B = 64° 22'$, $a = 13\cdot3$ cm, $b = 12\cdot0$ cm.

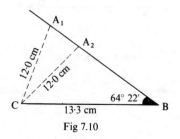

Fig 7.10

First sketch the triangle, as shown in fig 7.10. Since $b < a$, the ambiguous case is involved, and two triangles must be solved.

Now $\dfrac{\sin A}{a} = \dfrac{\sin B}{b}$

No.	Log.
13·3	1·1239
sin 64° 22′	$\overline{1}$·9550
	1·0789
12·0	1·0792
sin 88° 36′	$\overline{1}$·9997

$\therefore \dfrac{\sin A}{13\cdot3} = \dfrac{\sin 64° 22'}{12\cdot0}$

$\therefore \sin A = \dfrac{13\cdot3 \times \sin 64° 22'}{12\cdot0}$

$\therefore A = 88° 36'$ or $180° - 88° 36'$
$= 88° 36'$ or $91° 24'$

$A_1 = 88° 36'$
$\therefore C_1 = 180° - (64° 22' + 88° 36') = 27° 02'$

No.	Log.
12·0	1·0792
sin 27° 02′	$\overline{1}$·6575
	0·7367
sin 64° 22′	$\overline{1}$·9550
6·049	0·7817

$\therefore \dfrac{c_1}{\sin 27° 02'} = \dfrac{12\cdot0 \text{ cm}}{\sin 64° 22'}$

$c_1 = \dfrac{12\cdot0 \times \sin 27° 02'}{\sin 64° 22'}$ cm

$\approx 6\cdot05$ cm

$A_2 = 91° 24'$
$C_2 = 180° - (64° 22' + 91° 24') = 24° 14'$

No.	Log.
12·0	1·0792
sin 24° 14′	$\overline{1}$·6133
	0·6925
sin 64° 22′	$\overline{1}$·9550
5·464	0·7375

$\therefore \dfrac{c_2}{\sin 24° 14'} = \dfrac{12\cdot0 \text{ cm}}{\sin 64° 22'}$

$c_2 = \dfrac{12\cdot0 \times \sin 24° 14'}{\sin 64° 22'}$ cm

$$\approx 5{\cdot}46 \text{ cm.}$$

Answers   A = 88° 36'; C = 27° 02'; c = 6·05 cm.
          A = 91° 24'; C = 24° 14'; c = 5·46 cm.

Exercise 7.2

1 In triangle ABC, A = 77° 43', B = 60° 45', b = 4·31 cm. Find the
   length of the longest side.
2 In triangle DEF, D = 74°, E = 42° 16', DE = 2400 m. Solve the
   triangle to find the missing elements.
3 In triangle ABC, B = 122° 44', C = 15° 22', c = 2·5 cm. Solve the
   triangle.
4 In the triangle ABC, B = 31° 19', C = 109° 51', a = 28 m. Solve
   the triangle.
5 Examine the data for the following triangles and state whether the
   ambiguous case is involved or not, giving your reasons:
   (a) Triangle ABC: B = 63° 17', c = 14·2 cm, b = 10·1 cm.
   (b) Triangle ABC: C = 27° 38', a = 7·9 cm, c = 11·2 cm.
   (c) Triangle PQR: Q = 101° 33', p = 1·3 cm, q = 4·2 cm.
   (d) Triangle XYZ: Y = 69° 23', x = 8·2 cm, y = 9·5 cm.
   (e) Triangle PQR: P = 48° 32', q = 5·3 m, p = 4·1 m.
6 In the triangle ABC, B = 28° 05', a = 0·6 cm, b = 0·35 cm. Solve
   the triangle completely, giving both solutions if ambiguous.
7 In the triangle ABC, B = 42° 16', b = 1·8 cm, c = 2·4 cm. Solve
   the triangle for all possible solutions.
8 Solve completely the triangle in which A = 128° 17', a = 24·4 cm,
   c = 15·6 cm.
9 Solve completely the triangle in which B = 69° 14', b = 10·0 cm,
   c = 8·5 cm.
10 Give all possible solutions for the triangle in which B = 65° 08',
   a = 31·2 m, b = 29·2 m.
11 Solve completely the triangle in which B = 45° 07', a = 89·2 m,
   b = 67·4 m.
12 Solve completely the triangle in which C = 95° 44', a = 50 cm,
   c = 90 cm.
13 In the triangle ABC, A = 25° 25', a = 8 mm, b = 5 mm. Solve
   the triangle.
14 Solve the triangle ABC, where A = 140°, b = 2·4 cm, c = 4·5 cm.

The Area of a Triangle

The area of a triangle is found from the formula $\frac{1}{2}$ (base) × (altitude).
The base and altitude of a triangle are shown in fig 7.11.

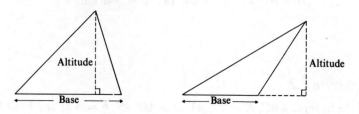

Fig 7.11

In fig 7.12, a triangle ABC has a perpendicular drawn from A to the
side BC.

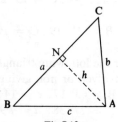

Fig 7.12

If h is the length of the perpendicular, then

$h = b \sin C$ or $c \sin B$ .... from the triangles ACN, ABN

Now area of the triangle $= \frac{1}{2} ah$

$= \frac{1}{2} ab \sin C$

$= \frac{1}{2} ac \sin B$

Similarly it can be shown that the area equals $\frac{1}{2} bc \sin A$. The area
of a triangle is usually represented by the symbol Δ.

Hero's formula

In a triangle ABC, where the sides are a, b, c, the perimeter is $(a + b + c)$.
Let the perimeter be $2s$, i.e. $2s = a + b + c$; s is called the semi-
perimeter.

Now   (i) $a + b - c = a + b + c - 2c = 2s - 2c = 2(s - c)$

(ii) $a + c - b = a + b + c - 2b = 2s - 2b = 2(s - b)$

(iii) $b + c - a = a + b + c - 2a = 2s - 2a = 2(s - a)$

In a triangle ABC, $\cos A = \dfrac{b^2 + c^2 - a^2}{2bc}$

and from Pythagoras' theorem $\sin^2 A = 1 - \cos^2 A$

$\therefore \sin^2 A = (1 - \cos A)(1 + \cos A)$

$$= \left[1 - \frac{b^2 + c^2 - a^2}{2bc}\right] \times \left[1 + \frac{b^2 + c^2 - a^2}{2bc}\right]$$

$$= \left[\frac{2bc - (b^2 + c^2 - a^2)}{2bc}\right] \times \left[\frac{2bc + (b^2 + c^2 - a^2)}{2bc}\right]$$

$$= \frac{a^2 - (b - c)^2}{2bc} \times \frac{(b + c)^2 - a^2}{2bc}$$

$$= \frac{(a - b + c)(a + b - c)}{2bc} \times \frac{(b + c - a)(b + c + a)}{2bc}$$

$$= \frac{2(s - b) \cdot 2(s - c) \cdot 2(s - a) \cdot 2s}{4b^2c^2}$$

$\therefore \sin A = \frac{\pm 2}{bc} \sqrt{s(s - a)(s - b)(s - c)}$

but $\Delta = \frac{1}{2} bc \sin A$

$\therefore \Delta = \sqrt{s(s - a)(s - b)(s - c)}$ (Hero's formula).

Using Hero's formula, the area of a triangle can be found from the three sides.

$$\boxed{\begin{aligned} \Delta &= \tfrac{1}{2} \times \textbf{(base)} \times \textbf{(altitude)} \\ &= \tfrac{1}{2} \, bc \sin A \\ &= \tfrac{1}{2} \, ac \sin B \\ &= \tfrac{1}{2} \, ab \sin C \\ &= \sqrt{s(s - a)(s - b)(s - c)}. \end{aligned}}$$

Example 5

The sides of a triangle are: $a = 12\cdot7$ cm, $b = 13\cdot9$ cm, $c = 8\cdot6$ cm.
Calculate the height of the perpendicular from A to side BC.

First the area of the triangle must be found, using Hero's formula.

	No.	Log.
$s = \frac{1}{2}(12\cdot7 + 13\cdot9 + 8\cdot6) = 17\cdot6$	17·6	1·2455
$s - a = 17\cdot6 - 12\cdot7 \qquad = 4\cdot9$	4·9	0·6902
$s - b = 17\cdot6 - 13\cdot9 \qquad = 3\cdot7$	3·7	0·5682
$s - c = 17\cdot6 - 8\cdot6 \qquad = 9\cdot0$	9·0	0·9542
		3·4581 ÷ 2
Now $\Delta = \frac{1}{2} ah$		
where h is the length of the perpendicular	Δ	1·7290(5)
	6·35	0·8028
$\therefore h = \dfrac{\Delta}{\dfrac{a}{2}} = \dfrac{\Delta}{6\cdot35}$	8·437	0·9262

$\approx 8\cdot44$ cm.

Example 6

The area of triangle ABC *is* $20\sqrt{3}$ cm². A $= 60°$ *and* $b = 8$ cm. *Find the length of side* a.

$$\Delta = \tfrac{1}{2}\, bc \sin A$$

$$\therefore c = \frac{2\Delta}{b \sin A} = \frac{2 \times 20\sqrt{3}}{8 \times \dfrac{\sqrt{3}}{2}} = 10 \text{ cm}$$

Now $a^2 = b^2 + c^2 - 2bc \cos A$ $\qquad\qquad$ $(\cos A = \cos 60° = \tfrac{1}{2})$
$\qquad\quad = 64 + 100 - 2.8.10.\tfrac{1}{2} = 164 - 80$
$\qquad\quad = 84$
$\therefore a = 9.16$ cm.

(Note: Calculation by logarithms is unnecessary, as only simple figures are involved in the working.)

Example 7

A triangle has the following measurements: $s - a = 5$ cm, $s - b = 3$ cm, $s - c = 2$ cm. *Calculate the area of the triangle.*

Add all three measurements: $(s - a) + (s - b) + (s - c)$
$\quad = 3s - (a + b + c) = 5 + 3 + 2 = 10$ cm
$\qquad\qquad\qquad\quad = s \qquad$ since $(a + b + c) = 2s$
$$\therefore \Delta = \sqrt{s(s - a)(s - b)(s - c)}$$
$$= \sqrt{10 \times 5 \times 3 \times 2}$$
$$= \sqrt{300} = 10\sqrt{3} \approx 17.3 \text{ cm}^2.$$

Exercise 7.3

1 In triangle PQR, P $= 30°$, PR $= 8$ cm, PQ $= 3.5$ cm. Find the area of the triangle.

2 In triangle ABC, A $= 47° \; 36'$, $b = 4.8$ cm, $c = 6.9$ cm. Find its area.

3 The sides of a triangle are 4 cm, 5 cm, and 7 cm long. Calculate its area to 2 sig. figs.

4 The sides of a triangle are 31.2 cm, 29.2 cm, 18.8 cm. Calculate the area of the triangle to 3 sig. figs.

5 In triangle ABC, $a = 5$ cm, $b = 4$ cm, and the area of the triangle is 8.6 cm². Calculate the size of angle C, given that it is obtuse.

6 The sides of a triangle are 6 cm, 15 cm, and 19 cm respectively. Calculate the altitude of the triangle, taking the longest side as the base. (Answer to 3 sig. figs.)

7 A triangle has the following measurements: $s - a = 1\cdot5$ cm, $s - b = 1\cdot8$ cm, $s - c = 2\cdot7$ cm. Calculate its area correct to 3 sig. figs.

8 The sides of a triangle are respectively $4p$, $7p$, and $5p$ units, and the area is 245 square units. Calculate the value of p.

9 The triangle ABC has an area of $\dfrac{25\sqrt{3}}{2}$ cm^2, C $= 60°$, and $a = 5$ cm. Find angles A and B.

10 In triangle ABC, B $= 62°$ 13', $c = 30$ cm, and the area is $53\cdot1$ cm^2; calculate side a.

11 Triangle ABC has C $= 40°$, $a = 160x$ cm, $b = 100x$ cm, and area $= 1290$ cm^2. Calculate the value of x.

12 The following measurements are given for a triangle ABC: $s = 30$ cm, $s - a = 8\cdot4$ cm, $s - b = 15\cdot6$ cm. Calculate the area of the triangle.

8 Coordinate Geometry: The Straight Line

Coordinates

Previous work in mathematics has used *Cartesian coordinates* to describe the position of a point; these are the coordinates used on squared paper for graphical work. Fig 8.1 shows a plane divided into four quadrants (as in trigonometry) and the signs of the x and y coordinates are shown for each quadrant. The x-coordinate of a point is sometimes called the **abscissa**, and the y-coordinate is called the **ordinate**.

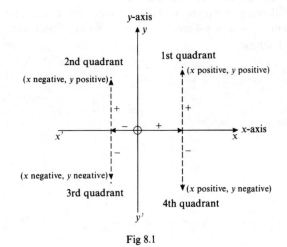

Fig 8.1

When giving the coordinates of a point, the value of x is written first; i.e. the point $(3, -6)$ has an x-coordinate of 3 and a y-coordinate of -6; the coordinates are enclosed in brackets. A fixed point, whose actual coordinates are not given, is usually referred to as a point (x_1, y_1); the *suffices* 2, 3, ... can be used in place of the *suffix* 1 to describe other points. If reference has to be made to a variable point which obeys a mathematical relationship, it is usually written as (x, y) without suffices.

When drawing graphs of mathematical relationships, *the x-axis is used for an independent variable, and the y-axis for the dependent variable.* For example, in the relationship $y = 3x^2 - 2x - 6$, the values of x are chosen independently, and the value of y is dependent on the value of x.

Functions

The function notation has been described in chapter 3. In the expression $f(x) = 2x^3 - x - 5$, suitable values of x are selected, and corresponding values of $f(x)$ calculated. Hence $f(x)$ is the dependent variable and gives the value of the ordinate of a point when a selected value of x is taken as the abscissa. The graphical work in exercise 8.1 supplements the work in chapters 3 and 5.

Exercise 8.1

1 Plot the following coordinates on a sheet of graph paper, taking 2 inches or 2 cm per 10 units on both the x and the y-axes, and labelling the coordinates with the capital letters as indicated:
    A$(4, 7)$; B$(7, 4)$; C$(3, -7)$; D$(-7, 3)$; E$(-4, -7)$; F$(-3, -2)$;
    G$(-4, 2)$; H$(7, -3)$.
2 Draw the graphs of (a) $y - 2x + 1 = 0$, and (b) $3y + x + 3 = 0$. At what point do these lines intersect?
3 On the same axes as you used for question 2, draw the lines $y = 5$ and $y = -3$. What are the coordinates of the points of intersection of these two lines with the straight line $y - 2x + 1 = 0$ (drawn in question 2)? What is the distance between the lines $y = 5$ and $y = -3$?
4 $f(x)$ is a linear function. $f(1) = -3$ and $f(3) = +3$. Draw the graph of the function for values of x between -2 and $+20$ and find the point at which it cuts the x-axis.
5 Draw the graph of $f(x) = x^2 - 4x - 5$ for values of x between -5 and $+10$. Find (a) the minimum value of the function, (b) the points at which it intersects the x-axis.
6 Draw the graph of $f(x) = 7 - 2x - x^2$. Does it have a maximum or minimum, and what is the value of the maximum or minimum?
7 Using a scale of 1 inch for 1 unit on the x-axis and 1 inch for 10 units on the y-axis, draw axes for values of x between -3 and $+3$ and values of y between -50 and $+20$. On these axes draw the graph of $f(x) = 2x^3 - 3x^2 - 12x + 6 = 0$.
    Determine (a) the maximum and minimum values of the curve, (b) the point at which the curve cuts the y-axis.
8 Draw the graph of $f(x) = 4/x$ for values of x between $+10$ and -10.

On the same axes, draw the graph of $f(x) = x$. At what point do these two graphs intersect?

9 Draw the graph of $f(x) = (x - 5)(11 - x)$ for values of x between 0 and 15. On the same axes draw the graphs of $f(x) = x - 5$ and $f(x) = 11 - x$.

(a) At what point do the two linear functions intersect?

(b) What are the coordinates of the point at the maximum of the quadratic function?

Distance between Two Points

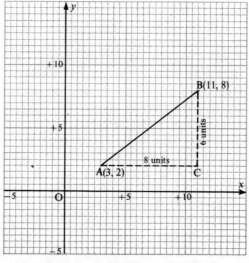

Fig 8.2

In fig 8.2, two points A and B have coordinates (3, 2) and (11, 8) respectively. The distance between A and B has to be found. Complete the triangle ABC by drawing BC parallel to the y-axis and AC parallel to the x-axis. The triangle ABC is right-angled at C. By Pythagoras' theorem $AB^2 = BC^2 + AC^2$. Now $AC = 11 - 3 = 8$ units and $BC = 8 - 2 = 6$ units. $\therefore AB^2 = (8)^2 + (6)^2 = 64 + 36 = 100$. $\therefore AB = \sqrt{100} = 10$ units.

The result either gives the distance between the points A and B, or the length of the line AB. The negative root is neglected, as no direction is given to the distance.

General formula
Let the two points, A and B, be (x_1, y_1) and (x_2, y_2) respectively (see fig 8.3). The distance between A and B is required. Complete the right-angled triangle ABC, as before. The length AC is $(x_2 - x_1)$ units, and the length BC is $(y_2 - y_1)$ units.

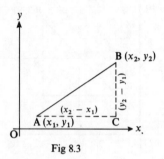

Fig 8.3

Using Pythagoras' theorem,

$$AB^2 = (x_2 - x_1)^2 + (y_2 - y_1)^2$$
$$\therefore \ AB = \sqrt{(x_2 - x_1)^2 + (y_2 - y_1)^2}$$

If d is the distance between A and B, then:

$$\boxed{d = \sqrt{(x_2 - x_1)^2 + (y_2 - y_1)^2}.}$$

Example 1
Find the distance between the points $(3, -2)$ *and* $(-4, 5)$.
In this example $(x_1, y_1) = (3, -2)$ and $(x_2, y_2) = (-4, 5)$.

$$\therefore d = \sqrt{(x_2 - x_1)^2 + (y_2 - y_1)^2}$$
$$= \sqrt{(-4 - 3)^2 + [5 - (-2)]^2}$$
$$= \sqrt{(-7)^2 + (7)^2}$$
$$= \sqrt{49 + 49}$$
$$= 7\sqrt{2}.$$

Mid-point of Two Given Points

Let the two given points, A and B, have coordinates of (2, 3) and (8, 9) respectively. Draw the right-angled triangle ABC with BC parallel to the y-axis and AC parallel to the x-axis, as shown in fig 8.4. Let P be the mid-point between A and B. Draw PN parallel to AC and PM parallel to BC.

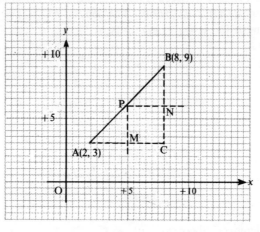

Fig 8.4

By the mid-point theorem in geometry, since $PM \parallel BC$ and $AP = PB$, then $AM = MC$. Let the coordinates of P be (p, q), then $MA = p - 2$ and $CM = 8 - p$.

$\therefore p - 2 = 8 - p$ i.e. $2p = 8 + 2$ i.e. $p = \dfrac{8 + 2}{2} = 5$

Similarly by the mid-point theorem, $BN = NC$

$\therefore q - 3 = 9 - q$ i.e. $2q = 9 + 3$ i.e. $q = \dfrac{9 + 3}{2} = 6$

$\therefore$ The point P is (5, 6).

General formula for the mid-point

Let the two given points, A and B, be (x_1, y_1) and (x_2, y_2) respectively. Let P be the mid-point and have coordinates (x, y). Complete the right-angled triangle ABC, as before, and draw the lines PM, PN respectively parallel to the sides BC and AC as in fig 8.5.

By the mid-point theorem, $AM = MC$

$$\therefore x - x_1 = x_2 - x$$

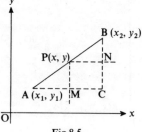

Fig 8.5

i.e. $$2x = x_2 + x_1$$

∴ $$x = \frac{x_2 + x_1}{2}$$

Similarly, $$y = \frac{y_2 + y_1}{2}$$

The mid-point of AB is $\left(\dfrac{x_2 + x_1}{2}, \dfrac{y_2 + y_1}{2}\right)$.

Example 2

Find the mid-point of the straight line joining the two given points $(-4, 2)$ *and* $(2, -9)$.

In this example $(x_1, y_1) = (-4, 2)$ and $(x_2, y_2) = (2, -9)$.

∴ the mid-point is $\left(\dfrac{(-4) + 2}{2}, \dfrac{2 + (-9)}{2}\right)$

$$= \left(\frac{-2}{2}, \frac{-7}{2}\right) = (-1, -3\tfrac{1}{2}).$$

Division of a Line in a given ratio

Let AB be the given line (see fig 8.6). Draw AC parallel to the x-axis and BC parallel to the y-axis, forming the right-angled triangle ABC. Let P be such a point that it divides AB in the ratio $m:n$. Let the co-ordinates of P be (x, y). Draw PM parallel to BC.

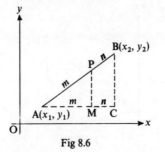

Fig 8.6

Then $\dfrac{AP}{PB} = \dfrac{AM}{MC}$ but $\dfrac{AP}{PB} = \dfrac{m}{n}$

$\therefore$ $\dfrac{m}{n} = \dfrac{AM}{MC} = \dfrac{x - x_1}{x_2 - x}$

$\therefore$ $mx_2 - mx = nx - nx_1$
$\therefore$ $mx_2 + nx_1 = mx + nx$
i.e. $x(m + n) = mx_2 + nx_1$
$\therefore$ $x = \dfrac{mx_2 + nx_1}{m + n}.$

Similarly, by drawing a line through P, parallel to AC, it can be shown that $y = \dfrac{my_2 + ny_1}{m + n}$.

The coordinates of the point (x, y) dividing a line joining (x_1, y_1) and (x_2, y_2) in the ratio of $m:n$ are therefore given as:

$$x = \frac{mx_2 + nx_1}{m + n}, \quad y = \frac{my_2 + ny_1}{m + n}$$

The mid-point is a special case of this formula, in which $m = n$; this reduces the formula to

$$\left(\frac{x_2 + x_1}{2}, \frac{y_2 + y_1}{2}\right) \text{ as before.}$$

Example 3

A line AB *joins two points with coordinates* $(2, -1)$ *and* $(3, 4)$. *A point* P *is chosen so that* AP:PB $= 2:1$. *Find the coordinates of* P.

In this example $(x_1, y_1) = (2, -1)$ and $(x_2, y_2) = (3, 4)$; $m = 2$ and $n = 1$.

Let the coordinates of P be (x, y), then

$$x = \frac{mx_2 + nx_1}{m + n} = \frac{(2 \times 3) + (1 \times 2)}{2 + 1} = \frac{6 + 2}{3} = 2\tfrac{2}{3}$$

$$y = \frac{my_2 + ny_1}{m + n} = \frac{(2 \times 4) + (1 \times -1)}{2 + 1} = \frac{8 - 1}{3} = 2\tfrac{1}{3}.$$

The required point is $(2\tfrac{2}{3}, 2\tfrac{1}{3})$.

Gradients

The gradient of a line is measured by: $\dfrac{\text{the increase in } y}{\text{the increase in } x}$ moving from

one point to another. In fig 8.7, the gradient of AB from A to B is measured by

$$\frac{\text{the increase in } y}{\text{the increase in } x} = \frac{20 \text{ units}}{10 \text{ units}} = 2.$$

If the gradient is measured from B to A, it is:

$$\frac{\text{the increase in } y}{\text{the increase in } x} = \frac{5 - 25}{5 - 15} = \frac{-20}{-10} = 2.$$

In whichever direction of AB the gradient is measured, it is found to be $+2$. Now concentrate on the line EF in fig 8.7. The gradient has

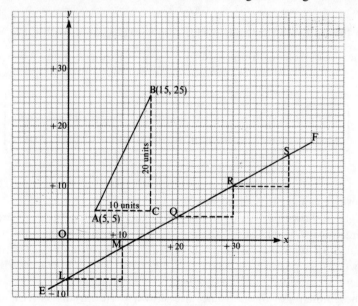

Fig 8.7

been determined between the pairs of points LM; QR and RS, by drawing the increases in x and the increases in y. Clearly all these triangles are similar, and hence the value of the gradient is the same. A straight line, then, possesses a constant gradient.

Angle of slope

In fig 8.8, a straight line AB is drawn, making an angle θ with the positive direction of the x-axis. P is the point where AB intersects the x-axis, and Q is any other point on the line. A line QR is drawn perpendicular to the x-axis. The gradient of AB, by definition, is:

$$\frac{\text{the increase in } y}{\text{the increase in } x} = \frac{QR}{PR}.$$

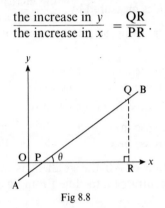

Fig 8.8

In the right-angled triangle PQR, right-angled at R, with $Q\hat{P}R = \theta$ (the angle of slope), we have $\tan \theta = \dfrac{QR}{PR} = $ gradient of line AB.

> The gradient of a straight line is the **tangent of the angle** which it makes with the positive direction of the x-axis.

Positive and negative gradients

Fig 8.9 shows two straight lines, AB and CD, sloping in different directions. AB makes an angle θ with the positive x-axis, and CD makes an angle ϕ with the positive x-axis. θ is acute and ϕ is obtuse, so $\tan \theta$ is positive, and $\tan \phi$ is negative. The gradient of AB is therefore positive, and the gradient of CD is negative.

The gradients of the two lines are shown by the triangles AEF and MNL respectively.

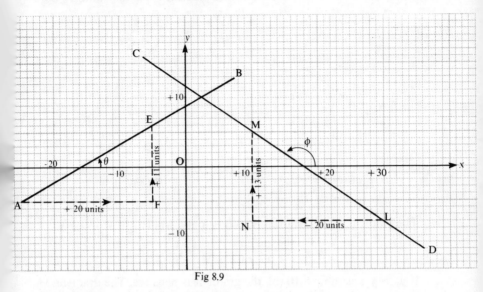

Fig 8.9

The gradient of AB from A to B is:

$$\frac{\text{increase in } y}{\text{increase in } x} = \frac{6 - (-5)}{-5 - (-25)} = \frac{11}{20} \text{ and it is positive.}$$

The gradient of CD from C to D is:

$$\frac{\text{increase in } y}{\text{increase in } x} = \frac{5 - (-8)}{10 - 30} = \frac{13}{-20} = \frac{-13}{20} \text{ and it is negative.}$$

The gradient of CD from D to C is:

$$\frac{-8 - 5}{30 - 10} = -\frac{13}{20} \text{ i.e. the same in either direction.}$$

Thus if a straight line rises from left to right, its gradient is positive, and if it rises from right to left, its gradient is negative. The gradient of a line it usually denoted by the letter m; in fig 8.9, we have

$$m_{AB} = \tan \theta \text{ and } m_{CD} = \tan \phi$$

The Gradient of the Line joining Two Points

Fig 8.10 shows two points, A and B, with coordinates (x_1, y_1) and (x_2, y_2) respectively; it is required to find the gradient of the line AB joining the two points.

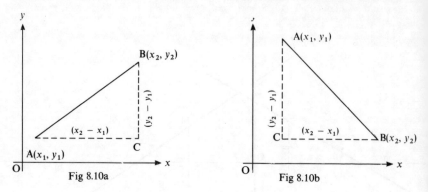

Fig 8.10a   Fig 8.10b

The gradient, m, of AB from A to B is found from:

$$m = \frac{\text{the increase in } y}{\text{the increase in } x} = \frac{y_2 - y_1}{x_2 - x_1}.$$

If $y_1 > y_2$, as in fig 8.10 (b), the gradient is negative. The gradient, m, is given in all cases by the formula:

$$\boxed{m = \frac{y_2 - y_1}{x_2 - x_1}.}$$

Example 4

Find the gradient of the line joining the point $(3, \ -2)$ *to* $(-5, \ -8)$.
It does not matter which point is selected as $(x_1, \ y_1)$ or $(x_2, \ y_2)$, as it has been shown that the slope measured in either direction of the line joining two points is the same.

Let $\qquad (x_1, y_1) = (3, \ -2)$ and $(x_2, y_2) = (-5, \ -8)$

Then $\qquad m = \dfrac{y_2 - y_1}{x_2 - x_1} = \dfrac{(-8) - (-2)}{(-5) - (3)} = \dfrac{-6}{-8} = \tfrac{3}{4}.$

Useful facts can be deduced from these formulae as follows.

Collinear points

If three, or more, points are collinear, then they lie on the same straight line. A straight line has a uniform, or constant, gradient. To test for collinearity, the gradients between adjacent pairs of points are found. If the gradients are the same, then the points lie on the same line, and hence are collinear.

Example 5

Are the points $(2, 2\frac{1}{4})$, $(5, 4\frac{1}{2})$, $(7, 6)$ *collinear?*

Let the points be A, B, C respectively.

$$\text{Gradient AB} = \frac{4\frac{1}{2} - 2\frac{1}{4}}{5 - 2} = \frac{2\frac{1}{4}}{3} = \frac{9}{12} = \frac{3}{4}$$

$$\text{Gradient BC} = \frac{6 - 4\frac{1}{2}}{7 - 5} = \frac{1\frac{1}{2}}{2} = \frac{3}{4}.$$

∴ the gradients are the same, hence the points are collinear.

Exercise 8.2

1 Find the distances between the following pairs of points:
   (a) $(2, 5)$ and $(5, 9)$;   (b) $(-1, 5)$ and $(-7, -3)$;
   (c) $(-1, -4)$ and $(-6, -16)$;   (d) $(-4, 5)$ and $(6, 4)$;
   (e) $(7, 2)$ and $(13, -10)$;   (f) $(ap^2, 2ap)$ and $(aq^2, 2aq)$.

2 The distance between the points $(16, k)$ and $(1, 1)$ is 17. Find the value of k.

3 If the distance between the points
$$(l + m, m + n) \text{ and } (l + n, l + m)$$
is $(l - m)$, prove that $(l - n)(m - n) = 0$.

4 Find the mid-point of the following pairs of points:
   (a) $(2, 5)$ and $(6, 13)$;   (b) $(5, -7)$ and $(-3, -1)$;
   (c) $(2a, 6)$ and $(4, 2b)$;   (d) $(ap^2, 2ap)$ and $(aq^2, 2aq)$.

5 The points A and B are $(k, 2)$ and $(-2, l)$ respectively. Point C, coordinates $(1, 1)$, is the mid-point of AB. Find the values of k and l.

6 Point A is $(11, 1)$ and point B is $(2, 6)$. Point P lies on the line AB and $2AP = PB$. Find the coordinates of P.

7 Find the gradients between the following pairs of points:
   (a) $(2, 5)$ and $(4, 9)$;   (b) $(3, -6)$ and $(-3, 2)$;
   (c) $(-4, -9)$ and $(-2, 1)$;   (d) $(5, -4)$ and $(-2, -3)$;
   (e) $(0, 0)$ and $(-3, -6)$;   (f) $(4, 2)$ and $(0, 0)$;
   (g) $(ap, \frac{a}{p})$ and $(aq, \frac{a}{q})$;   (h) $(at_1{}^2, 2at_1)$ and $(at_2{}^2, 2at_2)$.

8 Show that the four points: $(1, 18)$, $(5, 8)$, $(-1, -2)$, $(-5, 8)$ are the vertices of a parallelogram.

9 Test if the following points are collinear:
   (a) $(2, 5)$, $(-1, -1)$, $(-4, -7)$;
   (b) $(1, -1)$, $(-2, 4)$, $(0, 1)$;
   (c) $(7, -3)$, $(-2, 1)$, $(-6\frac{1}{2}, 3)$.

10 Point A is distant 5 units from point B. The coordinates of B are (2, 5) and the gradient of AB is $-\frac{3}{4}$. Calculate the coordinates of A.

Areas of Triangles and Quadrilaterals

Area of a triangle

Fig 8.11a shows a triangle, ABC, with the ordinates AK, BL, CM, drawn. The lines form three trapeziums, from which the area of the triangle can be found.

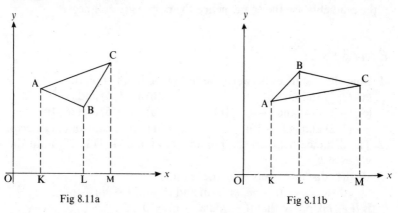

Fig 8.11a   Fig 8.11b

Area of triangle ABC = area of trapezium KACM – (area of trapezium KABL + area of trapezium LBCM). The area of a trapezium is given by:

$$(half \; sum \; of \; parallel \; sides) \; \times \; \begin{pmatrix} perpendicular \; distance \; between \\ parallel \; sides \end{pmatrix}$$

Let the vertices of the triangle be A(2, 4), B(6, 3), and C(9, 7); then AK = 4, BL = 3, CM = 7, KL = (6 – 2) = 4, LM = (9 – 6) = 3, KM = (9 – 2) = 7.

Area of trapezium KACM = $\dfrac{(4 + 7)}{2} \times 7 = 38\frac{1}{2}$ sq. units

Area of trapezium KABL = $\dfrac{(4 + 3)}{2} \times 4 = 14$ sq. units

Area of trapezium LBCM = $\dfrac{(3 + 7)}{2} \times 3 = 15$ sq. units

∴ area of ABC = $38\frac{1}{2} - (14 + 15) = 9\frac{1}{2}$ sq. units.

In fig 8.11b, the area of the triangle is found from:
ΔABC = trap. KABL + trap. LBCM – trap. KACM.

Area of a quadrilateral

Fig 8.12 shows a quadrilateral ABCD, with the ordinates drawn from the vertices. The figure is again divided into trapeziums, and the area of the quadrilateral is found from:

Quad. ABCD
= trap. KABL + trap. LBCM + trap. MCDN − trap. KADN

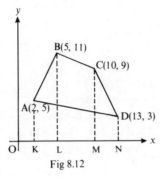

Fig 8.12

Let the coordinates of the vertices be $A(2, 5)$, $B(5, 11)$, $C(10, 9)$ and $D(13, 3)$.

Then $AK = 5$, $BL = 11$, $CM = 9$, $DN = 3$, $KL = 3$, $LM = 5$, $MN = 3$, $KN = 11$.

Area of
quad. ABCD $= \dfrac{(5 + 11)3}{2} + \dfrac{(11 + 9)5}{2} + \dfrac{(9 + 3)3}{2} - \dfrac{(5 + 3)11}{2}$

$= 24 + 50 + 18 - 44 = 48$ square units.

Notes on areas of figures

(1) Always make a sketch of the figure to locate the vertices.

(2) Mark in the ordinates to form trapeziums, triangles, or rectangles.

(3) Calculate the area of the figure from the numerical areas of the trapeziums, triangles, or rectangles.

(4) If the area enclosed by 3, or more, points is 0 square units, then the points are collinear.

Example 6

Find the area of the quadrilateral whose vertices are (3, 7), (5, −6), (7, 0) *and* (−4, 0).

Draw a diagram as in fig 8.13; note that the vertices were not given in order.

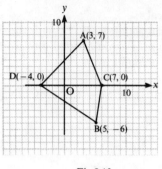

Fig 8.13

Area of quad. $= \Delta BCD + \Delta ACD$
Distance $DC = 7 - (-4) = 11$ units
Perp. height of $A = 7$ units; perp. height of $B = 6$ units.
Area of quad. $= \frac{1}{2} \times 11 \times 7 + \frac{1}{2} \times 11 \times 6$

$$= \frac{77}{2} + \frac{66}{2}$$

$$= \frac{143}{2}$$

$$= 71\frac{1}{2} \text{ square units.}$$

Exercise 8.3

1 Find the areas of the following triangles whose vertices are:
    (a) $(1, 3), (3, 7), (5, 1)$.
    (b) $(-2, -1), (0, 3), (2, -2)$.
    (c) $(-1, -7), (0, 0), (-3, 2)$.
    (d) $(0, 0), (p + q, p - q), (p - q, p + q)$.
    (e) $(0, 0), (3, 1), (k, 0)$.
2 Find the area enclosed by the following three points: $(1, 2), (4, 8)$, $(-2, -4)$. What do you deduce from the result?
3 Find the area of the triangle whose vertices are $(-1, t - 1), (t, t - 3)$, $(t - 6, 3)$. Deduce the values of t if the points are collinear.
4 Show that the area of the triangle whose vertices are $(-2, 2), (6, 2)$, (a, b) is independent of the value of a. If $a = 2b$, and the area of the triangle is 12, determine the coordinates of the third vertex.
5 The three vertices of a triangle are $(2, 4), (a, b), (-a, -b)$. If the area of the triangle is 4 sq. units, find two expressions relating a and b.
6 Find the areas of the quadrilaterals which have the following vertices:
    (a) $(6, 6), (3, 1), (1, 5), (4, 7)$.

(b) $(-2, 1)$, $(-3, 8)$, $(-6, 3)$, $(1, 2)$.

(c) $(0, 4)$, $(-2, 3)$, $(-3, -2)$, $(3, 0)$.

7 The three vertices, A, B, C, of a triangle are $(2, 5)$, $(-2, 3)$, $(-1, 4)$. D is the mid-point of BC. Verify that $AB^2 + AC^2 = 2AD^2 + \dfrac{BC^2}{2}$.

Find the area of the quadrilateral with vertices at the origin and points A, B, and C.

8 The vertices of a triangle are A $(-5, -5)$, B $(7, 0)$, C $(3, 10)$. Calculate (a) the lengths of AB and AC; (b) the area of the triangle; (c) the value of sin $B\widehat{A}C$ (expressed as a fraction).

The Equation of a Straight Line

Fig 8.14 shows a number of parallel straight lines. They all cut the x-axis. As they are parallel, the angle these lines make with the x-axis is the same, θ, in each case. Hence for each line tan θ has the same value, i.e. all the lines AB, CD, EF, GH, have the same gradient. Now look at the straight lines drawn in fig 8.15; they all pass through the same point P, but have different gradients.

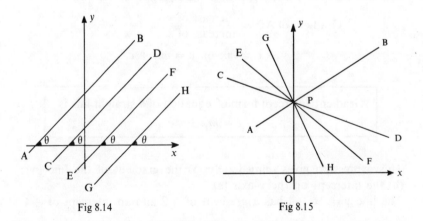

Fig 8.14   Fig 8.15

To fix the position of a straight line, two facts must be known, which can either be the gradient and the position of one point, or the positions of two points of the line.

In the first case, a useful point is the point where the line cuts the y-axis: from this, and the gradient, we can derive the **gradient/intercept** form of the equation of a line.

Gradient/Intercept equation of the straight line

In fig 8.16, a straight line, AB, cuts the y-axis at P, and has a gradient of m. Let Q be any point on the line, and let the coordinates of Q be

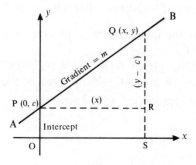

Fig 8.16

(x, y). Draw QS perpendicular to the x-axis, meeting it in S; draw PR perpendicular to QS, meeting it in R. Let the coordinates of P be $(0, c)$, i.e. c is the *intercept* of the line on the y-axis. If we now find a relationship between x and y for the point Q, this relationship will be true for all points on the straight line. The relationship will be an equation between y and x which is the equation of the line.

$$\text{Gradient of AB} = \frac{\text{increase of } y}{\text{increase of } x} = \frac{y - c}{x} = m$$

$$\therefore y - c = mx \text{ or } y = mx + c.$$

> Gradient/Intercept form of equation of a straight line is
> $$y = mx + c$$

This expression gives immediately **(a)** the gradient of the line (m); **(b)** the intercept on the y-axis (c).

The line $y = 2x + 4$ has a gradient of $+2$ and an intercept of $+4$, i.e. the line cuts the y-axis at $(0, 4)$. The line $y = -\dfrac{x}{3} - 4$ has a gradient of $-\frac{1}{3}$ and cuts the y-axis at $(0, -4)$.

Notice that the coefficient of y in the gradient form must be unity.

The lines $y = 3x + 4$ and $y = 3x - 6$ are parallel because in each case the gradient is $+3$, and lines with the same gradient are parallel.

We may however be given the gradient and the coordinates of some other fixed point of the line. From these two facts we derive the equation of the line.

Equation of a line, given its gradient and one point on the line
In fig 8.17, the line AB has a gradient m, and passes through the point Q of coordinates (x_1, y_1). The right-angled triangle PLQ, right-angled at L, is drawn, where P is any point on the straight line with coordinates (x, y).

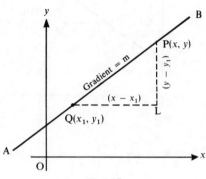

Fig 8.17

The gradient of AB $= \dfrac{\text{increase in } y}{\text{increase in } x} = \dfrac{PL}{QL} = \dfrac{y - y_1}{x - x_1} = m$ (given)

$\therefore \dfrac{y - y_1}{x - x_1} = m$ and from this, the equation of the line is found.

Example 7
A straight line has a gradient of $+3$ and passes through the point $(1, 4)$. Find its intercept on the y-axis.
In this case (x_1, y_1) is $(1, 4)$ and $m = +3$

Since $\dfrac{y - y_1}{x - x_1} = m,$

$\dfrac{y - 4}{x - 1} = +3$

$y - 4 = 3x - 3$ i.e. $y = 3x + 1$ which is the gradient form of the equation of the line. Hence the intercept on the y-axis is 1.

Thirdly, we can derive the equation of a line if we know the coordinates of *two* points on it.

Equation of a line, given two points on the line

The gradient of the line can be found from two points, as shown in fig 8.18, where Q and R are the two given points, with coordinates (x_1, y_1) and (x_2, y_2). Any point, P, is taken with coordinates (x, y). The relationship between x and y will give the equation of the straight line, as explained above.

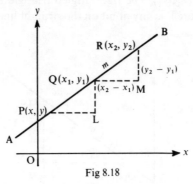

Fig 8.18

The gradient of AB $= m = \dfrac{y_2 - y_1}{x_2 - x_1}$ from triangle RMQ

In triangle QLP, the gradient of the line AB is

$$m = \frac{y_1 - y}{x_1 - x} = \frac{y - y_1}{x - x_1}.$$

But $$m = \frac{y_2 - y_1}{x_2 - x_1}$$

∴ $\dfrac{y - y_1}{x - x_1} = \dfrac{y_2 - y_1}{x_2 - x_1}$ This gives the equation of the straight line.

The formulae for the equation of a straight line through two points, or through one point with a given gradient, and the formula for the gradient of a line between two points, can all be summarised in the one statement:

$$\boxed{\frac{y - y_1}{x - x_1} = m = \frac{y_2 - y_1}{x_2 - x_1}}$$

Example 8
Find the equation of the straight line **AB**, *which passes through the*

points $(1, -1)$ *and* $(-2, -13)$; *hence find the equation of the straight line* CD *parallel to* AB, *which passes through the point* $(2, 11)$

On the line AB, $(x_1, y_1) = (1, -1)$ and $(x_2, y_2) = (-2, -13)$

Since $\qquad\qquad \dfrac{y - y_1}{x - x_1} = \dfrac{y_2 - y_1}{x_2 - x_1}$

$$\therefore \frac{y - (-1)}{x - 1} = \frac{-13 - (-1)}{-2 - 1} = \frac{-12}{-3} = 4$$

$$\therefore y + 1 = 4x - 4$$

$$\text{i.e. AB is } y = 4x - 5$$

$\therefore$ the gradient of AB is $+4$.

CD, the line parallel to AB, will also have a gradient of $+4$. CD passes through $(2, 11)$.

Since $\qquad\qquad \dfrac{y - y_1}{x - x_1} = m$

$$\therefore \frac{y - 11}{x - 2} = 4$$

$$\text{i.e. } y - 11 = 4x - 8$$

$\therefore$ the equation of the parallel line is $y = 4x + 3$.

Exercise 8.4

1 Find the gradient/intercept form of the equation of the following straight lines:
   (a) gradient $= +2$, intercept $= +3$;
   (b) gradient $= +4$, intercept $= -5$;
   (c) gradient $= -2$, intercept $= 6$;
   (d) gradient $= -1\frac{1}{2}$, intercept $= -\frac{3}{4}$.

2 Find the equation of the following straight lines:
   (a) the line has a gradient of 3, and passes through $(2, 1)$;
   (b) ,, ,, ,, ,,   ,, -2 ,,   ,,   ,, $(5, 2)$;
   (c) ,, ,, ,, ,,   ,, $\frac{3}{4}$,,   ,,   ,, $(-8, 3)$;
   (d) ,, ,, ,, ,,   ,, $-\frac{1}{2}$,,   ,,   ,, $(-4, -5)$.

3 Find the equation of the straight lines which pass through the following pairs of points:
   (a) $(2, 4)$ and $(6, 5)$;   **(b)** $(-4, 3)$ and $(2, -5)$;
   (c) $(2, -1)$ and $(3, 0)$;   **(d)** $(2, -3)$ and $(5, -3)$.

4 The three vertices of a triangle are $A(6, 3)$, $B(1, 2)$, and $C(7, -2)$. Find the coordinates of the mid-points of AB and BC; show that the line joining these mid-points is parallel to AC.

5 Find the equation of the straight line which passes through $(-4, 3)$ and is parallel to the line $y = 2x + 5$.

6 Write down the equations, in gradient form, of the following straight lines:

(**a**) A line that passes through the point $(2, 7)$ and has a gradient of m;

(**b**) A line that has a gradient of $+2$, and passes through the point $(t, \frac{1}{t})$;

(**c**) A line that is parallel to $y = -3x + 7$.

7 Find the equation of a straight line that is parallel to $y = -\frac{5}{2}x + 4$ and bisects the line joining $(3, 1)$ to $(-1, -5)$.

8 The coordinates of A are $(3, 1)$ and of B are $(-6, -5)$. P is a point on the line such that $AP = 2PB$. A straight line, of gradient -2, is drawn through P; calculate the intercept of this line on the y-axis.

9 A straight line AB of gradient m passes through the point $(3, 2)$ to form a triangle with the x and y-axes. Calculate (**i**) the intercepts the line AB makes on both of the axes, (**ii**) the area of the triangle formed by the axes and the line AB.

10 The straight line $y = mx + c$ passes through the points $(0, k)$ and $(h, 2k)$. Express m and c in terms of h and k.

Other Forms of Equation of the Straight Line

The general form

An equation of the first degree in both x and y represents a linear function. The general equation of the first degree is $ax + by + c = 0$, and this is the general form of the equation of a straight line. The general form and the gradient/intercept form of the straight line can be readily converted from one to the other. Here are some examples:

Gradient/Intercept	General form
(**a**) $y = 2x - 3$	$2x - y - 3 = 0$
(**b**) $y = \frac{1}{2}x + 4$	$x - 2y + 8 = 0$
(**c**) $y = -4x + 1$	$4x + y - 1 = 0$
(**d**) $y = -\frac{3}{8}x + \frac{1}{4}$	$3x + 8y - 2 = 0$

The steps of conversion in (**d**) are:

(**i**) $y = -\frac{3}{8}x + \frac{1}{4}$ i.e. $8y = -3x + 2$ i.e. $3x + 8y - 2 = 0$

(**ii**) $3x + 8y - 2 = 0$ i.e. $8y = -3x + 2$ i.e. $y = -\frac{3}{8}x + \frac{1}{4}$

When the general form of the equation is given, it is converted to the gradient/intercept form to deduce the gradient and the intercept on the y-axis.

The double-intercept form

Fig 8.19 shows a straight line, AB, which intersects the x-axis at $(a, 0)$ and the y-axis at $(0, b)$. The intercepts on the x and y-axes are a and

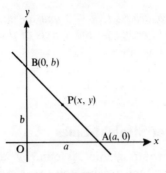

Fig 8.19

b respectively. Let P be any point on AB, with coordinates (x, y). As the three points A, P, B are collinear, then

gradient AP = gradient PB

i.e. $\dfrac{y - 0}{x - a} = \dfrac{b - y}{0 - x}$

i.e. $-xy = xb - xy - ab + ay$

i.e. $0 = xb - ab + ay$

$\therefore bx + ay = ab$ Now divide throughout by ab:

$\therefore \dfrac{x}{a} + \dfrac{y}{b} = 1.$

Since (x, y) was any point, this equation is true for every point on the line, and hence it is the equation of the line AB.

If the equation of a straight line is written in this manner, it gives, by inspection, the intercepts on the axes. Note that the Right Hand Side of the equation must be unity. Both the gradient/intercept and general forms can be converted to the double-intercept form by making the R.H.S. equal to unity. The sign and magnitude of the intercepts can thus be found by inspection.

General form	Double–intercept form	Intercepts on	
		x-axis	y-axis
$4x + 3y - 12 = 0$	$\dfrac{x}{3} + \dfrac{y}{4} = 1$	$+3$	$+4$

The intercepts can also be found quickly by setting $x = 0$ and $y = 0$ in turn in the equation of the line.

When $x = 0$, $3y - 12 = 0$ and $y = 4$; when $y = 0$, $4x - 12 = 0$ and $x = 3$, giving the intercepts on each axis.

Example 9

A straight line has a gradient of -3 and an intercept of -4 on the y-axis. What is the intercept on the x-axis of this line?

The equation of the line is $y = -3x - 4$ (from $y = mx + c$)

When $y = 0$, $0 = -3x - 4$ or $x = -\frac{4}{3}$.

Hence the intercept on the x-axis is $-\frac{4}{3}$.

The Intersection of Straight Lines

The intersection of two straight lines is found by putting their equations into the general form and solving the two simultaneous equations. For example, to find the intersection of

$$3x + 4y - 6 = 0 \text{ and } x - 2y + 3 = 0,$$

let (a, b) be the point of intersection. Thus (a, b) lies on both straight lines.

$\therefore 3a + 4b - 6 = 0$ and $a - 2b + 3 = 0$

The values of a and b are found by solving these equations simultaneously, obtaining $a = 0$, $b = \frac{3}{2}$.

$\therefore$ point of intersection is $(0, \frac{3}{2})$.

There is no need to use a point (a, b) when solving the problem. The equations can be solved for x and y directly.

Example 10

Find the point of intersection of the lines given by

$$y = 2x + \frac{3}{2} \text{ and } \frac{x}{3} + \frac{y}{2} = 1.$$

$y = 2x + \frac{3}{2}$ is converted to $4x - 2y + 3 = 0$ . . . . (i)

$\frac{x}{3} + \frac{y}{2} = 1$ is converted to $2x + 3y - 6 = 0$ . . . . (ii)

Hence, solving the equations (i) and (ii) simultaneously, we obtain:

$x = -\frac{27}{16}$, $y = -\frac{15}{8}$ and so the point of intersection of the lines is $(-\frac{27}{16}, -\frac{15}{8})$.

Exercise 8.5

1 Express the following equations in the general form:

(a) $y = 2x + 8$;   (b) $y = -\frac{3}{5}x + 1$;   (c) $y = +\frac{3}{4}x - \frac{1}{8}$;

(d) $\frac{x}{2} + \frac{y}{5} = 1$;   (e) $\frac{x}{-3} + \frac{y}{+1} = 1$;   (f) $\frac{x}{-6} + \frac{y}{-8} = 1$.

2 Express the following equations in the double-intercept form:
   (a) $3x + y - 6 = 0$;   (b) $4x - 3y + 24 = 0$;
   (c) $3x + 5y + 15 = 0$;   (d) $y = x + 6$;
   (e) $y = -2x + 8$;   (f) $y = -\frac{3}{4}x + \frac{1}{2}$.

3 Express the following equations in the gradient/intercept form:

   (a) $3x - 6y + 4 = 0$;   (b) $\frac{x}{2} + \frac{y}{1} = 1$;

   (c) $x + 3y - 7 = 0$;   (d) $\frac{x}{-2} + \frac{y}{-8} = 1$;

   (e) $\frac{x}{-4} + \frac{y}{5} = 1$;   (f) $3x + 7y + 2 = 0$.

4 Write down the equation of the straight line which has intercepts of 3 on the x-axis and -5 on the y-axis. What is the gradient of this line?

5 A straight line makes intercepts of p on the x-axis and q on the y-axis. If the line passes through the point $(3, -2)$, find the value of p in terms of q.

6 Find the point of intersection of the following pairs of lines:
   (a) $3x + 4y - 10 = 0$ and $2x - 3y - 1 = 0$;
   (b) $x + 2y + 5 = 0$ and $3x - 2y - 17 = 0$;
   (c) $2x + 3y + 13 = 0$ and $3x + y + 9 = 0$.

7 The straight line AB makes intercepts of $\frac{2}{3}$ on the x-axis and -1 on the y-axis. The straight line CD has a gradient of $+2$, and passes through the point $(1, -1)$. Calculate the point of intersection of AB and CD.

8 A straight line passes through the point $(1, -2)$ and has a gradient of m. Find the double-intercept form of the equation of this line and hence write down the intercepts on the axes. If the intercepts on the two axes are equal, determine the equation of the line.

9 A straight line makes intercepts of h on the x-axis and k on the y-axis. The line passes through the point $(1, \frac{3}{2})$ and the product $hk = -2$. Determine the equation of the line.

10 Point A is $(-3, 4)$ and point B is $(3, 4)$. Point C is $(3, 0)$. Point D is a variable point on the y-axis with coordinates $(0, a)$. Calculate the point of intersection, P, of the two lines AB and CD. Calculate the area of the triangle with vertices at P, A, and the origin, and show that the area is independent of the point B.

The Distance of a Point from a Straight Line

The distance of the origin from a straight line

In fig 8.20, the straight line AB has the equation $ax + by + c = 0$. It cuts the axes at A and B as shown. A perpendicular, of length p, is drawn from the origin to AB, meeting it at C. Let CO make an angle α with the x-axis.

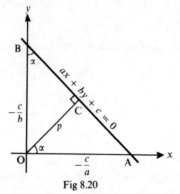

Fig 8.20

Then $\ O\hat{A}B = 90° - \alpha$ and $\ O\hat{A}B = 90° - O\hat{B}A \ \therefore \ O\hat{B}A = \alpha$

Rewrite the equation of the line AB in the double-intercept form:

$$\frac{x}{\dfrac{-c}{a}} + \frac{y}{\dfrac{-c}{b}} = 1$$

Hence the intercepts are $\dfrac{-c}{a}$ and $\dfrac{-c}{b}$ on the x and y-axes respectively.

The distance AB $= \pm \sqrt{\left(\dfrac{-c}{a}\right)^2 + \left(\dfrac{-c}{b}\right)^2} = \pm \sqrt{\dfrac{c^2}{a^2} + \dfrac{c^2}{b^2}}$

$$= \pm \frac{c}{ab}\sqrt{a^2 + b^2}$$

Now OC $=$ OA $\cos \alpha$ and $\cos \alpha = \dfrac{\text{OB}}{\text{AB}}$

$$\frac{\text{OB}}{\text{AB}} = \frac{\dfrac{-c}{b}}{\pm\dfrac{c}{ab}\sqrt{a^2 + b^2}} = \frac{\pm a}{\sqrt{a^2 + b^2}}$$

$$\therefore \ \text{OC} = \frac{-c}{a} \times \frac{\pm a}{\sqrt{a^2 + b^2}} = \frac{\pm c}{\sqrt{a^2 + b^2}}.$$

The perpendicular distance of the origin from a straight line is taken by convention as a positive distance. This is done by taking the sign of the square root to be the same as the sign of the constant term.

$$p = \frac{+c}{\sqrt{a^2 + b^2}}$$

The distance of a point from a straight line

In fig 8.21, the straight line, AB, has the equation $ax + by + c = 0$. It cuts the axes at A and B, and the perpendicular from the origin, OQ, makes an angle α with the x-axis. A point, P, of coordinates (x_1, y_1) has a perpendicular distance, d, indicated by PN. RP is drawn parallel to AB. PL is drawn perpendicular to the x-axis; LS is drawn perpendicular to OQ and NP produced meets SL in K.

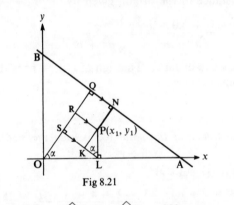

Fig 8.21

$\widehat{PLK} = 90° - \widehat{SLO}$, but $\widehat{SLO} = 90° - \alpha$

$\therefore \widehat{PLK} = \alpha$

In triangle PKL, $KP = PL \sin \alpha = y_1 \sin \alpha$

(y coordinate of P is y_1)

In triangle SLO, $OS = OL \cos \alpha = x_1 \cos \alpha$

(x coordinate of P is x_1)

Now PN = QR   (opp. sides of rectangle)

$\quad\quad = OQ - OR$

$\quad\quad = OQ - (OS + SR)$

$\quad\quad = OQ - OS - KP$ (KP and SR opp. sides of a rectangle)

$\quad\quad = p - x_1 \cos \alpha - y_1 \sin \alpha$ (p is length of perpendicular from origin to line AB)

but $\cos \alpha = \dfrac{-a}{\pm\sqrt{a^2 + b^2}}$ and $\sin \alpha = \dfrac{-b}{\pm\sqrt{a^2 + b^2}}$

$$\text{and } p = \frac{+c}{\pm\sqrt{a^2 + b^2}}$$

$$\therefore \text{ PN} = \frac{c}{\pm\sqrt{a^2 + b^2}} - \frac{x_1 \times (-a)}{\pm\sqrt{a^2 + b^2}} - \frac{y_1 \times (-b)}{\pm\sqrt{a^2 + b^2}}$$

$$= \frac{c + ax_1 + by_1}{\pm\sqrt{a^2 + b^2}}.$$

This formula is usually written in the same way as the original equation, i.e.

$$\boxed{d = \frac{ax_1 + by_1 + c}{\pm\sqrt{a^2 + b^2}}.}$$

The sign of the square root is taken in the formula, so that the perpendicular distance of the origin, given by

$$d = \frac{a.0 + b.0 + c}{\sqrt{a^2 + b^2}} = \frac{c}{\sqrt{a^2 + b^2}},$$

is always a positive distance. The sign of $\sqrt{a^2 + b^2}$ is the same as the sign of the constant term.

Example 11

What is the distance of the point (2, 3) *from the straight line*
$$3x + 4y + 7 = 0?$$
In this case $(x_1, y_1) = (2, 3)$

$$\therefore d = \frac{ax_1 + by_1 + c}{\pm\sqrt{a^2 + b^2}} = \frac{3 \times 2 + 4 \times 3 + 7}{+\sqrt{9 + 16}} = \frac{25}{5} = 5.$$

The sign of the perpendicular distance
Fig 8.22 shows the graph of the equation $y = \frac{1}{2}x + 5$. Two points, P (5, 5) and Q (5, 10), are marked, with their respective perpendiculars drawn in. The general form of the equation is $x - 2y + 10 = 0$. To find the perpendicular distances of P and Q, the calculation is:

Point P (5, 5)

$$d = \frac{ax_1 + by_1 + c}{\sqrt{a^2 + b^2}} = \frac{5 - 10 + 10}{+\sqrt{1 + 4}} = \frac{5}{\sqrt{5}} = \sqrt{5}$$

Note: The sign of $\sqrt{5}$ is positive because the constant term is $+10$.

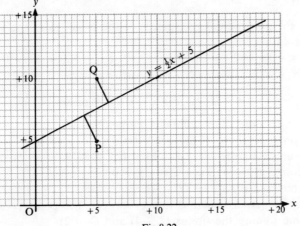

Fig 8.22

Point Q (5, 10)

$$d = \frac{ax_1 + by_1 + c}{\sqrt{a^2 + b^2}} = \frac{5 - 20 + 10}{+\sqrt{1 + 4}} = \frac{-5}{\sqrt{5}} = -\sqrt{5}.$$

Note the difference in sign. $d = +\sqrt{5}$ indicates the point P is on the same side of the line as the origin. $d = -\sqrt{5}$ indicates the point Q is on the opposite side of the line from the origin. The difference in sign indicates that the points lie on opposite sides of the straight line.

Example 12

Are the points (1, 2) *and* (2, −2) *on the same side of* $5x + 12y − 6 = 0$ *or not?*

Point (1, 2)

$$d = \frac{5 \times 1 + 12 \times 2 - 6}{-\sqrt{25 + 144}} = -\frac{23}{13}$$

Point (2, −2)

$$d = \frac{5 \times 2 + 12 \times (-2) - 6}{-\sqrt{25 + 144}}$$

$$= \frac{10 - 24 - 6}{-\sqrt{169}} = +\frac{20}{13}.$$

The points are on opposite sides of the straight line.

Note: If the distance is zero, the point lies on the line.

Exercise 8.6

1 Find the perpendicular distances of the origin from the following straight lines:

(a) $3x - 4y + 20 = 0$;   (b) $5x - 12y + 13 = 0$;

(c) $5x + 12y - 26 = 0$;   (d) $x + 2y - 3 = 0$;

(e) $\dfrac{x}{4} + \dfrac{y}{3} = 1$;   (f) $y = \dfrac{5}{12} x - 2$.

2 Find the lengths of the following perpendiculars:

(a) from $(1, 4)$ to $5x + 12y + 13 = 0$;

(b) from $(2, 3)$ to $5x + 12y - 13 = 0$;

(c) from $(-1, -3)$ to $x + 2y - 3 = 0$;

(d) from $(-2, 4)$ to $y = 3x - 4$.

3 Find the perpendicular distance between the point $(-3, 7)$ and the line joining the points $(2, -4)$ and $(-6, 2)$.

4 Find the distance between the two parallel lines $3x - 4y + 12 = 0$ and $3x - 4y + 2 = 0$. (Hint: find the difference between the distances from the origin.)

5 Find the equation of the straight line parallel to $y = \dfrac{5}{12} x + 2$ and distant 2 units from it on the opposite side to the origin. Express your answer in the gradient form.

6 The line forming the base of a triangle has the equation

$$6y - 8x + 3 = 0.$$

The other two sides are formed by the lines $3x + 2y - 10 = 0$ and $x + 6y - 10 = 0$. Calculate the altitude of the triangle.

7 Determine whether the points $(3, -1)$ and $(2, 4)$ are on the same side of the straight line $y = 2x - 6$, and determine their respective distances.

8 The coordinates of the points A, B, and C are $(2, 5)$, $(3, -2)$ and $(-1, 4)$ respectively. Find the distance of A from the line joining B and C.

Locus Problems

When the position of a variable point in a plane, or in three dimensional space, is governed by a geometrical law, then the set of all such points together is called the *locus* of the point. A simple example of a locus is that of a point which moves in a plane at a fixed distance from a given point. If such points are plotted and joined in a smooth curve, then the locus is a circle.

Loci formed by Parallel Lines

(**1**) The locus of a point which moves at a fixed distance from a given line is required. Let AB be the given line (see fig 8.23). Let P, co-

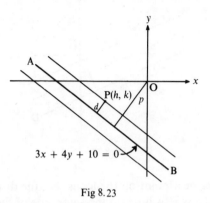

Fig 8.23

ordinates (h, k), be any point on the locus. The equation of AB is $3x + 4y + 10 = 0$ and let the fixed distance of P from AB be $\frac{1}{2}$. The distance of P from AB is given by:

$$d = \frac{3h + 4k + 10}{\sqrt{9 + 16}} = \frac{1}{2} \qquad \text{(fixed distance)}$$

$$\therefore \quad 3h + 4k + 10 = \frac{\pm\sqrt{25}}{2} \qquad \text{(the sign of the square root depends on which side of the line P is located)}$$

$$3h + 4k + 10 = \frac{\pm 5}{2} \qquad \text{(no side is specified)}$$

$$\therefore \quad \left. \begin{array}{l} 6h + 8k + 20 = +5 \\ 6h + 8k + 20 = -5 \end{array} \right\} \qquad \text{either solution can be taken}$$

These expressions relate the coordinates h and k of the point P; if they are transformed into x and y, they will form the equations of the locus.

The equations are $\left\{ \begin{array}{l} 6x + 8y + 20 = 5 \text{ i.e. } 6x + 8y + 15 = 0 \\ 6x + 8y + 20 = -5 \text{ i.e. } 6x + 8y + 25 = 0 \end{array} \right.$

The locus has two equations which satisfy the given conditions. They are both straight lines, parallel to the given line, at the fixed distance from the given line and on opposite sides of the line.

(2) The locus of a point which is equidistant from two parallel straight lines is required. Let AB and CD be the straight lines with equations $5x + 12y + 2 = 0$ and $5x + 12y + 18 = 0$ (see fig 8.24). Let P,

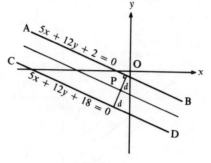

Fig 8.24

coordinates (h, k), be a point on the locus. Let the distance of P from both lines be d. Now P will be on the same side of the origin for one of the lines, but on the other side from the origin for the other line; hence the signs of the distances will be opposite. (The distance of the origin is given by $\dfrac{+2}{+13}$ and $\dfrac{+18}{+13}$, i.e. since both square roots have the same sign, the origin lies on the same side of both lines.)

$$d_1 = \frac{5h + 12k + 2}{\sqrt{25 + 144}} \qquad d_2 = \frac{5h + 12k + 18}{\sqrt{25 + 144}}$$

and $d_1 = -d_2$

$$\therefore \frac{5h + 12k + 2}{\sqrt{169}} = -\frac{5h + 12k + 18}{\sqrt{169}}$$

$\therefore 10h + 24k + 20 = 0$ i.e. $5h + 12k + 10 = 0$.

Replace h and k by x and y to obtain the equation of the locus: $5x + 12y + 10 = 0$ i.e. the locus is parallel to the given parallel lines. Also note that the constant term of the locus is the average of the two constant terms in the given parallel lines.

The Perpendicular Bisector between Two Points

Let A and B be two given points with coordinates $(5, 2)$ and $(1, 8)$ respectively. Let P, coordinates (x, y), be a point on the perpendicular bisector. Now, by geometry, AP = PB, so the perpendicular bisector is the locus of a point which moves so that it is equidistant from two given points (see fig 8.25).

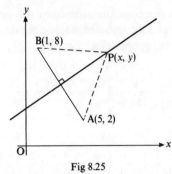

Fig 8.25

Now write down this condition:
$$AP = PB$$
i.e. $\sqrt{(x - 5)^2 + (y - 2)^2} = \sqrt{(x - 1)^2 + (y - 8)^2}$

Squaring and multiplying out,
$$x^2 - 10x + 25 + y^2 - 4y + 4 = x^2 - 2x + 1 + y^2 - 16y + 64$$
$$-8x + 12y - 36 = 0$$
$$\therefore 2x - 3y + 9 = 0 \text{ which is the required locus.}$$

The Bisector of Two Intersecting Lines

Let AB and CD be the intersecting straight lines (see fig 8.26) with equations $4x + 3y - 21 = 0$ and $5x - 12y + 12 = 0$ respectively. Let P be a point on the locus with coordinates (h, k). Then the perpendicular distances of P from each of the lines are equal (geometrical theorem). P can have two positions, P and P_1, shown in fig 8.26, so there will be two loci. The point P is on the same side of one line as the origin, but on the opposite side from the origin for the other line.

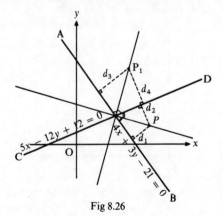

Fig 8.26

The point P_1 is on the opposite side from the origin for both lines. Hence d_2 is positive and d_1 negative, while both d_3 and d_4 are negative. Now write down the geometrical condition:

$$d_1 = -d_2$$

i.e. $$\frac{4h + 3k - 21}{-\sqrt{16 + 9}} = -\frac{5h - 12k + 12}{\sqrt{25 + 144}}$$

$$d_3 = d_4$$

i.e. $$\frac{4h + 3k - 21}{-\sqrt{16 + 9}} = +\frac{5h - 12k + 12}{+\sqrt{25 + 144}}$$

The condition can be summarised as:

$$\frac{4h + 3k - 21}{\sqrt{25}} = \pm \frac{5h - 12k + 12}{\sqrt{169}}$$

i.e. $$13(4h + 3k - 21) = \pm 5(5h - 12k + 12)$$
$$52h + 39k - 273 = \pm (25h - 60k + 60)$$

The solutions are $27h + 99k - 333 = 0$ or $77h - 21k - 213 = 0$.

Since (h, k) is a point on the locus, replacing h and k by x and y will give the equations of the locus.

$\therefore$ the loci are $27x + 99y - 333 = 0$ or $77x - 21y - 213 = 0$ which reduce to $3x + 11y - 37 = 0$ or $77x - 21y - 213 = 0$.

Lines through a Point of Intersection

Fig 8.27 shows two straight lines,
$$3x - y - 6 = 0 \text{ and } 2x + y - 14 = 0,$$
intersecting at the point P, coordinates (h, k). The coordinates h and k satisfy both equations, as P lies on both straight lines.

$\therefore$ $3h - k - 6 = 0$ and $2h + k - 14 = 0$.

Hence $(3h - k - 6) + \lambda(2h + k - 14) = 0$

                          where λ is a numerical coefficient

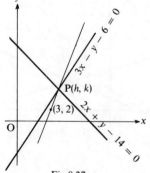

Fig 8.27

i.e. $h(3 + 2\lambda) + k(\lambda - 1) + (-6 - 14\lambda) = 0.$

This establishes a relationship between h and k, which depends on
the value of λ. If h and k are replaced by x and y, we have:
$$x(3 + 2\lambda) + y(\lambda - 1) + (-6 - 14\lambda) = 0$$
or alternatively $3x - y - 6 + \lambda(2x + y - 14) = 0.$

This is a linear equation, and hence represents any straight line which
passes through (h, k), i.e. the point of intersection.

The gradient of the line is: $-\dfrac{3 + 2\lambda}{\lambda - 1}$

i.e. it depends solely on λ. The line can be fixed in the plane by (1) giving
its gradient or (2) giving another point through which the line passes.

Let the line pass through $(3, 2)$; substitute this point in the equation
of the line:
$$3x - y - 6 + \lambda(2x + y - 14) = 0 \quad \text{where } x = 3 \text{ and } y = 2$$
$\therefore$ $9 - 2 - 6 + \lambda(6 + 2 - 14) = 0$ i.e. $1 - 6\lambda = 0$

$\therefore$ $\lambda = \frac{1}{6}$

Now substitute in the equation of the line:
$$3x - y - 6 + \tfrac{1}{6}(2x + y - 14) = 0$$
$$18x - 6y - 36 + 2x + y - 14 = 0$$
$$20x - 5y - 50 = 0$$
$$\therefore 4x - y - 10 = 0.$$

This is the equation of the straight line which passes through the
intersection of the given lines and the point $(3, 2)$.

The converse

Show that the line $4x - y = 10$ passes through the intersection of
the two lines $3x - y - 6 = 0$ and $2x + y - 14 = 0.$

Any line passing through the given intersection has the equation:
$$3x - y - 6 + \lambda(2x + y - 14) = 0$$
i.e. $x(3 + 2\lambda) + y(\lambda - 1) - (6 + 14\lambda) = 0$ . . . . (i)

The given line for test is $4x - y = 10$ with gradient of $+4$.

The gradient of any line passing through the intersection is $\dfrac{3 + 2\lambda}{1 - \lambda}$.

$\therefore$ $\dfrac{3 + 2\lambda}{1 - \lambda} = 4$ for a line parallel to the given line

i.e. $3 + 2\lambda = 4 - 4\lambda$ whence $\lambda = \frac{1}{6}$

Substituting in (i),

$$x(3 + \frac{2}{6}) + y(\frac{1}{6} - 1) - (6 + \frac{14}{6}) = 0$$

$$\frac{20x}{6} - \frac{5y}{6} - \frac{50}{6} = 0$$

Multiply throughout by 6, and divide throughout by 5.

$\therefore 4x - y - 10 = 0.$

Hence this line does pass through the point of intersection.

Other Locus Problems

Always take a point P and give it coordinates (x, y) or (h, k) and then write down the geometrical condition as an algebraic expression. A rough diagram or sketch should also be made to check the facts used.

Example 13

The coordinates of A *are* $(2, -4)$ *and of* B *are* $(0, 3)$. *Point* P *is a variable point, such that the area of the triangle* ABP *is 10 sq. units. Show that the locus of* P *is two parallel straight lines, and find the distance between these lines.*

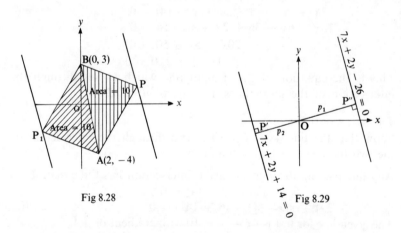

Fig 8.28   Fig 8.29

Firstly, make a sketch of the problem, as in fig 8.28. Secondly, write down the conditions as an equation. Let the coordinates of P be (h, k). The area of the triangle is found from $\frac{1}{2} \times$ (base) $\times$ (perp. height).

Base $=$ distance AB $= \sqrt{2^2 + 7^2} = \sqrt{53}$

Perpendicular height $=$ distance from point P to the line through AB.

AB is:
$$\frac{y + 4}{3 + 4} = \frac{x - 2}{0 - 2}$$
$$7x + 2y - 6 = 0$$

$$\therefore \text{ Perpendicular height of } \mathbf{P} = \frac{7h + 2k - 6}{\pm\sqrt{49 + 4}} = \frac{7h + 2k - 6}{\pm\sqrt{53}}$$

$$\text{Area of triangle} = 10\,\text{sq. units} = \tfrac{1}{2} \times \sqrt{53} \times \frac{7h + 2k - 6}{\pm\sqrt{53}}$$

i.e. $\qquad 20 = \pm(7h + 2k - 6)$

Replacing (h, k) by (x, y) to give the equation of the line,

$\qquad 7x + 2y - 6 = \pm 20$

This gives two lines which are:

$\left. \begin{array}{l} 7x + 2y - 26 = 0 \\ 7x + 2y + 14 = 0 \end{array} \right\}$ i.e. two parallel lines as each has a gradient of $-3\tfrac{1}{2}$.

To find the distance between the two parallel lines, consider a line, perpendicular to both straight lines, and which passes through the origin, i.e. the line P′ P″ in fig 8.29. The length OP″ is measured in the opposite direction to the line OP′. If the line $7x + 2y + 14 = 0$ were on the other side of the origin, the perpendicular distance between the two lines would be $p_1 - p_2$. This formula still holds for the lines shown in fig 8.29, as p_2 is in a negative direction with regard to p_1, and the actual distance is, arithmetically, $p_1 + p_2$, i.e. $[p_1 - (-p_2)]$. Writing the length of the perpendicular drawn, we have:

distance between the parallel lines $= p_1 - p_2$

$$p_1 - p_2 = \frac{7 \times 0 + 2 \times 0 - 26}{-\sqrt{49 + 4}} - \frac{7 \times 0 + 2 \times 0 + 14}{-\sqrt{49 + 4}}$$

i.e. giving both square roots the same sign

$$\therefore d = p_1 - p_2 = \frac{26 - (-14)}{\sqrt{53}} = \frac{40\sqrt{53}}{53}.$$

Example 14

Write down the equation of a line passing through the mid-point of A, *(1, −1,), and* B, *(5, 3). If the gradient of the line is +2, show that the line passes through the point (4, 3), and find its equation in the gradient form. Prove that* A *and* B *are equidistant from, and on opposite sides of, the line.*

$$\text{Mid-point of AB is given by} \left(\frac{x_1 + x_2}{2}, \frac{y_1 + y_2}{2} \right)$$

$$\therefore \text{mid-point is} \left(\frac{1 + 5}{2}, \frac{-1 + 3}{2} \right) \quad \text{i.e. (3, 1).}$$

Assuming a gradient of m, the equation of a line through this point is: $\qquad \dfrac{y - 1}{x - 3} = m$

i.e. $y = mx - 3m + 1$

If $m = +2$, the line is $y = 2x - 6 + 1$ or $y = 2x - 5$

Substituting the point $(4, 3)$ in this line:
$$3 = 2 \times 4 - 5 = 3$$
Hence the point lies on the line.

The perpendicular distance of a point from the line is given by:

$$\frac{2x_1 - y_1 - 5}{-\sqrt{4 + 1}}$$

Distance of point A = $\dfrac{2 \times 1 - 1 \times (-1) - 5}{-\sqrt{5}} = \dfrac{-2}{-\sqrt{5}} = \dfrac{2}{5}\sqrt{5}$
from line

Distance of point B = $\dfrac{2 \times 5 - 1 \times 3 - 5}{-\sqrt{5}} = \dfrac{+2}{-\sqrt{5}} = -\dfrac{2}{5}\sqrt{5}.$
from line

The distances are numerically equal, and as the signs are opposite, the points lie on opposite sides of the line.

Exercise 8.7

1 Find the equation of the locus of a point which moves so that it is always at a distance of 2 units from the line $5x + 12y - 2 = 0$.

2 Find the equation of a line which is parallel to $x + 2y - 4 = 0$ and passes through the point $(4, -3)$. What is the perpendicular distance between these two lines?

3 Find the locus of a point P which moves so that it is equidistant from A $(3, -6)$, and B $(-5, 2)$. Are the points $(-2, -1)$ and $(3, -2)$ on the locus?

4 A point P moves so that it is equidistant from the lines
$$3x + 2y + 6 = 0 \text{ and } 3x + 2y + 8 = 0.$$
What is the equation of the locus of P, and what is the distance between the two given straight lines?

5 Find the equation of the locus of points which are equidistant from the straight lines $3x + 4y - 5 = 0$ and $8x - 15y + 3 = 0$.

6 The straight line $y = 6x + 4$ intersects the line joining the points $(-3, -\frac{1}{2})$ and $(7, -8)$ at the point P. Find the equation of the straight line joining P to the origin.

7 Find the equation of the straight line which passes through the intersection of $2x - y + 7 = 0$ and $3x + 2y - 4 = 0$ and has a gradient of $+3$.

8 Show that the line $2x - 5y + 6 = 0$ passes through the intersection of the two straight lines

$$x - 3y + 4 = 0 \text{ and } 3x + 5y - 16 = 0.$$

9 The base of a triangle is the line joining A and B, of coordinates
(0, 4) and (−1, −3) respectively. The apex of the triangle is a variable point P. If the area of the triangle is 2 square units, show that
P lies on the straight line $y - 7x + 8 = 0$.

10 A is a variable point on the x-axis with coordinates $(t, 0)$. B is
a variable point on the y-axis with coordinates $(0, \dfrac{4}{t})$. A triangle
is formed by the line AB and the two axes in the first quadrant.
Prove the area of the triangle AOB is constant and find its area.
Show that the locus of the mid-point of AB has the equation
$xy = 1$.

Exercise 8.8   *Miscellaneous*

1 The vertices of a triangle, P, Q, R, are (1, 7), (−2, 5), (2, 3) respectively.
S is the mid-point of QR. Verify that
$$PQ^2 + PR^2 = 2PS^2 + 2SR^2.$$
Calculate the area of the quadrilateral OQPR where O is the origin.
(C)

2 The coordinates of A, B and C are (3, 1), (1, 5) and (4, 2) respectively.
P is the mid-point of BC.
Q lies on AC and is such that CQ : QA = 3 :1.
R lies on AB and is such that AR : RB = 1 :3.
Find the equations of the lines AP, BQ and CR and prove that
the lines are concurrent. (C)

3 A is the point (2, 0) and B the point (6, 0). P is the point where the
line A to $(h, 4)$ meets the line joining B to $(-h, 4)$. Find the coordinates of P in terms of h. Verify that P lies on the line $x + y = 4$.
(C)

4 The points A $(-3, -2)$, B $(9, 3)$ and C $(5, 13)$ are the vertices of
a triangle. Calculate:
(a) the lengths of AB and AC;
(b) the area of the triangle BAC;
(c) the angle BAC. (C)

5 **(i)** Prove that the distance from the point (1, 1) to the point
$$\left[\frac{2t^2}{1 + t^2}, \frac{(1 - t)^2}{1 + t^2} \right] \text{ is the same for all values of } t.$$

(ii) Find the area of the triangle whose vertices are $(a, 0)$, $(0, b)$
and (x, y). (C)

Other Uses of Straight Line Graphs

In science when two variables are believed to be connected, a set of corresponding measurements is made. Such results can then be used to discover if there is a mathematical law relating the two variables. If the mathematical law is thus found, it can be used to predict further values and these in turn used to test the law experimentally.

Usually the results obtained from the experiments are plotted as a graph and an attempt made to deduce the law from this graph. Now if the graph obtained is a straight line the mathematical relationship is easily deduced. If the quantities involved are denoted as x and y, the law will be of the form $y = mx + c$, where m is the gradient of the line and c the intercept on the y-axis of the graph.

However the law will rarely be of this simple form. More often it will be one of the two types $y = ax^b$ or $y = ab^x$, where a and b are constants, which have to be found. Using logarithms we can convert each of these two types to a straight line graphical form.

$$\boxed{y = ax^b}$$

Then $\log y = \log a + \log x^b = \log a + b \log x$, using the rules for logarithms explained in Chapter 1.

Now write $\log y = Y$ and $\log x = X$.

Then $Y = \log a + bX$ or $Y = bX + \log a$ which is of the form $Y = mX + c$ where $m = b$ and $\log a = c$.

Hence if we plot the values of $\log y (Y)$ against those of $\log x (X)$ as an (X, Y) graph and the result is a straight line, we can say that the law connecting the variables x and y is of the form $y = ax^b$. The values of a and b can now be deduced from the gradient and intercept of the line. If a straight line graph is not obtained the variables must be related in some other way.

Example 15

The following pairs of values of x and y were obtained by experiment:

x	2	3	4	5
y	113	312	640	1118

It is suspected that y is a function of x of the form $y = ax^b$. By plotting log y against log x, test this theory and if correct, find approximate values for a and b. From your graph find the value of x for which $y = 215$.

If $y = ax^b$, then as above $Y = \log y = \log a + b \log x = bX + \log a$ where $X = \log x$. A table of values is made:

x	2	3	4	5
$X = \log x$	0·3010	0·4771	0·6021	0·6990
y	113	312	640	1118
$Y = \log y$	2·0531	2·4942	2·8062	3·0483

The values of X and Y are plotted as pairs on graph paper (fig 8.30), choosing the largest scales possible on the paper available, as accurately as possible. It then appears that the points very nearly lie on a straight line and this confirms that the law is of the given form. The fact that they do not lie exactly on a line can reasonably be said to be due to experimental errors and we draw the line which best fits the points. This is called the **line of best fit**. (There will of course be some difference of opinion as to the position of this line. Hence any results can only be approximate).

From the graph the gradient is approximately 2·5. Hence $b = 2·5$.

The intercept cannot be obtained directly from the graph but using the value already found for the gradient we can estimate the intercept.

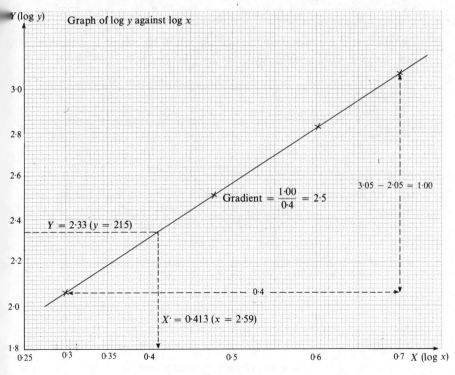

Fig 8.30

From fig 8.31, if the line were continued backwards with the same gradient, let c be the further drop in Y, to reach $X = 0$.

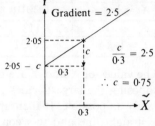

Then $\dfrac{c}{0\cdot3} = 2\cdot5$ and $c = 0\cdot75$

Hence when $X = 0$,
$Y = 2\cdot05 - 0\cdot75 = 1\cdot30$.
The intercept is thus $1\cdot30$ and
$\log a = 1\cdot30$ giving $a \approx 20$.

Fig 8.31

The required law is therefore $y = 20x^{2\cdot5}$.
If $y = 215$, $Y = \log y = 2\cdot33$. Reading from the graph we obtain $X = \log x = 0\cdot413$ giving $x = 2\cdot59$ approximately.

$$y = ab^x$$

Then $\log y = \log a + \log b^x = \log a + x \log b$.
So $Y = (\log b)x + \log a$ which is of the form $Y = mx + c$ where $m = \log b$ and $c = \log a$.
In this case if we plot $Y(\log y)$ against x and the result is a straight line, the law will be confirmed. Then a and b can be found from the graph.

Example 16

The following experimental results were obtained for two variables x and y:

x	1	2	3	5	7
y	57	217	2630	11 900	17 200

Test the assumption that these variables are connected by a law of the form $y = ab^x$, where a and b are constants, by drawing a graph. If the assumption is justified, find approximate values for a and b. Find the value of y corresponding to $x = 4\cdot5$.

If $y = ab^x$, then $Y = \log y = (\log b)x + \log a$.
We therefore plot values of $\log y$ against those of x.

x	1	2	3	5	$\dot{7}$
y	57	217	2630	11 900	17 200
$Y = \log y$	1·76	2·34	3·42	4·08	5·24
	(correct to two decimal places)				

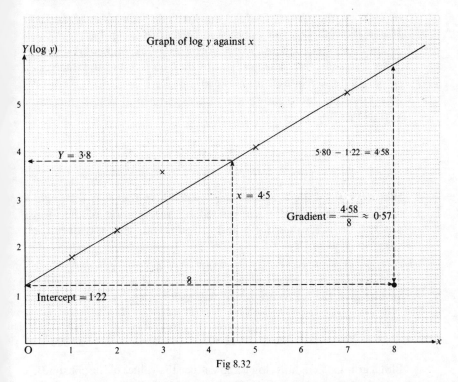

Fig 8.32

The resulting graph is shown in fig 8.32. All but one of the points lie very well on a straight line. It is reasonable to say therefore that the law is confirmed and that the value $y = 2630$ when $x = 3$ was incorrectly determined and can therefore be ignored. (From the graph, when $x = 3$, $Y = \log y = 2.93$ and hence it would appear that the correct value of y in this case is about 851).

From the graph, we find that the gradient ≈ 0.57 and the intercept on the Y-axis ≈ 1.22.

Then $\log b = 0.57$ or $b = 3.7$ and $\log a = 1.22$ giving $a = 17$.

The law is of the form $y = 17 \times (3.7)^x$.

When $x = 4.5$, reading from the graph gives $Y = \log y = 3.8$ and thus $y = 6310$.

Exercise 8.9

1 An equation has the form $y = mn^x$, where y and x are variables and m and n are constants. A set of values of x and y are plotted with $\log y$ on the y-axis and x on the x-axis. A straight line of best fit is plotted and it is found to have a gradient of 0·38 and an intercept of 1·48 on the y-axis. Deduce approximate values for m and n. If $y = 190\,000$, find the value of x, correct to 2 sig. figs.

2 A graph of log y against log x, for a set of experimental results, gives a straight line. The gradient of the line is 2 and the intercept is 1·7. Find the relationship between x and y.

3 The following table shows a set of experimental results for two variables, x and y:

x	2	4	8	12
y	50	70	99	121

Plot a graph of log x against log y, and show that it is a straight line. Assuming x and y are related by an equation of the form $y = px^n$, deduce values for p and n. From your graph, deduce the value of y when $x = 10$.

4 A scientific law obeys an equation of the form $y = mn^x$ where x and y are variables and m and n are constants. Experimental results give the following set of values for x and y:

x	1	2	3	4	5
y	264	1740	11 500	75 900	500 000

Plot a graph of x against log y and deduce the values of the constants, m and n. For what value of x is $y = 4500$? (Use your graph to determine the value of x correct to 2 significant figures).

5 It is thought that two variables, λ and V, are connected by the equation $\lambda V^2 = bV + S$, where b and S are constants. The following series of experimental results was taken:

V	1	2	5	8
λ	12	3·5	0·8	0·406

Plot values of λV^2 on the y-axis against values of V on the x-axis to verify that the variables do obey the equation. From your graph deduce the approximate value of the constants, b and S, and also find the value of λ when $V = 4$.

6 The following values of x and y were obtained in an experiment:

x	2	3	4	5	6
y	45	100	370	710	2293

Plot log y against x. Assuming that the relation between x and

y is of the form $y = an^x$, estimate *from your graph* approximate values for a and n.   (C)

7 The following pairs of values for x and y have been found by experiment:

y	5·00	6·23	7·36	7·85	8·61	9·37
x	10	20	40	50	80	100

It is thought that a law of the form $y = a \log x + b$, where a and b are constants, connects x and y. By plotting y against $\log x$, show that such a law is approximately valid. Estimate, from your graph, the value of a.   (C)

8 The following pairs of values for p and μ have been found by experiment, and some corresponding values of $\mu \sqrt{p}$ have been calculated:

p	1	2	3	4	5	8
μ	5·9	5·6	5·8	6·0	6·2	7·0
$\mu \sqrt{p}$		7·9	10·1		13·9	

It is thought that a law of the form $\mu\sqrt{p} = ap + b$, where a and b are constants, connects p and μ. By plotting $\mu\sqrt{p}$ against p, show that such a law is approximately valid and find suitable values for a and b.   (C)

156

9 Calculus (1): Differentiation

This chapter will introduce a new branch of Mathematics called **Calculus** which was largely established by Newton and Leibnitz in the seventeenth century. It involves important ideas and techniques. A simple starting point is to consider the problem of finding the gradient at a point of a curve.

Gradients

The gradient of a straight line is the ratio $\dfrac{\text{increase in } y}{\text{increase in } x}$ in going from one point to another on the line (fig 9.1). You have also seen that if (x_1, y_1), (x_2, y_2) are two points on a line, the gradient

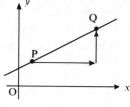

Fig 9.1

equals $\dfrac{y_2 - y_1}{x_2 - x_1}$ which is equivalent to the definition given above. This gradient will of course be constant all along the line and equal to the tangent of the angle made with the positive direction of the x-axis.

For a curve, the idea of a gradient is more complicated. A curve changes direction from one point to another and we therefore define the gradient at a point to be the gradient of the *tangent line at that point* (fig 9.2). The problem is to find this gradient, if we know the equation of the curve.

A rough way would be to draw the graph as accurately as possible and draw in the tangent by eye, but this is obviously an

Fig 9.2

arbitrary method. We now develop a method of finding a **gradient function** associated with the given function represented by the curve.

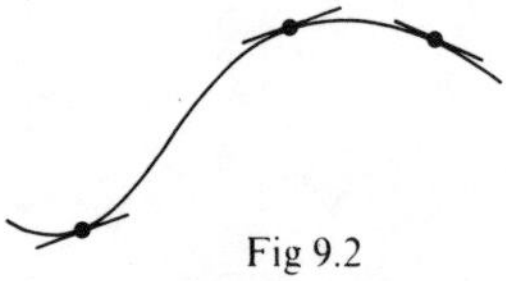

Example 1

Consider the curve $y = x^2$ *(fig 9.3).* We try to find the gradient at the point P $(3, 9)$ where $x = 3$ i.e., the gradient of the tangent at that point. For a start, take another point Q_1 on the curve where Q_1 is $(4, 16)$. Draw PR parallel to the x-axis (fig 9.4).

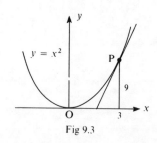

Fig 9.3

Then the gradient of $PQ_1 = \dfrac{Q_1R}{PR}$

$$= \dfrac{16 - 9}{1}$$

$$= 7$$

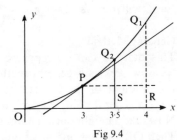

Fig 9.4

This is not the gradient of the tangent at P but it is a rough approximation. It is in fact the average gradient over the stretch PQ_1.

We can improve on this by taking a second point Q_2, closer to P on the curve, where Q_2 is $(3·5, 12·25)$. Then the gradient of PQ_2 is

$$\dfrac{Q_2S}{PS} = \dfrac{12·25 - 9}{0·5} = 6·5$$

Clearly this is a better approximation to the gradient of the tangent at P. It seems hopeful therefore to continue with this method, bringing Q closer and closer to P on the curve. The following table shows the successive values of the gradient of PQ at each new position.

Coordinates of Q	QR	PR	Gradient
(3·4, 11·56)	2·56	0·4	6·4
(3·3, 10·89)	1·89	0·3	6·3
(3·2, 10·24)	1·24	0·2	6·2
(3·1, 9·61)	0·61	0·1	6·1
(3·01, 9·060 1)	0·060 1	0·01	6·01
(3·001, 9·006 001)	0·006 001	0·001	6·001
(3·000 1, 9·000 600 01)	0·000 600 01	0·0001	6·0001

It would appear that the closer Q gets to P, the closer the gradient approaches the value 6. The pattern of the above results suggests that as we continue with the process, the gradient becomes as near 6 as we please, and the process can continue indefinitely. We say, as Q tends to P $(Q \to P)$ the gradient of PQ $\to 6$. We cannot directly find the gradient of the tangent as Q must always be a neighbouring point to P on the curve (however close) and PQ is never the actual tangent at P. The above sequence is an example of a **limiting** process. As $Q \to P$ so the gradient of PQ tends to a limiting value of 6 and this limiting value is the gradient of the tangent at P.

Repeat the process for the same curve at the points where $x = 2$, $x = 5$ and $x = -3$. (The limiting values will be $4, 10$ and -6 and these will be the gradients of the respective tangents).

General Method
This method can be repeated for every point on the curve. What we now seek is a general method which will produce a formula for the gradient of the tangent at any given point.
Take a general point P (x, x^2) on the curve $y = x^2$.
Now take a neighbouring point Q whose x-coordinate is $x + h$. (fig 9.5).
Then $QR = (x + h)^2 - x^2$ and $PR = h$.
The size of h is not fixed and we can vary it as we please, except that it must not equal 0.

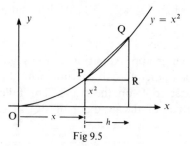

Fig 9.5

$$\text{Gradient of PQ} = \frac{QR}{PR} = \frac{(x + h)^2 - x^2}{h}$$

$$= \frac{x^2 + 2xh + h^2 - x^2}{h}$$

$$= \frac{2xh + h^2}{h}$$

$$= 2x + h.$$

Clearly so long as h is positive, this value will be greater than $2x$, but as $h \to 0$ the gradient of PQ will tend to $2x$. Hence the limiting value will be $2x$ and this will be the gradient of the tangent at P. When $x = 3$ for instance, the gradient of the tangent will equal 6 (as we found above) and so on.

Gradient Function

For the curve $y = x^2$, the gradient of the tangent at any point is given by the value of $2x$ and $2x$ is the **gradient function** for this curve. Every curve will have its own special gradient function which can (at present) only be found by a limiting process as above.

Example 2
Find the gradient function for the curve $y = 3x^2 - 2$.

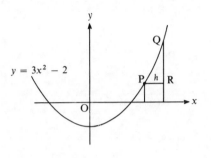

Fig 9. 6

Take a point P, coordinates (x, y), i.e. $(x, 3x^2 - 2)$ on the curve. (fig. 9.6) Now take a neighbouring point Q whose x-coordinate is $x + h$. Then the y coordinate of Q is $3(x + h)^2 - 2$.
Then QR $= 3(x + h)^2 - 2 - (3x^2 - 2)$ and PR $= h$.

$$\text{Gradient of PQ} = \frac{3(x + h)^2 - 2 - (3x^2 - 2)}{h}$$

$$= \frac{3x^2 + 6xh + 3h^2 - 2 - 3x^2 + 2}{h}$$

$$= 6x + 3h.$$

Now as Q gets closer to P, $h \to 0$ and $3h \to 0$.
Hence the limiting value of the gradient of PQ will be $6x$ and so the gradient function of $3x^2 - 2$ is $6x$. For example the gradient at the point where $x = 3$ is 18; where $x = -1$ the gradient is -6 and so on.

Example 3

Find the gradient function for $\dfrac{1}{x}$.

The curve is a rectangular hyperbola (fig 9.7).

P is the point $(x, \dfrac{1}{x})$ and Q is $(x + h, \dfrac{1}{x + h})$

RQ $= h$ and the y-step from P to Q is negative.

$$PR = \frac{1}{x+h} - \frac{1}{x}$$

$$= \frac{x - (x+h)}{x(x+h)}$$

$$= \frac{-h}{x(x+h)}$$

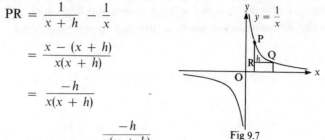

Fig 9.7

Hence the gradient of PQ $= \dfrac{\dfrac{-h}{x(x+h)}}{h}$

$$= -\frac{h}{x(x+h)} \times \frac{1}{h}$$

$$= \frac{-1}{x(x+h)}.$$

Now as Q → P, $h \to 0$ and $\dfrac{-1}{x(x+h)} \to -\dfrac{1}{x^2}$. The gradient function

of $\dfrac{1}{x}$ is $-\dfrac{1}{x^2}$. Note that this is always negative. The gradient at all

points of the curve $y = \dfrac{1}{x}$ is negative. For example, when $x = 2$,

the gradient is $-\frac{1}{4}$ and so on.

Notation

If a curve is given by the equation $y = f(x)$ the gradient function is
written $f'(x)$. An alternative notation which is also widely used will
be given later.

Exercise 9.1

For the curves given by the following functions find the gradient
function by the above limiting process:

1 $y = 2x^2$ **2** $y = 2x^2 - 5$ **3** $y = 2x^2 + 7$
4 $y = x^2 + x - 1$
5 $y = x^3$ (NB: $(x+h)^3 = x^3 + 3x^2h + 3xh^2 + h^3$)

6 $y = x^3 + x^2 - 2x$ **7** $y = \dfrac{2}{x}$ **8** $y = 2x^3$

Find the gradients of the tangents to the following curves at the points
specified. (Use the gradient functions already found).

9 $y = 2x^2$ when $x = 3$ **10** $y = 2x^2 - 5$ when $x = -2$
11 $y = 2x^2 + 7$ when $x = 3$ **12** $y = x^2 + x - 1$ when $x = 2$

13 $y = x^3$ when $x = -2$ **14** $y = \dfrac{2}{x}$ when $x = -4$

An alternative notation for the gradient function

Once again take the curve $y = x^2$ and the point $P(x, x^2)$ (fig 9.8).
Now consider the neighbouring point Q whose x-coordinate is $x + \delta x$.
δx (read delta x) is the *increment* or increase in the value of x. Note
carefully that δx is one symbol, like sin x
or log x and does not denote $\delta \times x$. The
ordinate of Q will be written $y + \delta y$,
where δy is the corresponding increment
in y.

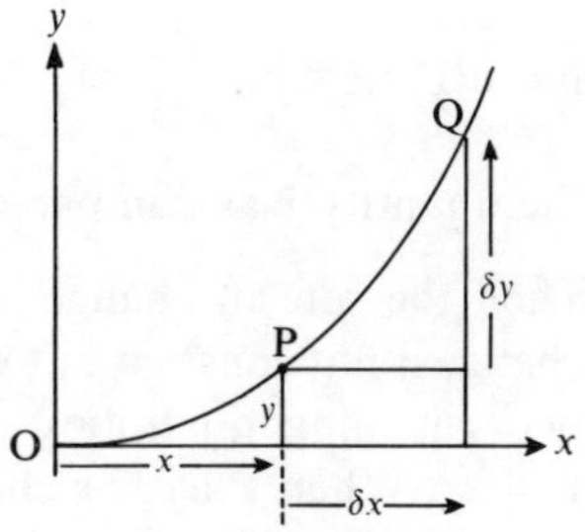

Now at P $y = x^2$ and at Q

$$y + \delta y = (x + \delta x)^2$$

since both points lie on the

curve $y = x^2$.

Therefore $\delta y = (x + \delta x)^2 - x^2$

$$= 2x\delta x + (\delta x)^2$$

Hence the gradient of PQ $= \dfrac{\delta y}{\delta x}$

$$= 2x + \delta x.$$

Fig 9.8

Now as $Q \to P$, $\delta x \to 0$ and the limiting value of $\dfrac{\delta y}{\delta x}$ will be $2x$.

To signify that this is the limiting value of $\dfrac{\delta y}{\delta x}$ we rewrite it in the

form $\dfrac{dy}{dx}$ (read *dee y* by *dee x*) and we have $\dfrac{dy}{dx} = 2x$.

The method is the same as before but the notation $\dfrac{dy}{dx}$ (due to Leibnitz)
instead of f$'(x)$ for the gradient function is extremely useful, as it suggests
how the result was obtained by taking the ratio $\dfrac{\delta y}{\delta x}$ to the limit. However the student must be careful. $\dfrac{dy}{dx}$ is not a fraction but a single
symbol which signifies the gradient of the tangent on the curve y
$= f(x)$.

Gradient function of $y = f(x)$ is f$'(x) = \dfrac{dy}{dx}$

where $\dfrac{dy}{dx}$ is the limiting value of $\dfrac{\delta y}{\delta x}$ as $\delta x \to 0$.

Differentiation: Rate of Change

This process of finding $\dfrac{dy}{dx}$ or $f'(x)$ is called **differentiation** and $\dfrac{dy}{dx}$ is also called the **derivative** of y with respect to x. (A further name is the **differential coefficient**). When we are considering the graphical relationship between y and x, as given by the curve $y = f(x)$ then $\dfrac{dy}{dx}$ is of course the gradient function for this curve. A wider meaning however is that $\dfrac{dy}{dx}$ measures the instantaneous rate of change of the quantity y as compared with x. For example, if $y = x^2$, $\dfrac{dy}{dx} = 2x$.

Then the rate of change of y (compared with x) is $2x$. This rate of change is not constant but varies with the value of x. You must imagine two changing quantities, y and x, bound together by the relation $y = x^2$. When x has reached the value 3 (say) the value of y is 9 but y is changing (at that point) 6 times as much as x is changing. If the relation between x and y is shown by a graph, the rate of change (y compared with x) is given by the gradient of that graph.

The notation can be generally applied. For example if s is a function of t, $s = f(t)$, then $\dfrac{ds}{dt}$ is the derivative of s wrt t (wrt is a useful abbreviation for "with respect to"). $\dfrac{ds}{dt}$ is the rate of change of s wrt t.

Again, if $A = r^2$, $\dfrac{dA}{dr} = 2r$. A is changing $2r$ times as much as r is changing.

Example 4

If $y = 3x^2 - x + 5$, differentiate y wrt x. (We could also say find $\dfrac{dy}{dx}$).

We work from first principles without the graphical illustration. We have $y = 3x^2 - x + 5$. An increase in x to $x + \delta x$ produces a corresponding increase in y to $y + \delta y$. (In this context, "increase" is meant algebraically and a negative increase is possible).

Then $y + \delta y = 3(x + \delta x)^2 - (x + \delta x) + 5$

and so $\delta y = 3(x + \delta x)^2 - (x + \delta x) + 5 - (3x^2 - x + 5)$
$= 3x^2 + 6x\delta x + 3(\delta x)^2 - x - \delta x + 5 - 3x^2 + x - 5$
$= 6x\delta x + 3(\delta x)^2 - \delta x$

Therefore $\dfrac{\delta y}{\delta x} = 6x + 3(\delta x) - 1$.

If we now consider $\delta x \to 0$ the limiting value of $\dfrac{\delta y}{\delta x}$ would be $6x - 1$

and so $\dfrac{dy}{dx} = 6x - 1$

Hence the gradient of the tangent to the curve $y = 3x^2 - x + 5$ at $x = 2$ would be 11.

Exercise 9.2

Taking an increment δx find the corresponding change δy in y and then find the limiting value of $\dfrac{\delta y}{\delta x}$ for each of the following:

1 $y = 6x^2 - 4$ **2** $y = \frac{1}{2}x^2 + 1$ **3** $y = 2x^3$
4 $y = x^2 - x + 3$ **5** $y = 2x^2 + 3x - 7$
6 For each of the above curves, find the gradient of the tangents where $x = -1, 0, 2$.

Summary

In general, if $y = f(x)$ then $\dfrac{dy}{dx}$ is the limiting value of $\dfrac{f(x + \delta x) - f(x)}{\delta x}$ as $\delta x \to 0$. In our work we assume that this limiting value will exist and can be found by this process.

We write $\dfrac{dy}{dx} = \lim_{\delta x \to 0} \dfrac{f(x + \delta x) - f(x)}{\delta x}$

where $\lim_{\delta x \to 0}$ means the limiting value as $\delta x \to 0$.

Differentiation of the general term: ax^n

You will have noticed a certain common pattern in the derivatives so far obtained and have perhaps guessed that the derivative has some connection with the power of the variable (x) used. Look at these results already obtained:

Function	Derivative
x^2	$2x$
$3x^2 - 2$	$6x$ (i.e. $3 \times 2x$)
$\dfrac{1}{x}$ (i.e. x^{-1})	$-\dfrac{1}{x^2}$ (i.e. $-x^{-2}$)
x^3	$3x^2$
$x^2 + x - 1$	$2x + 1$
$6x^2 - 4$	$12x$ (i.e. $6 \times 2x$)

From these results it would appear that the derivative of x^n is nx^{n-1}. For example the derivative of x^2 is $2x$ and of x^3 is $3x^2$. The derivative of x^{-1} is $(-1)x^{-1-1}$ i.e. $-x^{-2}$.

The derivative of $3x^2$ is $3 \times$ the derivative of x^2, i.e. $3 \times 2x = 6x$. The derivative of $-2x$ is -2: this is correct as the gradient of the line $y = -2x$ is -2 (constant). Or we can say, the derivative of x is $1x^{1-1} = x^0 = 1$.

The derivative of a constant is zero: the gradient of the line $y = 5$ (for example) is always zero.

So we can say:

$$\boxed{\text{If } y = ax^n, \frac{dy}{dx} = nax^{n-1} \quad \text{where } a \text{ is a constant}}$$

(This result will be proved when we have studied the binomial theorem).

Example 5

If $y = ax^2 + bx + c$, find $\dfrac{dy}{dx}$.

Take an increment δx in x and a corresponding increment δy in y.

Then
$$y + \delta y = a(x + \delta x)^2 + b(x + \delta x) + c$$
$$= ax^2 + 2ax\delta x + a(\delta x)^2 + bx + b\delta x + c$$

So
$$\delta y = 2ax\delta x + a(\delta x)^2 + b\delta x$$

and
$$\frac{\delta y}{\delta x} = 2ax + a\delta x + b \to 2ax + b \text{ as } \delta x \to 0$$

Hence
$$\frac{dy}{dx} = 2ax + b.$$

The derivatives of the separate terms ax^2, bx, c are $2ax$, b and 0 respectively. Hence the derivative of y is the sum of the separate derivatives.

To differentiate single terms, express them in index form first.

Example 6

If $y = \sqrt{x}$, find $\dfrac{dy}{dx}$.

Write $y = x^{\frac{1}{2}}$, then $\dfrac{dy}{dx} = \frac{1}{2}x^{\frac{1}{2}-1} = \frac{1}{2}x^{-\frac{1}{2}} = \dfrac{1}{2\sqrt{x}}$.

Example 7

If $y = \dfrac{3}{x^2}$, find $\dfrac{dy}{dx}$.

Write y as $3x^{-2}$.

Then $\dfrac{dy}{dx} = 3 \times (-2) \times x^{-2-1}$

$\qquad = -\dfrac{6}{x^3}.$

Example 8

If $s = 4t^3 - 2t^2 - \dfrac{1}{t}$, find $\dfrac{ds}{dt}$.

$s = 4t^3 - 2t^2 - t^{-1}$

and so $\dfrac{ds}{dt} = 12t^2 - 4t - (-1)t^{-1-1}$

$\qquad = 12t^2 - 4t + \dfrac{1}{t^2}.$

Exercise 9.3

Find the derivatives of the following wrt x:

1 $5x^2$ **2** $3x^3$ **3** $10x^4$ **4** $8x$

5 8 **6** $x^2 + 5x$ **7** $x - \dfrac{1}{x}$ **8** $9x + \dfrac{5}{x}$

9 $4x^3 - 2x + 3$ **10** $\sqrt[3]{x}$ **11** $\dfrac{1}{x^3}$ **12** $x^3 - \dfrac{1}{x} + \dfrac{3}{x^2}$

13 $\dfrac{1}{\sqrt{x}}$ **14** $4\sqrt{x} - 5$ **15** $4x^3 - \dfrac{3}{x^4}$

16 If $y = 2t^2 - \dfrac{1}{t}$, find $\dfrac{dy}{dt}$.

17 If $s = 3t^3 + 2t^2 - 7t + 3$, find $\dfrac{ds}{dt}$.

18 Find the gradient at the point where $x = 1$ on the curve
$$y = x^3 + x^2 - 3.$$

19 Find the gradient of the tangent where $x = 9$ on the curve $y = 3\sqrt{x} - 4$ ($\sqrt{x}$ meaning the positive root).

20 If $A = \frac{1}{2}x^2 - 3x + 4$, what is the rate of change of A wrt x when $x = 5$?

10 Calculus (2): Techniques in Differentiation

Polynomials

As we have seen, if $y = ax^2 + bx + c$, then $\dfrac{dy}{dx} = 2ax + b$. The derivative of y is the sum of the derivatives of the separate terms. This result is true for any polynomial or collection of terms involving powers of a variable. Hence expressions containing brackets and fractional terms can be differentiated only if simplified first.

Example 1

Differentiate wrt x **(a)** $(2x - 1)^2$, **(b)** $(3x - 1)(2x + 4)$.

(a) Expand first: $\quad y = (2x - 1)^2$

$$= 4x^2 - 4x + 1$$

and now $\quad \dfrac{dy}{dx} = 8x - 4.$

(b) Again expand first: $\quad y = (3x - 1)(2x + 4)$

$$= 6x^2 + 10x - 4$$

and $\quad \dfrac{dy}{dx} = 12x + 10.$

However this method would not be practicable for expressions such as $(3x - 2)^{10}$ or $(3x^2 - 1)^5 (x + 4)^3$ nor possible for $\sqrt{3x - 4}$. More general methods will be seen later to handle such expressions.

Example 2

Differentiate $\dfrac{3x^4 + 2x^2 - 1}{2x^2}$ *wrt x.*

Express as a polynomial by division:

$$y = \frac{3x^4 + 2x^2 - 1}{2x^2}$$

$$= \frac{3x^4}{2x^2} + \frac{2x^2}{2x^2} - \frac{1}{2x^2}$$

$$= \frac{3}{2}x^2 + 1 - \frac{1}{2}x^{-2}$$

Hence $\quad \dfrac{dy}{dx} = 3x + x^{-3} = 3x + \dfrac{1}{x^3}.$

Note however that an expression such as $\dfrac{x^4 + x^2 - 1}{x^3 + 3}$ cannot be dealt with in this way, as the denominator is not a monomial (single term).

Exercise 10.1

Differentiate wrt x:

1 $5x^4 - 3x^2 + 7x - 7$ **2** $(x + 1)^2$ **3** $(2x - 3)^2$

4 $(x - 1)(2x + 1)$ **5** $\dfrac{x + 1}{x}$ **6** $\dfrac{x^2 + 5}{x}$

7 $\dfrac{3x^3 - x + 4}{2x}$ **8** $(x - 2)^3$ **9** $\dfrac{(x - 3)^2}{2x}$

10 $(3x - 2)(x + 1)^2$

Composite Functions

A function such as $y = \dfrac{1}{(3x - 4)^2} = (3x - 4)^{-2}$ poses a problem for differentiation. We cannot express it as a polynomial. To find $\dfrac{dy}{dx}$ we treat it as a **composite** function, built up in two stages from a core function $(3x - 4)$ which we will call u and then taking $y = u^{-2}$. So we have $u = 3x - 4$ and $y = u^{-2}$.

Consider an increment δx in x. This will produce an increment δu in u which in turn will change y by an amount δy.

Now by ordinary fractions, $\dfrac{\delta y}{\delta x} = \dfrac{\delta y}{\delta u} \times \dfrac{\delta u}{\delta x}$ as δu cancels.

The important point now is that we shall assume that

$$\lim_{\delta x \to 0} \frac{\delta y}{\delta x} = \left(\lim_{\delta u \to 0} \frac{\delta y}{\delta u} \right) \times \left(\lim_{\delta x \to 0} \frac{\delta u}{\delta x} \right)$$

i.e. that $\qquad \dfrac{dy}{dx} = \dfrac{dy}{du} \times \dfrac{du}{dx}$

Then in the above problem, as $\dfrac{du}{dx} = 3$ and $\dfrac{dy}{du} = -\dfrac{2}{u^3}$ we have

$\dfrac{dy}{dx} = -\dfrac{2}{u^3} \times 3 = -\dfrac{6}{(3x - 4)^3}$, substituting for u.

Interpreting this important method, the total rate of change of y wrt x is the product of the rates of change of y wrt u and u wrt x, and this rule will apply to all composite functions. Note that we could not directly say that $\dfrac{dy}{dx} = \dfrac{dy}{du} \times \dfrac{du}{dx}$ as these are not fractions which can be manipulated. It must be admitted however that the notation is helpful in suggesting the result achieved.

This method is also called the **function of a function** rule or sometimes the **chain rule**. With practice the student will be able to apply the rule readily and write down the result.

Example 3

Differentiate **(a)** $\sqrt{2x + 5}$, **(b)** $2(3x^2 - x + 7)^3$ *wrt x.*

(a) Take $u = 2x + 5$ and $y = u^{\frac{1}{2}}$.

Now $\qquad\qquad\qquad \dfrac{dy}{dx} = \dfrac{dy}{du} \times \dfrac{du}{dx}$

and $\qquad\qquad\qquad \dfrac{du}{dx} = 2,\ \dfrac{dy}{du} = \tfrac{1}{2}u^{-\frac{1}{2}}.$

Hence $\qquad\qquad \dfrac{dy}{dx} = \tfrac{1}{2}u^{-\frac{1}{2}} \times 2 = u^{-\frac{1}{2}}$

$$= \frac{1}{\sqrt{2x + 5}}.$$

Now bearing the rule in mind, this result can be obtained directly as follows: $\qquad\qquad y = (2x + 5)^{\frac{1}{2}}$

Take $2x + 5$ as the core of this composite function.

Differentiate y as a power, wrt to the core.

Then multiply by the derivative of the core wrt x.

So $\qquad\qquad\qquad \dfrac{dy}{dx} = \tfrac{1}{2}(2x + 5)^{-\frac{1}{2}} \times 2$

$$= \frac{1}{\sqrt{2x + 5}} \text{ as before.}$$

(b) $y = 2(3x^2 - x + 7)^3$

The core function is $3x^2 - x + 7$.

Then $\qquad\qquad \dfrac{dy}{dx} = 2 \times 3(\text{core})^2 \times (\text{derivative of the core})$

$$= 6(3x^2 - x + 7)^2\,(6x - 1).$$

Exercise 10.2

Differentiate wrt x:

1 $(2x + 5)^3$	**2** $(2x + 5)^5$	**3** $(3x - 1)^3$	**4** $(3x - 1)^7$
5 $(3x - 1)^9$	**6** $(4x - 3)^4$	**7** $(4x - 3)^7$	**8** $\sqrt{4x - 3}$

9 $(7x + 1)^3$ **10** $(ax + b)^4$ a, b constants   **11** $(x^2 + 1)^4$

12 $(x^2 - 1)^5$ **13** $(3x^2 - 1)^2$ **14** $(2x^3 - 1)^3$ **15** $(2x^3 - 1)^4$

16 $(x^2 + x + 1)^3$ **17** $(x^2 + 3x - 1)^2$ **18** $(x^2 - 2x + 5)^4$

19 $(4x^2 - x + 5)^3$ **20** $(2x^3 - x^2 + 1)^4$ **21** $(x^3 + x^2 + x + 1)^2$

22 $\sqrt{3x - 1}$ **23** $\sqrt{4x + 3}$ **24** $\sqrt{x^2 - x + 1}$

25 $\sqrt{3x^2 - 2x + 1}$ **26** $\dfrac{1}{4x - 3}$ **27** $\dfrac{1}{3x + 5}$

28 $2(5x^3 - 4)^4$ **29** $\sqrt[3]{3x + 4}$ **30** $\sqrt{x - \dfrac{1}{x}}$

31 $\sqrt[3]{2x^3 - 5}$ **32** $\dfrac{1}{x^2 + x - 3}$ **33** $(4x^4 - x^2 + 2)^3$

34 $\dfrac{1}{\sqrt{3x - 5}}$ **35** $\dfrac{1}{x^3 - 6}$ **36** $\dfrac{1}{\sqrt{x^2 + 2x - 4}}$

37 $\dfrac{1}{(2x - 3)^2}$ **38** $\dfrac{1}{(2x^2 - x - 6)^2}$

We can now extend our technique in differentiating to deal with complicated expressions, those which are products of two or more expressions and those which are in the form of a quotient.

The Product Rule

$y = (3x - 1)^3 (x^2 + 5)^2$ is the product of two expressions, $(3x - 1)^3$ and $(x^2 + 5)^2$. Each of these can be differentiated wrt x as composite functions. How is $\dfrac{dy}{dx}$ found from these two derivatives?

Let $y = uv$ where u and v are each functions of x. (In the above, u would be $(3x - 1)^3$ and v would be $(x^2 + 5)^2$.

Take an increment δx in x which will in turn produce increments δu in u and δv in v, finally producing a change δy in y.

Then $y + \delta y = (u + \delta u)(v + \delta v) = uv + v\,\delta u + u\,\delta v + \delta u \delta v$

$\therefore \qquad \delta y = v\,\delta u + u\,\delta v + \delta u \delta v$

and $\qquad \dfrac{\delta y}{\delta x} = v\dfrac{\delta u}{\delta x} + u\dfrac{\delta v}{\delta x} + \dfrac{\delta u}{\delta x}\delta v$

Now let $\delta x \to 0$, then $\delta u \to 0$, $\delta v \to 0$.

$\therefore \dfrac{\delta u}{\delta x} \to \dfrac{du}{dx},\ \dfrac{\delta v}{\delta x} \to \dfrac{dv}{dx}$ and $\dfrac{\delta y}{\delta x} \to \dfrac{dy}{dx}.$

Then in the limit, $\dfrac{dy}{dx} = v\dfrac{du}{dx} + u\dfrac{dv}{dx}$ (the third term $\to 0$
as $\delta v \to 0$)

So, if $y = uv$, we have the product rule

$$\frac{dy}{dx} = v\frac{du}{dx} + u\frac{dv}{dx}$$

Example 4

Differentiate $(3x - 2)(x^2 + 3)$ *wrt x.*

Take u as $3x - 2$ and v as $x^2 + 3$.

Then $\dfrac{du}{dx} = 3$ and $\dfrac{dv}{dx} = 2x$.

Hence $\dfrac{dy}{dx} = (x^2 + 3) \times 3 + (3x - 2) \times (2x)$

$$= 3x^2 + 9 + 6x^2 - 4x$$

$$= 9x^2 - 4x + 9.$$

Check by expanding the original expression and differentiating.

Example 5

Differentiate $y = x^2(2x - 5)^4$ *wrt x*

Take u as x^2 and v as $(2x - 5)^4$

$$\frac{dy}{dx} = \underset{\underset{\dfrac{du}{dx}}{\downarrow}}{2x}\underset{v}{(2x - 5)^4} + \underset{u}{x^2} \times \underset{\dfrac{dv}{dx}}{\underline{4(2x - 5)^3 \times 2}}$$

Now simplify as far as possible

$$\frac{dy}{dx} = 2x(2x - 5)^3 (2x - 5 + 4x)$$

$$= 2x(2x - 5)^3 (6x - 5).$$

Example 6

Differentiate $y = (3x - 1)^3 (x^2 + 5)^2$

$$\frac{dy}{dx} = 3(3x - 1)^2 \times 3 \times (x^2 + 5)^2 + (3x - 1)^3 \times 2(x^2 + 5) \times 2x$$

$$= (3x - 1)^2 (x^2 + 5) [9(x^2 + 5) + 4x(3x - 1)]$$

$$= (3x - 1)^2 (x^2 + 5) (21x^2 - 4x + 45).$$

Do not use the product rule if one of the elements is a constant, as for example in $y = 2\pi(x^2 - 3x + 1)^3$. Here 2π is a multiplying constant and $(x^2 - 3x + 1)^3$ is differentiated as a composite function.

The rule can be extended to cover triple products such as $y = uvw$ where u, v and w are each functions of x. Each factor is differentiated wrt x while the other two remain untouched. The three results are then added together:

If $y = uvw$, then $\dfrac{dy}{dx} = \dfrac{du}{dx}vw + \dfrac{dv}{dx}uw + \dfrac{dw}{dx}uv.$

Exercise 10.3

Differentiate these functions wrt x. Simplify as far as possible, leaving your answers in factor form.

1 $(2x - 1)(x + 4)^2$ **2** $x(x^2 - 1)^5$

3 $(x^2 - 1)(x^3 + 1)^2$ **4** $(3x - 1)^2(x^2 - 5)$

5 $3x^3(x + 5)^4$ **6** $(3x - 2)^2(2x - 3)^3$

7 $(4x - 3)(x^2 - 2x - 8)^2$ **8** $(x^2 + x - 5)(x^2 - x + 3)$

9 $\sqrt{x}(x + 3)^2$ **10** $x(x + 1)(x - 3)$

11 $x^2(x - 1)(x + 2)^2$ **12** $(x + 1)(x - 1)^2(x + 3)$

13 Prove the product rule for three functions u, v, w by taking uvw as a product of u and (vw).

The Quotient Rule

A method can also be found to obtain the derivative of a function expressed in the form of a quotient $\dfrac{u}{v}$, such as $\dfrac{x^2 + 1}{3x - 2}$. (A quotient such as $\dfrac{x^2 + 1}{3x}$ should be divided out first and then differentiated).

To save space we sometimes write quotients such as $\dfrac{u}{v}$ in the form

u/v e.g. $\dfrac{3x^2 + 1}{2x - 5}$ as $(2x^2 + 1)/(2x - 5)$.

Consider $y = \dfrac{u}{v}$ where u and v are each functions of x. Let δx be an increment in x, thus causing corresponding increments δu, δv, δy in u, v and y respectively.

Then $y + \delta y = \dfrac{u + \delta u}{v + \delta v}$ and hence $\delta y = \dfrac{u + \delta u}{v + \delta v} - \dfrac{u}{v} = \dfrac{v\delta u - u\delta v}{v(v + \delta v)}$

Therefore $\dfrac{\delta y}{\delta x} = \dfrac{v\dfrac{\delta u}{\delta x} - u\dfrac{\delta v}{\delta x}}{v(v + \delta v)}$ dividing both sides by δx.

Now to find the limiting value of $\frac{\delta y}{\delta x}$, let $\delta x \to 0$, then δu and δv each $\to 0$, $\frac{\delta y}{\delta x} \to \frac{dy}{dx}$, $\frac{\delta u}{\delta x} \to \frac{du}{dx}$ and $\frac{\delta v}{\delta x} \to \frac{dv}{dx}$.

Hence
$$\boxed{\frac{dy}{dx} = \frac{v\dfrac{du}{dx} - u\dfrac{dv}{dx}}{v^2}}.$$
This is the quotient rule.

(Since the order of the terms in the numerator is important, the rule can easily be remembered by the phrase — "bottom dee top minus top dee bottom all over bottom squared" — not to be quoted in examinations!).

Example 7

Find $\dfrac{dy}{dx}$ *if* $y = \dfrac{(x-3)}{(x^2+2)}$

Here $u = x - 3$ and $v = x^2 + 2$

Then
$$\frac{dy}{dx} = \frac{(x^2+2)(1) - (x-3)(2x)}{(x^2+2)^2}$$
$$= \frac{x^2 + 2 - 2x^2 + 6x}{(x^2+2)^2}$$
$$= \frac{-x^2 + 6x + 2}{(x^2+2)^2}$$

Example 8

Differentiate $\dfrac{x-3}{\sqrt{x+1}}$ *wrt x.*

$u = x - 3$, $v = (x+1)^{\frac{1}{2}}$

Then
$$\frac{dy}{dx} = \frac{(x+1)^{\frac{1}{2}}(1) - (x-3)\frac{1}{2}(x+1)^{-\frac{1}{2}}}{(x+1)}.$$

To simplify, multiply numerator and denominator by $2(x+1)^{\frac{1}{2}}$

Then
$$\frac{dy}{dx} = \frac{2(x+1) - (x-3)}{2(x+1)^{\frac{3}{2}}}$$
$$= \frac{x+5}{2(x+1)^{\frac{3}{2}}}$$

Exercise 10.4

Differentiate wrt x, simplifying your answers as much as possible:

1 $\dfrac{(x + 1)}{(x + 2)}$ **2** $\dfrac{(2x - 1)}{(x + 1)}$ **3** $\dfrac{(3x - 5)}{(x - 5)}$

4 $\dfrac{x}{(x - 4)}$ **5** $\dfrac{(x^2 - 1)}{(x + 2)}$ **6** $\dfrac{x^2}{(2x - 3)}$

7 $\dfrac{(x^2 + x - 1)}{(x - 1)}$ **8** $\dfrac{(x^2 + x - 1)}{(x^2 - x - 1)}$ **9** $\dfrac{(2x^2 - 2)}{(x^3 + 4)}$

10 $\dfrac{x}{\sqrt{x - 2}}$ **11** $\dfrac{(3x - 1)}{\sqrt{2x + 1}}$ **12** $\dfrac{(x^3 - 1)}{(x^3 + 1)}$

13 Prove the quotient rule by differentiating $y = \dfrac{u}{v}$ as a product uv^{-1}.

Further Derivatives

If the function $y = 3x^2 - 2x + 1$ is differentiated wrt x we obtain $\dfrac{dy}{dx} = 6x - 2$.

This means that the rate of change of y as compared with x, $(6x - 2)$, is itself dependent on x and is therefore another function of x. We can therefore find *its* derivative wrt x, i.e. the rate of change of $\dfrac{dy}{dx}$ wrt x.

This second derivative should be written $\dfrac{d(\frac{dy}{dx})}{dx}$ but as it can be in-

terpreted as an *operator* $\dfrac{d}{dx}$ being used twice $\dfrac{d}{dx}\left(\dfrac{d}{dx}\right)y$, a less clumsy symbol is $\dfrac{d^2 y}{dx^2}$. (Read dee two y by dee x squared). Note that this is a single symbol and also note the difference between the second derivative $\dfrac{d^2 y}{dx^2}$ and the square of $\dfrac{dy}{dx}$, i.e. $\left(\dfrac{dy}{dx}\right)^2$.

Then for the above function, $\dfrac{d^2 y}{dx^2} = 6$ which shows that the rate of change of $\dfrac{dy}{dx}$ wrt x is constant $(=6)$ (fig 10.1).

The second derivative of $y = f(x)$ is sometimes also written as $f''(x)$.

$y = 3x^2 - 2x + 1$ shows the relationship between y and x. The gradient is given by $\dfrac{dy}{dx}$.

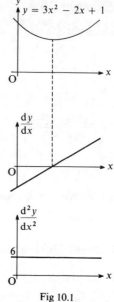

$\dfrac{dy}{dx} = 6x - 2$. Values of $\dfrac{dy}{dx}$ on this graph correspond to the gradients of the first graph. The gradient of this graph is given by $\dfrac{d^2y}{dx^2}$.

$\dfrac{d^2y}{dx^2} = 6$, showing that the rate of change of $\dfrac{dy}{dx}$ wrt x is constant ($=6$), i.e. the gradient of the previous curve is always 6.

Fig 10.1

We shall see later that the second derivative can be given a physical meaning in certain problems. Higher derivatives such as the third derivative $\dfrac{d^3y}{dx^3}$ etc. can be found but will not be used at this stage.

Exercise 10.5

Find $\dfrac{dy}{dx}$ and $\dfrac{d^2y}{dx^2}$ for the following:

1 $y = 3x^3$ **2** $y = 4x - 6$ **3** $y = x^4 + 2x^3 - 3x^2 + 1$

4 $y = \dfrac{1}{x}$ **5** $y = \sqrt{x}$ **6** $y = \dfrac{1}{x + 1}$ **7** $y = \dfrac{x + 1}{x - 1}$

8 If $s = 4t^3 - 3t^2 + t - 5$, find $\dfrac{ds}{dt}$ and $\dfrac{d^2s}{dt^2}$.

9 If $s = 3t^2 - \dfrac{1}{t}$, find $\dfrac{ds}{dt}$ and $\dfrac{d^2s}{dt^2}$.

10 If $y = ax^n$, where a is a constant, find $\dfrac{dy}{dx}$ and $\dfrac{d^2y}{dx^2}$.

11 If $y = 2x^5 + 4x^4 - 3x^3 + 2x^2 - 7$, find $\dfrac{dy}{dx}, \dfrac{d^2y}{dx^2}, \dfrac{d^3y}{dx^3}$.

12 If $y = \dfrac{1}{x}$, find $\dfrac{dy}{dx}, \dfrac{d^2y}{dx^2}, \dfrac{d^3y}{dx^3}, \dfrac{d^4y}{dx^4}, \dfrac{d^5y}{dx^5}$. Can you predict $\dfrac{d^7y}{dx^7}$, $\dfrac{d^{10}y}{dx^{10}}$?

13 If $y = \dfrac{x}{1+x}$, find $\dfrac{dy}{dx}$ and $\dfrac{d^2y}{dx^2}$ and show that

$$(1+x)\frac{d^2y}{dx^2} + 2\frac{dy}{dx} = 0.$$

14 If $y = \dfrac{1}{2-x}$ prove that $y\dfrac{d^2y}{dx^2} = 2\left(\dfrac{dy}{dx}\right)^2$.

Exercise 10.6   Miscellaneous

Differentiate wrt x using any appropriate method. Simplify your your answers wherever possible.

1 $3x^5 - 2x^2 + \dfrac{4}{x}$ **2** $(x-1)(2x^2 - 3x + 1)$

3 $x\sqrt{x+2}$ **4** $\dfrac{(x-1)}{(2x-3)}$ **5** $(3x^2 - 1)^3$

6 $\dfrac{x^3}{1+x^2}$ **7** $\sqrt{2x^2 + x - 1}$ **8** $\dfrac{1}{\sqrt{4x-3}}$

9 $x - \dfrac{1}{\sqrt{x}}$ **10** $\dfrac{x^3 - 2x + 1}{x^2}$ **11** $(x - \dfrac{2}{x})^2$

12 $\sqrt{x} + \dfrac{1}{\sqrt{x}}$ **13** $3\sqrt[3]{x} - \dfrac{1}{\sqrt[3]{x}}$ **14** $\sqrt{x^3 + 1}$

15 $\sqrt{x^2 + 2x}$ **16** $\dfrac{1}{x\sqrt{x}}$ **17** $(1 + \sqrt{x})^3$

18 $\sqrt[3]{3 - x^4}$ **19** $\dfrac{x^4 + 5x^2 + 3}{3x^3}$ **20** $(x-1)^3(3x-2)^2$

21 $(1 + ax)^n$ where a is a constant   **22** $\dfrac{4x^3}{5} - \dfrac{1}{2x^3}$

23 $(3x^2 - 5x + 7)^4$ **24** $\dfrac{(2x^2 - 1 + 3x)}{(x-3)^2}$

25 If $y = x - \dfrac{3}{x}$, find the value of $\dfrac{dy}{dx}$ when $x = 2$.

26 If $y = \dfrac{(x-1)}{(x+2)}$, find the value of $\dfrac{dy}{dx}$ when $x = -4$.

27 If $s = 3t^2 - \dfrac{1}{2t}$, find $\dfrac{ds}{dt}$ when $t = 2$.

28 Given that $y(1 - x^2)^{\frac{1}{2}} = 10$, prove that $(1 - x^2)\dfrac{dy}{dx} = xy$.   (C)

29 Given that $y = 243(4x + 5)^{-2}$, calculate the value of $\dfrac{dy}{dx}$ when $x = 1$.   (C)

30 If $y = 3x^2$, prove that $\dfrac{\dfrac{d^2y}{dx^2}}{\left[1 + \left(\dfrac{dy}{dx}\right)^2\right]^{\frac{3}{2}}} = \dfrac{6}{\left[(1 + 36x^2)\right]^{\frac{3}{2}}}$.

31 If $y = \dfrac{1}{\sqrt{4x - 3}}$, show that the value of $\dfrac{\dfrac{d^2y}{dx^2}}{\left[1 + \left(\dfrac{dy}{dx}\right)^2\right]^{\frac{3}{2}}}$ when $x = 1$ is $2\cdot4$.

32 If $y = x^2(10 - 2x)$, find the values of $\dfrac{d^2y}{dx^2}$ when $\dfrac{dy}{dx} = 0$.

33 If $y = x^2$, find $\dfrac{dy}{dx}$. Now writing $x = \sqrt{y}$, find $\dfrac{dx}{dy}$ and substitute for y in your answer. Compare your two answers. (This is an example of an important result, that $\dfrac{dy}{dx} = 1\left/\dfrac{dx}{dy}\right.$.)

34 If $y = \dfrac{1}{\sqrt{x}}$, find $\dfrac{dy}{dx}$ and $\dfrac{dx}{dy}$ (both in terms of x). Hence confirm the result in No. 33.

35 If $y = 2x^2 - 5x$, prove that $x^2\dfrac{d^2y}{dx^2} - 2x\dfrac{dy}{dx} + 2y = 0$.

11 Calculus (3): Applications of Differentiation

Tangents

For the curve given by $y = f(x)$, $\frac{dy}{dx}$ gives the gradient of the tangent at any point. Given the coordinates of a point on the curve and the value of $\frac{dy}{dx}$ at that point, the equation of the tangent can be easily obtained.

Example 1
Find the equation of the tangent to the curve $y = x^3 - 3x + 2$ where $x = 2$.

$$\frac{dy}{dx} = 3x^2 - 3$$

So when $x = 2$ the gradient of the tangent $= 9$.
When $x = 2$, $y = 4$.
The equation of the tangent is $y - 4 = 9(x - 2)$ i.e. $y = 9x - 14$.

Example 2
Find the coordinates of the points on the curve $y = x^3 - 4x^2 - 3x - 2$ where the tangent is parallel to the x-axis.
A line parallel to the x-axis has gradient 0. Therefore if the tangent is parallel to the x-axis, $\frac{dy}{dx}$ must be zero at the point of contact.

$$\frac{dy}{dx} = 3x^2 - 8x - 3$$

$$= (3x + 1)(x - 3) \text{ and when this equals } 0,$$

we have $x = -\frac{1}{3}$ or 3.
These are the only values of x which give points on the curve where the gradient is 0. Their coordinates are $(-\frac{1}{3}, -\frac{64}{27})$ and $(3, -20)$.

Exercise 11.1

Find the equations of the tangents to the following curves at the points given:

1 $y = 6x^2 - 4$, $x = 3$ 2 $y = \frac{1}{2}x^2 + 1$, $x = -2$
3 $y = 2x^3$, $x = 1$ 4 $y = x^2 - x + 3$, $x = -1$

5 $y = \dfrac{(x - 1)}{(x + 3)}$, $x = 2$ **6** $y = (x^2 + x - 4)^2$, $x = -2$

7 $y = \dfrac{1}{\sqrt{x}}$, $x = 9$ **8** $y = \dfrac{(x^2 + x - 1)}{(x^2 - x + 1)}$, $x = 1$

9 The tangent to the curve $y = x^2 - 5x + 3$ at a certain point is parallel to the line $y = x$. Find its equation and find where it cuts the x-axis.

10 Find the coordinates of the points on the curve
$$y = x^3 - 2x^2 - x + 5$$
where the tangents are parallel to the line $y = -2x + 3$. Then find the equations of the tangents.

11 Find the equation of the tangent to the curve $x^3y = 2 + x$ at the point where $x = -1$.

12 Find the equation of the tangent to the curve $4y = x^2$ at the point $(2, 1)$.

13 Find the equations of the tangents to the curve
$$y = x^3 - x^2 - x + 1$$
at the points $(1, 0)$ and $(-1, 0)$. Find also the coordinates of the point at which the tangent at the point $(-1, 0)$ meets the curve again.
                   (L)

14 Find the equation of the tangent to the curve $y = x^3 - x^2 - 2$ at the point where $x = 2$. Find the coordinates of the point where this tangent meets the curve again.

15 Find the gradient of the curve whose equation is $y = \dfrac{1}{x}$ at the point where $x = t$.

State whether the gradient is large or small and positive or negative when (**a**) t is large and positive, (**b**) t is small and positive, (**c**) t is large and negative, (**d**) t is small and negative.

Obtain the equation of the tangent to the curve at the point where $x = t$ and find the coordinates of the points L and M where this tangent cuts the x and y-axes respectively.

Calculate the area of the triangle LOM where O is the origin.   (L)

Stationary Points

A point on a curve at which $\dfrac{dy}{dx} = 0$ is called a stationary point and the value of the function at that point a stationary value. At such points the tangent is parallel to the x-axis. To find the stationary points, set $\dfrac{dy}{dx} = 0$ and solve the resulting equation.

Example 3
Find the stationary points of the function $4x^3 + 15x^2 - 18x + 7$.
For this function $f'(x) = 12x^2 + 30x - 18$
$$= 6(2x^2 + 5x - 3)$$
$$= 6(2x - 1)(x + 3)$$
Hence $f'(x) = 0$ when $x = \frac{1}{2}$ or -3.
The stationary points are therefore at $x = \frac{1}{2}$ and $x = -3$ and the stationary values of the function at these points are $\frac{9}{4}$ and 88 respectively.

Turning points: maximum and minimum points
Fig 11.1 shows a curve passing through a stationary point and reaching a **maximum** value at that point. As x increases (from left to right) the gradient of the curve decreases from a $+$ value through 0 to a $-$ value.

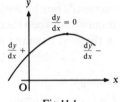

Fig 11.1

Fig 11.2 shows a curve reaching a **minimum** at a stationary point. As x increases the gradient increases from a $-$ value through 0 to a $+$ value.
Maximum and minimum points are also referred to as **turning points**, as the tangent turns over at such a point.

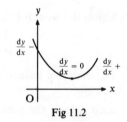

Fig 11.2

Points of inflexion
A third type of stationary point is shown in fig 11.3, where the curve has neither a maximum nor a minimum value. This is called **a point of inflexion**, not a turning point.

For all stationary points, the necessary condition is that $\dfrac{dy}{dx} = 0$.

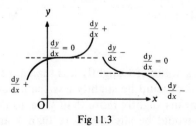

Fig 11.3

A maximum or a minimum value is not meant in an absolute sense. Fig 11.4 shows a curve which has two maximum and one minimum values though one of the 'maximum' values is greater than the other, and the minimum value is not the lowest value possible for the function.

The terms maximum and minimum apply only in a local sense near the stationary point. A function can have more than one of each.

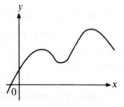

Fig 11.4

For a maximum or minimum, $\frac{dy}{dx} = 0$ is necessary but by itself is not sufficient to distinguish between them. A simple test is shown in the next example which is sufficient for our purpose.

Example 4
Find the maximum and minimum values of the function
$$4x^3 + 15x^2 - 18x + 7.$$

Using the working done in Example 3, $\frac{dy}{dx} = 0$ when $x = \frac{1}{2}$ or -3.

$$\frac{dy}{dx} = 6(2x - 1)(x + 3)$$

To settle the nature of the turning points consider the *sign* of $\frac{dy}{dx}$ on each side of the turning point. (The actual *value* of $\frac{dy}{dx}$ is irrelevant).

x	$\frac{1}{2}^-$	$\frac{1}{2}$	$\frac{1}{2}^+$	-3^-	-3	-3^+
	$x + 3$ is $+$		$x + 3$ is $+$	$x + 3$ is $-$		$x + 3$ is $+$
	$2x - 1$ is $-$		$2x - 1$ is $+$	$2x - 1$ is $-$		$2x - 1$ is $-$
$\frac{dy}{dx}$	$-$	0	$+$	$+$	0	$-$
Outline	╲	—	╱	╱	—	╲
	Minimum			Maximum		

Note: $\frac{1}{2}^-$ means a value of x slightly less than $\frac{1}{2}$, i.e. to the left along the x-axis. Then $2x$ would be slightly less than 1 and $2x - 1$ would be negative. $\frac{1}{2}^+$ means slightly more than $\frac{1}{2}$, i.e. to the right along the x-axis. Then $2x$ would be slightly more than 1 and $2x - 1$ would be positive. Similarly -3^- means slightly less than -3 (to the left of -3 in fact). Think perhaps of $-3\frac{1}{2}$: then $x + 3$ would be negative, and so on.

Hence at $x = \frac{1}{2}$ the function has a minimum value of $\frac{9}{4}$ and at $x = -3$ the function has a maximum value of 88.

Example 5

Find the maximum and minimum values of the function $\dfrac{2x - 1}{x^2 + 2}$.

$$f'(x) = \frac{(x^2 + 2)(2) - (2x - 1)(2x)}{(x^2 + 2)^2}$$

$$= \frac{-2x^2 + 2x + 4}{(x^2 + 2)^2}$$

$$= \frac{-2(x - 2)(x + 1)}{(x^2 + 2)^2}$$

$f'(x) = 0$ when $x = -1$ or 2. Hence the turning points are where $x = -1$ and $x = 2$. The sign of $f'(x)$ is now tested on each side of the turning values of x. Notice that the denominator of $f'(x)$ is always positive and may therefore be ignored in the test.

x	-1^-	-1	-1^+	2^-	2	2^+
$f'(x)$	$-$	0	$+$	$+$	0	$-$
outline	$\searrow$	minimum	$\nearrow$	$\nearrow$	maximum	$\searrow$

Note: When $x = -1^-$ (slightly less than -1) the numerator will have the signs $(-2) \times (-) \times (-)$ i.e. negative. When $x = -1^+$ (slightly more than -1) the signs will be $(-2) \times (-) \times (+)$ i.e. positive, and so on.

Hence at $x = -1$ the function has a minimum value of -1 and when $x = 2$ the function has a maximum value of $\frac{1}{2}$.

Example 6

Find the stationary points of $y = x^3 - 6x^2 + 12x + 3$ *and distinguish between them.*

$$\frac{dy}{dx} = 3x^2 - 12x + 12 = 3(x^2 - 4x + 4) = 3(x - 2)^2$$

The only value of x making $\dfrac{dy}{dx} = 0$ is $x = 2$ (a double value).

Testing the sign of $\dfrac{dy}{dx}$ on either side of $x = 2$ we have:

x	2^-	2	2^+
$\dfrac{dy}{dx}$	$+$	0	$+$
Outline	╱	$-$	╱

Hence the function has a point of inflexion when $x = 2$.

Notes on stationary points

A function of the second degree (a quadratic function) will give a gradient function of the first degree and hence can have only one turning point, which will either be a maximum or a minimum. The graph of such a function is a parabola.

A curve representing a function of the third degree (a cubic), as in Example 4, will have f'(x) of the second degree. f'(x) = 0 is then a quadratic equation, which may have two real roots (two turning points – one maximum and one minimum) or two equal roots (a double stationary point – a point of inflexion) or no real roots (no stationary points).

Such considerations will not apply to less straightforward functions and to those in the form of a quotient. In all cases use the test given above.

Exercise 11.2

For each of the following functions find **(a)** the position of the stationary points (if any), **(b)** the nature of these points (maximum, minimum or point of inflexion) and **(c)** the maximum or minimum values of the function where they occur.

1 $x^2 - 2x$ **2** $5 + 4x - x^2$

3 $\sqrt{x}$ **4** $x^3 - 3x - 2$

5 $3x - x^2$ **6** $2x^3 + 3x^2 - 36x + 5$

7 $x^3 - 3x^2 + 3x + 2$ **8** $x + \dfrac{1}{x} + 4$

9 $x^3 - 7$ **10** $x^3 - 6x^2 + 9x - 1$

11 $3x^4 - 4x^3 + 2$

12 Find $\dfrac{\mathrm{d}y}{\mathrm{d}x}$ for the two curves $y = x - \dfrac{4}{x}$ and $y = \dfrac{4x^2}{3} - \dfrac{1}{x^2}$ and prove that the curves have the same gradient when $x = \frac{1}{2}$.
Find the range of values of x for which the gradient of the curve $y = x - \dfrac{4}{x}$ is greater than 5. Determine whether the curve has any turning points. (C)

13 Find the values of x for which the gradient of the curve $y = \dfrac{2 - x}{x^2 + 1}$ is zero, giving your answers in surd form. (C)

14 Find the maximum and minimum values of $\dfrac{4x - 3}{x^2 + 1}$.

Curve Sketching

It is highly useful to be able to produce a quick sketch of the graph of a function and the position and nature of the turning points helps in this. To sketch a curve, a fairly general outline is required with the position of the turning points and other special points marked but no table of values is required.

The main guides will be: points where $x = 0$ or $y = 0$ (if readily found); the position and nature of the turning points; the trend of the curve as x and y tend to infinity (positive or negative); the gradient as given by $\dfrac{\mathrm{d}y}{\mathrm{d}x}$; symmetry about the axes; any other special points which can be spotted from the equation.

Example 7
Sketch the curve given by $y = 4x^3 + 15x^2 - 18x + 7$ *(previously met in Examples 3 and 4).*

$$\frac{\mathrm{d}y}{\mathrm{d}x} = 6(2x - 1)(x + 3)$$

(1) When $x = 0$, $y = 7$ $\qquad\Big\}$ point A
(2) When $x = 0$, gradient $= -18$ $\Big\}$ (fig 11.5)
(3) Turning point when $x = \frac{1}{2}$, $y = 2\frac{1}{4}$ is a minimum (point B).
(4) Turning point when $x = -3$, $y = 88$ is a maximum (point C).

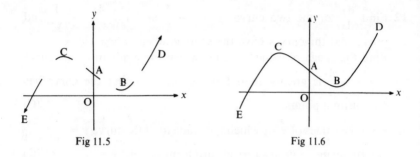

Fig 11.5   Fig 11.6

(5) When x is very large the value of the function is dominated by the term $4x^3$.

If $x \to +\infty$, $4x^3 \to +\infty$ and the graph is suggested as at D.

If $x \to -\infty$, $4x^3 \to -\infty$ and the graph is suggested as at E.

(6) It is not possible to find where $y = 0$ and there is no apparent symmetry about the axes.

These features are now joined up to produce the sketch (fig 11.6) which is not of course drawn to scale.

Example 8

Sketch the curve $y^2 = 6x - 2$

Differentiating wrt x, $2y\dfrac{\mathrm{d}y}{\mathrm{d}x} = 6$

$$\text{or } \frac{\mathrm{d}y}{\mathrm{d}x} = \frac{3}{y} = \frac{3}{\pm\sqrt{6x - 2}}.$$

(1) From the equation of the curve, as $y^2 \geqslant 0$, $6x - 2 \geqslant 0$ and so $x \geqslant \frac{1}{3}$. Hence no part of the curve can lie to the left of $x = \frac{1}{3}$.

(2) When $x = \frac{1}{3}$, $y = 0$.

(3) When $x = \frac{1}{3}$, $\dfrac{\mathrm{d}y}{\mathrm{d}x}$ is infinite. The tangent at this point is parallel to the y-axis.

(4) The curve has no stationary points as $\dfrac{\mathrm{d}y}{\mathrm{d}x}$ is never zero.

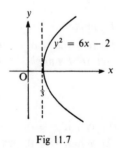

(5) The curve is symmetrical about the x-axis, as there are two numerically equal values of y [the two square roots of $(6x - 2)$ for each value of x].

Fig 11.7

(6) As $x \to \pm\infty$, $y \to \pm\infty$.

The curve (a parabola) is sketched in fig 11.7 from these details.

Example 9

Sketch the function $y = x + \dfrac{9}{x}$.

$$\frac{dy}{dx} = 1 - \frac{9}{x^2}$$

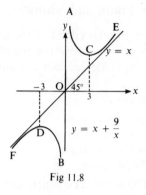

Fig 11.8

(**1**) As $x \to 0$, $y \to \infty$ and $\dfrac{dy}{dx} \to \infty$.

If x is on the positive side of zero, $y \to + \infty$ (as at A) (fig 11.8), and if x is on the negative side of zero, $y \to - \infty$ (as at B). The y-axis is an **asymptote**.

(**2**) Turning points: $\dfrac{dy}{dx} = 0$ when $1 - \dfrac{9}{x^2} = 0$ i.e. when $x^2 = 9$ or $x = \pm 3$. Testing these values, $x = +3$ gives a minimum point (C, $y = 6$) and $x = -3$ gives a maximum point (D, $y = -6$). There are no other stationary points.

(**3**) As $x \to \pm\infty$, the term $\dfrac{9}{x}$ becomes relatively insignificant and the graph tends to become the line $y = x$. $\dfrac{dy}{dx} \to 1$, the gradient of the line $y = x$, which is another asymptote. When x is positive, y is greater than x (as at E) and when x is negative, y is less than x (as at F).

(**4**) The curve cannot cross either the x- or y-axes. The curve is a hyperbola, sketched in fig 11.8.

Exercise 11.3

Sketch the graphs of the following functions:

1 x^2 **2** x^3 **3** $x^2 - 2x$ **4** $x^2 - 3x - 4$

5 $x - \dfrac{2}{x}$ **6** $2x^3 - 3x^2 - 12x + 1$ **7** $(x - 3)(2 - x)$

8 $x^3 - 3x^2 + 3x$ **9** $\sqrt{x}$ (positive root only)

10 $x(x - 3)$ **11** $x - \sqrt{x}$ (positive root)

12 $x + \dfrac{4}{x}$ **13** $x(x - 3)^2$

14 Find the maximum and minimum values of the function
$$3x^3 - 2x^2 - 5x - 5,$$
distinguishing between them. Sketch the graph of the function. (L)

Maximum and Minimum Problems

The method used to find the turning values of a function can be applied in problems where the maximum or minimum value of a quantity varying under certain conditions is required.

Example 10

A rectangular area is to be enclosed using an existing wall as one side and 100 m of fencing are available for the other three sides. It is desired to make the area as large as possible. Find the necessary dimensions of the enclosure.

ABCD is the rectangular area required, area A m^2 (fig 11.9).

Fig 11.9

The method of solution is to set up a function giving the quantity to be maximised (the area A) in terms of *one* variable and then to differentiate to find the turning points.

Take the length of AB as the variable.
Let AB = CD = x m and then we know that BC = $100 - 2x$.

Hence $A = x(100 - 2x)$
                               $= 100x - 2x^2$. This is the function required.

Then $\dfrac{dA}{dx} = 100 - 4x$.

For a turning point, $\dfrac{dA}{dx} = 0$ and so this gives
$$100 - 4x = 0 \text{ or } x = 25$$

Only one turning value is obtained and physically it is clear that this must give a maximum value. (The minimum area would be obtained by taking $x = 0$). It is always wise to test the nature of the turning value however, as it may not be so obvious.

x	25^-	25	25^+
$\dfrac{dA}{dx}$	+	0	−
	╱	Max.	╲

Hence the maximum area is obtained by taking $x = 25$ m and so BC $= 50$ m and the area is $25 \times 50 = 1250$ m^2.

Example 11

A closed rectangular tank is to be made to contain 9 m^3 of water. The length must be twice the breadth and the total surface area must be a minimum. Find the dimensions of the tank.

Take a variable x m as the breadth. The length is then $2x$ m. It is not obvious what the height must be, so take a second variable (temporarily).

Let the height $= h$ m (fig 11.10).

The quantity to be minimised is the surface area (A).

$$A = 2(2x^2 + 2xh + xh)$$
$$= 4x^2 + 6xh$$

We must now eliminate h. The relation between x and h is given by the condition that the volume is 9 m³.

Then $\qquad 9 = x \times 2x \times h = 2x^2h$

which gives $\qquad h = \dfrac{9}{2x^2}.$

Hence $\qquad A = 4x^2 + \dfrac{27}{x}.$

Differentiating wrt x, $\quad \dfrac{\mathrm{d}A}{\mathrm{d}x} = 8x - \dfrac{27}{x^2}$

and for a turning point, $\dfrac{\mathrm{d}A}{\mathrm{d}x} = 0.$

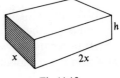

Fig 11.10

This gives $8x = \dfrac{27}{x^2}$ or $x^3 = \dfrac{27}{8}$

i.e. $\qquad x = \tfrac{3}{2} = 1{\cdot}5$ m.

Testing this value gives a definite minimum for A when $x = 1{\cdot}5$. Hence the minimum surface area is obtained when the dimensions are $1{\cdot}5$ m $\times$ 3 m $\times$ 2 m and the surface area is then

x	$1{\cdot}5^-$	$1{\cdot}5$	$1{\cdot}5^+$
$\dfrac{\mathrm{d}A}{\mathrm{d}x}$	$-$	0	$+$
	$\searrow$	$-$	$\nearrow$

$$4 \times (1{\cdot}5)^2 + \frac{27}{1{\cdot}5} = 27 \text{ m}^2.$$

Exercise 11.4

1 If $x + y = 6$, find the maximum value of $P = xy$.

2 If $x + 2y = 6$, find the greatest value of $P = xy$.

3 If 60 m of fencing are available, what are the dimensions of the rectangular enclosure which has the greatest possible area and what is this area?

4 If $xy = 8$, find the minimum value of $V = x + 2y$, where x and y are positive.

5 The total length of wire used in making this framework (fig 11.11) is 96 cm. Find the maximum volume of the outline of the framework.

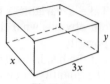

Fig 11.11

6 A stone is thrown up. Its height h units from the ground after t s is given by $h = 96t - 16t^2$. Find the greatest height the stone reaches.

7 A box is made from thin metal, with no top and with vertical sides. The base is a rectangle whose length is twice the breadth. If the box can hold 288 cm^3, what are its dimensions if the minimum amount of material is to be used in its construction? (Ignore any overlaps).

8 A cylinder is designed to hold 54π dm^3 of liquid but must be made so as to have minimum surface area. Taking r dm as the radius of the cylinder, find an expression for the surface area (A) in terms of r and hence find the value of r required.

9 From the rectangular piece of metal shown in fig 11.12, the shaded squares (each of side x cm) are removed and the remainder folded along the dotted lines to make a tray. Find an expression for the volume contained by the tray in terms of x and hence find the value of x which will make this volume a maximum.

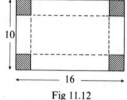

Fig 11.12

10 The side AB of a rectangle ABCD is 12 cm and the side AD is 9 cm. P is a point in the side CD and Q a point in the side BC such that angle APQ is a right angle. Taking DP as x cm, express BQ in terms of x.

P now moves along CD and Q along BC so that the angle APQ remains a right angle. Calculate the least value of BQ. (C)

11 MQRS is a rectangular field. MQ is 150 m long, MS is 60 m long and a road runs along the side MQ. A man starting from M, who wishes to reach the opposite corner R in the shortest possible time, can walk along the road at 100 m per minute and across the field at 60 m per minute. Find an expression for the time, in minutes, he will take if he walks along the road to P, a point x m from Q, and then across the field from P to R.

Find the numerical value of x for the time taken to be a minimum and find this minimum time. (C)

12 A frame, made of thin metal rods, consists of a rectangle of length h cm and width x cm with a rod joining the middle points of the sides of length h cm. If the total length of rod available is 24 cm, calculate the greatest area which the outside of the whole frame can enclose.

13 ABCD is a rectangle in which the side AB is greater than the side BC. The sides AB and BC vary in length in such a way that the perimeter of the rectangle is always 16 cm. Points P and Q are taken on AB and CD respectively so that PBCQ is a square. Denoting BC by x, express the area of the rectangle APQD in terms of x. Hence find the maximum area of the rectangle APQD as x varies, showing clearly that the area obtained is a maximum and not a minimum. Also find the value of x which makes the length of PD a minimum.   (O)

14 A circular cylindrical can, open at the top and of capacity 1 m³, is to be made of thin sheet metal. The bottom is a circle of radius x m. Find the value of x for which the total area of the bottom and the curved surface is a minimum. (Leave π in your answer).

15 Sketch the curve $y = 4x - x^2$.

Find the area of the largest rectangle that can be drawn between the curve and the x-axis, with two corners on the curve and two on the x-axis.   (L)

Small Changes: Approximations

Suppose $y = f(x)$ and x is changed by a small amount δx. There will consequently be a corresponding change δy in the value of y, and it is desired to find this change in a simple way.

In fig 11.13 AD represents a portion of the graph of the function $y = f(x)$. The point A represents the value of y with the given value of x. $AB = \delta x$ and correspondingly $BD = \delta y$. However if δx is small we may take $BC \approx \delta y$ ($\approx$ meaning approximately equal to) where C lies on the tangent at A.

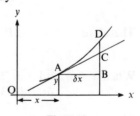

Fig 11.13

Now $\dfrac{BC}{AB} = \dfrac{dy}{dx}$ and therefore $BC = \dfrac{dy}{dx} \times AB = \dfrac{dy}{dx} \times \delta x$.

Hence

$$\boxed{\delta y \approx \dfrac{dy}{dx} \times \delta x \text{ or } \delta y \approx f'(x) \times \delta x.}$$

Example 12

If $P = \dfrac{54}{x^2}$ *and* x *is decreased from* 3 *to* 2·98 *what is the approximate change in the value of P?*

We have $\qquad\qquad \delta P \approx \dfrac{\mathrm{d}P}{\mathrm{d}x} \times \delta x$

and when $x = 3,$ $\quad \delta x = -0·02$ $\qquad\qquad$ (**NB** – a *decrease* in x).

Also $\qquad\qquad\quad \dfrac{\mathrm{d}P}{\mathrm{d}x} = -\dfrac{2 \times 54}{x^3}$

Therefore $\qquad\quad \delta P \approx \left(-\dfrac{2 \times 54}{3^3} \right) \times (-0·02)$

$\qquad\qquad\qquad\qquad\quad = 0·08$

i.e. an *increase* in the value of **P**.

Example 13

If $P = \sqrt[3]{6t^2 - 1}$ *and* t *is increased by a small amount* δt, *find an approximation for the corresponding increase in P. Calculate the approximate percentage increase in P if there is a* 3% *increase in t when* $t = 4$.

We have $\qquad\qquad \delta P \approx \dfrac{\mathrm{d}P}{\mathrm{d}t} \times \delta t$

and $\qquad\qquad\quad \dfrac{\mathrm{d}P}{\mathrm{d}t} = \tfrac{1}{3}(6t^2 - 1)^{-\frac{2}{3}} \times 12t$

$\qquad\qquad\qquad\qquad = \dfrac{4t}{(6t^2 - 1)^{\frac{2}{3}}} = \dfrac{4t}{P^2}$

Therefore $\qquad\quad \delta P \approx \dfrac{4t}{P^2} \times \delta t$

When $t = 4$ and t is increased by 3%,

$\qquad\qquad \delta t = 0·03 \times 4 = 0·12 \text{ and } P = \sqrt[3]{95}$

The percentage increase in P is $\dfrac{\delta P}{P} \times 100$

$\qquad\qquad\qquad\qquad \approx \dfrac{4t}{P^3} \times \delta t \times 100$

$\qquad\qquad\qquad\qquad = \dfrac{16}{95} \times 0·12 \times 100$

$\qquad\qquad\qquad\qquad = 2·02\%.$

Exercise 11.5

1 If $k = 3\sqrt{x}$, find the approximate increase in the value of k if x is increased from 4 to 4·01.

2 If $z = 2\sqrt{(9 + x^2)}$, find the approximate change in z when x is decreased from 4 to 3·99.

3 Find the approximate change in the area of a circle if the radius is increased from 2 to 2·01 cm.

4 If $T = \dfrac{5}{s}$, find the change in T when s is increased from 5 to 5·03.

5 If the volume of a sphere of radius r is given by $V = \dfrac{4}{3}\pi r^3$, find an expression for the approximate change in the volume corresponding to a change δr in the radius.
When the radius is 5 cm, it is increased by 1%. Find the approximate percentage change in the volume.

6 $P = (3x^2 - 1)^3$. When $x = 2$, it is decreased by 3%. Find the approximate percentage change in P.

7 If $y = 8x^{-\frac{2}{3}}$, find the change in y when x is increased from 8 to 8·3.

8 If $y = \frac{1}{2}ax^2$, where a is a constant, find the approximate percentage change in y when x is increased by 5% from the value $x = 3$.

9 The surface area (A) of a sphere of radius r is given by the formula $A = 4\pi r^2$. When $r = 5$ cm it is decreased by 2%. Find the approximate percentage decrease in the surface area.

10 Find the approximate increase in the circumference of a circle when the radius is increased from 10 cm to 10·03 cm.

11 The height (h) of a cone remains constant at 4 cm. The radius (r) of the circular base is 3 cm and this is increased to 3·15 cm. Find the approximate increase in the volume (V) of the cone.

12 Given that $y = \sqrt{(1 + x^3)}$, find $\dfrac{dy}{dx}$. Use your answer to calculate an approximation for the increase in y when x increases from 2 to 2·015.   (C)

13 If $P = Av^{\frac{2}{3}}$, where A is a constant, find the approximate percentage error in the value of P if there is an error of 3% in the measurement of v.

14 Find an approximate value for the radius of a sphere if its volume is increased by 20 cm³ when the radius is increased by 0·02 cm.

15 $T = ar^{\frac{4}{3}}$ where a is a constant. What percentage increase in r (approximately) will cause a 4% increase in T?

Rates of Change: Velocity and Acceleration

For the function $y = f(x)$, $\dfrac{dy}{dx}$ measures the rate of change of y as compared with x, (actually, the rate of *increase* of y wrt x). An important rate of change is the **speed** of a moving object, which is the rate of change of distance wrt time at a given instant. When the direction of motion is also relevant, we use the term **velocity**.

Now the *average* speed of a body is the rate $\dfrac{\text{distance travelled}}{\text{time taken}}$.

If the time taken $\to 0$ we obtain the limiting value of the average speed, i.e., the instantaneous speed or the speed at a given instant. In calculus terms, if s^* represents distance (measured from some fixed point) travelled in time t then the velocity v at any instant will be given by $\dfrac{ds}{dt}$. (An instrument which gives the numerical value of this derivative, i.e., the speed, is the speedometer of a car). The standard unit of velocity is the *metre per second*, written either as m/s or m s^{-1}. Other units are cm/s or cm s^{-1} (centimetre per second) and km/h or km h^{-1} (kilometre per hour). [The principal British unit still used is the mile per hour, written mile/h or m.p.h.] If the velocity itself changes, then we have the rate of change of velocity wrt time, called the **acceleration** (a). Then $a = \dfrac{dv}{dt} = \dfrac{d^2s}{dt^2}$.

Acceleration is the rate of *increase* of velocity compared with time, and hence the standard unit of acceleration will be metre per second per second, written m s^{-2}. A negative acceleration is sometimes called a **retardation**, i.e., 'slowing down'.

If s is a known function of t, we can find the velocity and the acceleration at any time.

Example 14

Discuss the motion of a particle travelling along a straight line such that t s after the start its distance s m from a fixed point on the line is given by $s = t^2 - t$. Illustrate graphically.

$$s = t^2 - t = t(t - 1).$$
$$s = 0 \text{ when } t = 0 \text{ (start) or when } t = 1.$$
$$v = \frac{ds}{dt} = 2t - 1.$$

---

*Note carefully: s represents the quantity 'distance' and s the unit symbol 'second(s)'.

When $t = 0$, $v = -1$ ($\leftarrow$ direction) and $v = 0$ when $t = \frac{1}{2}$.

$$a = \frac{dv}{dt} = 2 \text{ (constant in direction} \quad \rightarrow \quad)$$

The following table summarises the facts of the motion:

t s	0	$\frac{1}{2}$	1	2	3	......
s m	0	$-\frac{1}{4}$	0	2	6	......
v m s^{-1}	-1	0	$+1$	$+3$	$+5$	......
a m s^{-2}	$+2$	$+2$	$+2$	$+2$	$+2$	......
Direction of motion	$\leftarrow$ retarding	Momentary rest	$\rightarrow$	$\rightarrow$ accelerating	$\rightarrow$	

Take the positive direction (increasing s) along the line to be from left to right.

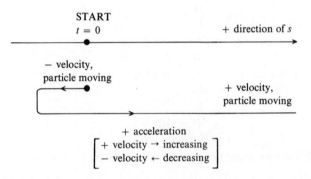

The particle starts from 0 with velocity -1 m s^{-1} ($\leftarrow$ direction), slows down to momentary rest ($t = \frac{1}{2}$, $s = -\frac{1}{4}$ m), reverses and returns to 0 ($t = 1$) and then continues in the direction $\rightarrow$ with increasing velocity. The acceleration is $+2$ m s^{-2} throughout the motion.
Fig 11.14 is a sketch of the s-t (distance-time) graph and fig 11.15 of the v-t (velocity-time graph), illustrating the above facts.

s-t (distance-time) graph. Distance from start ($+$ or $-$) related to time.

Gradient of curve $= \dfrac{ds}{dt} = v$.

(Turning point where $v = 0$ at $t = \frac{1}{2}$, $s = -\frac{1}{4}$).

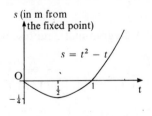

Fig 11.14

v-t (velocity-time) graph. Velocity
related to time.

$$\text{Gradient} = \frac{dv}{dt} = a = 2 \text{ (constant)}.$$

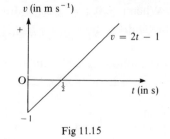

Fig 11.15

Example 15

Consider a particle moving as in Example 14 *but with*
$$s = t^3 - 12t^2 + 36t + 2.$$

As before take → as the + direction.

$$t = 0 \text{ gives } s = 2.$$

The particle starts 2 m from O.

$$v = \frac{ds}{dt} = 3t^2 - 24t + 36$$
$$= 3(t^2 - 8t + 12)$$
$$= 3(t - 2)(t - 6)$$

When $t = 0$, $v = +36$ and $v = 0$ when $t = 2$ or 6.

$$a = \frac{dv}{dt} = 6t - 24$$
$$= 6(t - 4).$$
$$a = 0 \text{ when } t = 4.$$

The table shows some key points in the motion:

t	0	1	2	3	4	5	6	7
s	+ 2	+27	+34	+29	+18	+7	+ 2	+ 9
v	+36	+15	0	− 9	−12	−9	0	+15
a	−24	−18	−12	− 6	0	+6	−12	+18
	→	→	rest	←	←	←	rest	→

A diagrammatic representation of the motion:

From these figures we can sketch the s-t, v-t and a-t graphs (figs 11.16–11.18).

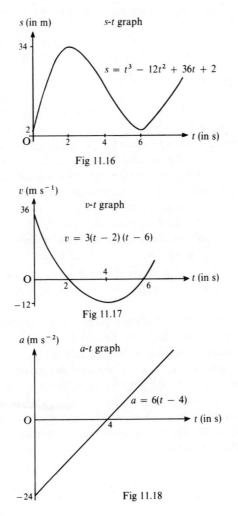

s (in m) s-t graph

34

$s = t^3 - 12t^2 + 36t + 2$

2

O 2 4 6 t (in s)

Fig 11.16

v (m s^{-1})

36   v-t graph

$v = 3(t - 2)(t - 6)$

O 2 4 6 t (in s)

−12

Fig 11.17

a (m s^{-2}) a-t graph

$a = 6(t - 4)$

O 4 t (in s)

−24   Fig 11.18

Exercise 11.6

1 The motion of a particle on a straight line is given by $s = t^2 + 2t$ where s is measured in m from a certain fixed point of the line and t (in s) is timed from the start.

Find the distance of the particle and its velocity after 2, 4, 6, 8 s, its acceleration and sketch the s-t and v-t graphs.

2 After t s the distance s m of a moving body along a straight line is given by $s = t(t - 4)$. Copy and complete this table:

t	0	1	2	3	4	5	6
s							
v							

When and where is the object momentarily at rest? Sketch the s-t and v-t graphs and describe the motion briefly.

3 The motion of an object in a straight line is given by $s = 6t - t^2$ (s in m, t in s).

Compile a table showing values of s, v and a for $t = 0, 1, 2, 3, 4, 5, 6, 7$ s and sketch the s-t and v-t graphs.

4 For a particle moving in a straight line its distance s m is measured from a certain point of the line and is given by $s = t^2 - 3t + 2$ where t is timed in s from the start. When is the particle momentarily at rest? What is its distance from the point at 1, $1\frac{1}{2}$, 2, 3 s? What is its acceleration? Make a diagrammatic representation of the motion and sketch the s-t and v-t graphs.

5 For a body moving in a straight line its distance s m from a fixed point at time t (in s) is given by $s = t^2 - 5t + 6$. Sketch the s-t and v-t graphs.

6 s m is the distance from a certain point 0 on a straight line at time t s of a body moving along the line and $s = 2t^3 - 9t^2 + 12t - 4$. Analyse the motion, making a table of values of s, v and a for $t = 0$, 1, $1\frac{1}{2}$, 2, 3 s and sketch the s-t, v-t and a-t graphs.

7 A particle moves in a straight line through O so that at time t s, its distance s m from O is given by $s = t^3 - 15t^2 + 63t - 49$. Calculate the distances from O at which the velocity of the particle is zero. Calculate also the accelerations of the particle at the instants when its velocity is zero.

(C)

Comparing Rates of Change

Example 16

Suppose the radius (r) of a circle is 3 cm *at a certain instant and is increasing at the rate of* 0·5 cm s^{-1}. *At what rate will the area (A) be increasing at that instant?*

We are given that $\dfrac{dr}{dt} = 0\cdot5$. We have to find $\dfrac{dA}{dt}$. We can compare the two rates of change by using the 'chain' rule: $\dfrac{\mathrm{d}A}{\mathrm{d}t} = \dfrac{\mathrm{d}A}{\mathrm{d}r} \times \dfrac{\mathrm{d}r}{\mathrm{d}t}$ as

it is possible to find $\dfrac{\mathrm{d}A}{\mathrm{d}r}$ from the relation $A = \pi r^2$.

Then
$$\begin{aligned}
\frac{\mathrm{d}A}{\mathrm{d}t} &= 2\pi r \times \frac{\mathrm{d}r}{\mathrm{d}t} \\
&= 2\pi \times 3 \times 0\cdot5 \\
&= 3\pi \approx 9\cdot42.
\end{aligned}$$

At this instant therefore the area is increasing at the rate $9\cdot42$ $\text{cm}^2 \text{ s}^{-1}$ (i.e. square cm per second) approximately.

This method of comparing rates of change is quite general. Suppose a quantity y can be expressed as a function of a single variable x, then the rate of change of y wrt $t, \dfrac{\mathrm{d}y}{\mathrm{d}t}$, and the rate of change of x wrt $t, \dfrac{\mathrm{d}x}{\mathrm{d}t}$, are related by the equation $\dfrac{\mathrm{d}y}{\mathrm{d}t} = \dfrac{\mathrm{d}y}{\mathrm{d}x} \times \dfrac{\mathrm{d}x}{\mathrm{d}t}$.

$\dfrac{\mathrm{d}y}{\mathrm{d}x}$ is found from the function $y = f(x)$.

Example 17
A hollow right circular cone has base radius 4 cm and vertical height 20 cm. It is held upside down with its axis vertical, and contains water. Water is being added at the constant rate of $1\cdot5$ cm^3 s^{-1} and leaks away through a small hole in the vertex at the constant rate of 2 cm^3 s^{-1}. At what rate is the depth of the water changing when the depth is 12 cm?

Fig 11.19 shows the facts. When the volume of water in the cone is V let its depth be x and the radius of its circular face be r. V is to be expressed in terms of x alone.

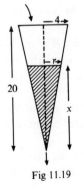

By similar triangles, $\dfrac{x}{20} = \dfrac{r}{4}$ or $r = \dfrac{x}{5}$.

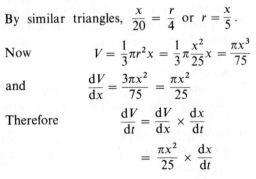

Now
$$V = \frac{1}{3}\pi r^2 x = \frac{1}{3}\pi \frac{x^2}{25} x = \frac{\pi x^3}{75}$$

and
$$\frac{\mathrm{d}V}{\mathrm{d}x} = \frac{3\pi x^2}{75} = \frac{\pi x^2}{25}$$

Therefore
$$\frac{\mathrm{d}V}{\mathrm{d}t} = \frac{\mathrm{d}V}{\mathrm{d}x} \times \frac{\mathrm{d}x}{\mathrm{d}t}$$
$$= \frac{\pi x^2}{25} \times \frac{\mathrm{d}x}{\mathrm{d}t}$$

Fig 11.19

But $\dfrac{\mathrm{d}V}{\mathrm{d}t}$ is the rate of increase of the volume of water.

$$\therefore \qquad \frac{\mathrm{d}V}{\mathrm{d}t} = 1 \cdot 5 - 2 = -0 \cdot 5$$

$$\therefore \qquad -0 \cdot 5 = \frac{\pi \times 12^2}{25} \times \frac{\mathrm{d}x}{\mathrm{d}t} \text{ since } x = 12$$

Therefore $\qquad \dfrac{\mathrm{d}x}{\mathrm{d}t} = -\dfrac{0 \cdot 5 \times 25}{\pi \times 144} = -0 \cdot 027$

The water level is falling at the rate of $0 \cdot 027$ cm s^{-1}.

Exercise 11.7

[Leave answers in terms of π]

1 At what rate is the area of a circle decreasing when its radius is 4 cm and decreasing at the rate of $0 \cdot 2$ cm s^{-1}?

2 When the circumference of a circle is 250 mm the radius is increasing at the rate of 5 mm s^{-1}. Find the rate at which the area of the circle is increasing.

3 An ink blot has radius of 12 mm and its area is increasing at the rate of 3π mm^2 s^{-1}. At what rate is the radius increasing?

4 The radius (r) of a sphere is 4 cm and is increasing at the rate of $0 \cdot 5$ cm s^{-1}. At what rate is the volume (V) increasing? ($V = \frac{4}{3}\pi r^3$)

5 A circular cylinder has a radius of $0 \cdot 3$ m and it is being filled with water at the rate of $0 \cdot 009$ m^3 s^{-1}. Find the rate at which the water level is rising.

6 A hollow circular cone, radius of base 75 mm and height 250 mm, is held vertex down with axis vertical and contains water. The water is leaking away through a small hole in the vertex at the rate of 900 mm^3 s^{-1}. Find the rate at which the water level is falling when it is 100 mm above the vertex.

7 A is the point $(0, 4)$ on the y-axis and another point B is travelling along the positive x-axis at a speed of 2 units/s. Find the rate at which the distance AB is increasing when B passes through the position $(3, 0)$.

8 A balloon is being inflated by blowing in 10^4 cm^3 s^{-1} of air. At what rate is the radius increasing when the diameter of the balloon is 45 cm?

9 The length of each side of a cube is expanding at the rate of $0 \cdot 02$ m s^{-1}. At what rate is the volume increasing when each side is $0 \cdot 25$ m long?

10 The height of a cone remains constant at 30 cm. The radius of the base is increasing at the rate of 0·1 cm s^{-1}. Find the rate at which the surface area of the cone is increasing, when $r = 40$ cm. (Surface area $= \pi r \sqrt{r^2 + h^2}$ where $r =$ radius of base, $h =$ height.)

11 The area, A cm^2, of the image of a rocket on a radar screen is given by the formula $A = \dfrac{12}{r^2}$ where r km is the distance of rocket from the screen. The rocket is approaching at 0·5 km s^{-1}. When the rocket is 10 km away, at what rate is the area of the image changing?

When A is changing at 0·096 cm^2 s^{-1}, how far away is the rocket?

(C)

12 A container is such that when the depth of liquid in it is x cm, the volume is V cm^3, where $V = \frac{1}{5}(x^5 + 10x^3 + 45x)$. Find the area of the surface of the liquid when the depth is 3 cm.

Water is poured into the container at a constant rate of 720 cm^3 s^{-1}. Find the rate at which the depth is increasing when the area of the surface of the water is 540 cm^2. (C)

13 Car A is travelling along a straight road X'OX at a speed of 10 m s^{-1} while a second car B is travelling along a perpendicular road Y'OY at a speed of 15 m s^{-1}. At a certain time, A and B are each 1 km from the intersection O of the two roads and moving towards O. Find an expression for the distance between the two cars t s later and hence find **(i)** the rate at which this distance is decreasing when $t = 10$ and **(ii)** the time when this distance is a minimum.

14 A spherical balloon is being inflated by pumping in 10^2 cm^3 s^{-1} of air. At what rate is the surface area of the balloon increasing when the radius of the balloon is 10 cm? (Surface area of sphere of radius $r = 4\pi r^2$.)

12 Permutations, Combinations and Probability

Arrangements or Permutations

There are four different books on a shelf. In how many ways could they be arranged in order?

If we label the books *a*, *b*, *c*, *d* for convenience, list all the possible orders of these letters. How many can you find?

Working systematically, first write down all the arrangements in which *a* comes first:

$$a\ b\ c\ d$$
$$a\ b\ d\ c$$
$$a\ c\ b\ d \qquad \text{i.e., 6 arrangements}$$
$$a\ c\ d\ b$$
$$a\ d\ b\ c$$
$$a\ d\ c\ b$$

Clearly if we now take book *b* first there will also be 6 arrangements. Similarly, if book *c* is first there will also be 6 arrangements and 6 more with book *d* first. Hence there are a total of 24 possible arrangements of the four books.

Now consider this alternative method.

Let us have four boxes into each of which one book can be put. Into the the first box we can put any one of the 4 books. There are 4 possible ways of doing this. Three books are now left. There are therefore 3 ways of putting a book into the second box and these 3 ways follow *any one* of the previous 4. So there are $4 \times 3 = 12$ ways of filling the first two boxes.

Box 1	Box 2	Box 3	Box 4
Any one of 4	Any one of 3	Any one of 2	No choice

Two books are now left, so there are 2 ways of filling box 3 taken with the 12 ways for boxes 1 and 2. Hence there are now $4 \times 3 \times 2$ ways of filling the first three boxes. Only one book is left to put in box 4 and there is only one way of doing this.

Altogether there are $4 \times 3 \times 2 \times 1$ ways of filling the boxes, i.e., of arranging the four books in order. We call this the number of arrangements or **permutations** of the four books.

This approach is an example of a very useful method, which we can generalize as follows. If there are *A* ways of doing one operation and *B* ways of doing a second operation, then there are $A \times B$ ways of doing both (one after the other). This principle can be extended (as above) for more than two operations.

Example 1

In how many ways can 3 books be arranged in order if 7 different books are available?

Box 1	Box 2	Box 3
7	6	5
ways	ways	ways

The first space or box can be filled by any one of the seven books – i.e. in 7 ways. The second space can be filled by any one of the 6 books left – i.e. in 6 ways. The third space can be filled in 5 ways as there are now only 5 books left to choose from.

Hence the number of arrangements $= 7 \times 6 \times 5 = 210$.

This is the number of arrangements or permutations of 3 books out of 7 or of 7 books taken 3 at a time and is written $^7P_3$. Then $^7P_3 = 210$. If there were 4 books and all four have to be permuted, then we have found that $^4P_4 = 4 \times 3 \times 2 \times 1 = 24$.

Example 2

In how many ways can the 1st, 2nd and 3rd prizes be awarded in a race if there are 10 competitors?

The 1st prize can be awarded in 10 ways, the 2nd in 9 ways and the 3rd in 8 ways.

Then $^{10}P_3 = 10 \times 9 \times 8 = 720$.

Example 3

In how many ways can the letters of the word MEASURING be arranged or permuted?

Answer: $^9P_9 = 9 \times 8 \times 7 \times 6 \times 5 \times 4 \times 3 \times 2 \times 1 = 362\,880$.

Factorials

A product such as $1 \times 2 \times 3 \times 4 \times 5$ is called **factorial** 5 and is written 5!. The answer to Example 3 is thus 9! and could be left in this factorial form. Factorial n, i.e. $n!$

$$= 1 \times 2 \times 3 \times \ldots \times (n - 2) \times (n - 1) \times n$$

where the ... indicate the missing integers 4 to $(n - 3)$.

Hence the number of arrangements or permutations of n different objects is $^nP_n = n \times (n - 1) \times (n - 2) \times \ldots \ldots \times 3 \times 2 \times 1 = n!$

The number of arrangements of 3 objects taken from n different ones is $^nP_3 = n \times (n - 1) \times (n - 2)$.

Similarly $^nP_4 = n(n - 1)(n - 2)(n - 3)$
and $^nP_5 = n(n - 1)(n - 2)(n - 3)(n - 4)$
and so the number of arrangements of r objects taken from n different
ones is $^nP_r = n(n - 1)(n - 2) \ldots\ldots [n - (r - 1)]$.
∴ $^nP_r = n(n - 1)(n - 2) \ldots\ldots (n - r + 1)$
This general formula should not be used in simple problems but a
direct approach made as in the above Examples.

Exercise 12.1

1 In how many ways can 5 different books be arranged in order?

2 In how many ways can 7 different objects be placed in a line?

3 Find the number of possible permutations of the letters of the
word NOTABLE.

4 Find the number of arrangements using any 3 letters of the word
CHEMISTRY.

5 In how many ways can the 1st, 2nd and 3rd prizes be awarded
in a class of 20?

6 How many 4 digit numbers can be formed using the digits 1, 2,
3, 4, 5 (no digit being repeated)?

7 There are three different ways of going from A to B and 4 different
ways of going from B to C. How many possible routes are there
from A to C via B?

8 In how many ways can 1 boy and 1 girl be selected from 10 boys
and 8 girls?

9 How many numbers greater than 100 can be formed using the
digits 1, 2, 3, 4, 5 if no digit may be repeated?

10 A boy has 6 double-sided pop records. In how many different
orders can he play them?

11 An examination paper has 6 questions but only 4 are to be answered.
In how many different ways can the answers be arranged?

12 There are six seats in a railway carriage. In how many different
ways can 4 people occupy them?

13 How many numbers can be formed using the three digits 1, 2, 3
if (a) no digit can be repeated, (b) if any digit can be used more
than once?

14 In how many ways can the letters of the word MEDIAN be
arranged? How many of these ways begin with M? How many
end with N? How many begin with M and end with N?

15 How many four and five digit numbers can be formed using the
digits 1, 2, 3, 4, 5 (no repetitions allowed)? How many will be
greater than 5000? How many will be even numbers?

16 What are the values of 3!, 4!, 5!, 6!?

17 Prove that $\dfrac{10!}{8!} = 90.$ **18** Find the value of $\dfrac{12!}{9!}.$

19 Find the values of $\dfrac{12!}{10! \times 2!}, \dfrac{8!}{5! \times 3!}$ and $\dfrac{15!}{10! \times 5!}.$

20 Simplify $\dfrac{n!}{(n-2)!}.$

Arrangements including Identical Objects

Example 4

In how many ways can the letters of the word MAXIMUM be arranged in order?

We must allow for the fact that in any arrangement the 3 Ms will be indistinguishable.

First write the Ms in different type to distinguish between them:

$$MAXIMUm$$

Then the number of arrangements is now $^7P_7 = 7!$

Now consider any one of these arrangements. e.g., MUAImXM. The 3 Ms can themselves be arranged in 3! ways without upsetting the other letters. We could have

M	UAI	m	X	M	3! arrangements in-
M	UAI	M	X	m	terchanging the Ms,
M	UAI	M	X	m	only leaving the
M	UAI	m	X	M	other letters in fixed
m	UAI	M	X	M	positions.
m	UAI	M	X	M	

If the Ms are indistinguishable however all these 6 arrangements will be the same. As this would be true for all the 7! possible arrangements with different Ms, the actual number of arrangements with identical Ms is $\dfrac{7!}{3!} = \dfrac{1 \times 2 \times 3 \times 4 \times 5 \times 6 \times 7}{1 \times 2 \times 3} = 840.$

In general the number of arrangements of n objects of which r are identical is $\dfrac{n!}{r!}.$

Example 5

How many different numbers of 5 digits can be made using the digits 2, 2, 3, 3, 3?

If the five digits were all different we could form 5! numbers.
But the two 2 s are indistinguishable and so are the three 3 s. The

two 2 s can be permuted in 2! ways by themselves and the three 3 s in 3! ways, and these would count as only one way in any sequence. Hence the number of different numbers that could be formed is

$$\frac{5!}{2! \times 3!} = \frac{1 \times 2 \times 3 \times 4 \times 5}{1 \times 2 \times 1 \times 2 \times 3} = 10.$$

Conditional Permutations

Example 6

Find the number of ways in which the letters of the word SHALLOW *can be arranged* (**a**) *if the two* L s *must not come together* (**b**) *if the two* L s *must always be together.*

Leaving out the two Ls, the letters SHAOW can be arranged in 5! ways.

$_\uparrow \mathrm{S} _\uparrow \mathrm{H} _\uparrow \mathrm{A} _\uparrow \mathrm{O} _\uparrow \mathrm{W} _\uparrow$ (**a**) With each of these ways the first L can be inserted in any one of 6 places:

When this is done there are then 5 possible places for the second L not next to the first. Hence the total number of arrangements with the two L s separated is 5! × 6 × 5 provided the L s can be distinguished. They cannot and so the number of arrangements is $\dfrac{5! \times 6 \times 5}{2} = 5! \times 15 = 1800.$

(**b**) In this case take the two L s (LL) as one object. There are then 6 places for it in each of the 5! arrangements of the letters SHAOW. Hence the number of arrangements is 6 × 5! = 6! = 720.

Example 7

How many numbers greater than 2000 *can be formed using the digits* 1, 3, 4, 5, 6? *How many of these will be even?* (*No repetitions allowed*).

If all five digits are used every number will be > 2000. Then there will be 5! = 120 possible numbers.

If only 4 digits are used, the first (left hand) digit cannot be 1. There are 4 choices for this first digit (3 or 4 or 5 or 6). Whichever is chosen there will be 4 choices for the second digit, 3 for the next and 2 for the last.

Thus there are 4 × 4 × 3 × 2 = 96 possible numbers.

Hence the total number possible is 120 + 96 = 216.

If all five digits are used and the number is to be even, the last (right hand) digit must be 4 or 6 which gives 2 choices. When this is done, there are 4 choices for the first digit, 3 for the second, 2 for the third and 1 for the fourth digit. Hence there are $4 \times 3 \times 2 \times 1 \times 2 = 48$ even numbers using all 5 digits.

If four digits are used the fourth digit must be 4 or 6 (two choices). The first digit cannot be 1 and either 4 or 6 has already been used: therefore there are 3 choices for the first digit. For the second digit there are now 3 choices (as only two have so far been used) and for the third digit there are 2 choices left.
Hence the number of arrangements $= 3 \times 3 \times 2 \times 2 = 36$.
Thus the total number of even numbers is $48 + 36 = 84$.

Exercise 12.2

1 How many numbers greater than 3000 can be made from the digits 1, 2, 3, 4, 5 without repeating any of them?

2 How many arrangements are there of the letters of the word LETTERS?

3 The letters of the word DIGIT are to be permuted. In how many of these permutations (a) do the I s come together, (b) are the I s separated?

4 How many numbers greater than 150 can be formed from the digits 1, 2, 3, 4, 5 if no repetitions are allowed?

5 How many even numbers greater than 5000 can be formed using the digits 1, 2, 4, 5, 6, no digit being used more than once?

6 In how many ways can 5 boys and 3 girls sit in a line if the girls do not sit next to each other?

7 In how many ways can 4 boys and 3 girls stand in a line if the boys and girls must alternate?

8 How many arrangements are there of the letters of the word CALCULUS?

9 In how many ways can I place 3 white balls, 2 green balls and 1 black ball in a line?

10 There are four seats in a car (including the driver's seat). In how many ways can 4 people sit in the car if only two of them can drive?

11 How many three digit odd numbers can be formed from the digits 1, 2, 3, 4, 5 if it is permitted to use any of the digits more than once?

12 Find the number of arrangements of the letters of the word PERCENTAGE if the E s must be placed next to each other. (Leave the result in terms of factorials).

13 A system of car registration numbers is such that each car must have two letters, the first a consonant, the second a vowel, followed by three digits. How many cars can be registered under this system?

14 Find the number of (i) five letter, (ii) four-letter arrangements of the letters of the word SHOOT. How many of the latter contain only one O?   (C)

15 Find how many numbers greater than 100 can be formed from the digits 1, 2, 3, 4, 5, no digit being used more than once in any particular number.   (C)

16 In how many ways can 4 men and 3 women be seated in a straight line if men and women are placed alternately? How many arrangements are possible if one man always occupies a particular place?
(C)

17 How many of the numbers from 1000 to 9999 (both inclusive) do *not* have four different digits?   (C)

18 In how many different ways can the letters of the word SALOON be arranged (**a**) if the two O s must not come together, (**b**) if the consonants and vowels must occupy alternate places?   (C)

Selections or Combinations

If we are given the five letters a, b, c, d, e, in how many ways can three of them be permuted?

As we found above, the number of arrangements is $5 \times 4 \times 3 = 60$, and each of these is a different *order*.

Suppose however we are not interested in the actual order of the letters but only in the number of *selections* of three letters. Then for example the 6 permutations abc, acb, bca, bac, cab, cba would all count as one selection of three letters from the five and this would apply to each set of three letters chosen.

Thus the number of selections or combinations of three letters chosen from the given five would be $\dfrac{5 \times 4 \times 3}{3!} = 10$ and we write this as $^5C_3$.

It is important to be clear about the difference between permutations and combinations and the following will illustrate this difference. Given five objects, three are chosen:

Permutations

$^5P_3$

$= 5 \times 4 \times 3$

Order is important.
abc is a different arrangement
from *acb* and so on.

Combinations

$^5C_3$

$= \dfrac{5 \times 4 \times 3}{3!}$

Order is irrelevant: *abc*, *acb*, etc
are the same selections. Each set
of three can be permuted in 3!
ways but these count as 1 combination.

$$\text{Hence } {}^5C_3 = \frac{{}^5P_3}{3!}$$

Similarly, the number of selections of four objects from seven is

$$^7C_4 = \frac{{}^7P_4}{4!} = \frac{7 \times 6 \times 5 \times 4}{1 \times 2 \times 3 \times 4} = 35.$$

$$\text{Hence } {}^nC_r = \frac{{}^nP_r}{r!} = \frac{n(n-1)(n-2)\ldots\ldots(n-r+1)}{r!}$$

$^nC_1 = n$ as this is the number of ways of selecting 1 from n objects;

$^nC_2 = \dfrac{n(n-1)}{2!}$, $^nC_3 = \dfrac{n(n-1)(n-2)}{3!}$ and so on.

Example 8

In how many ways can 4 boys be chosen from 6?

The number of permutations of 4 boys from 6 is $^6P_4 = 6 \times 5 \times 4 \times 3$.

Hence the number of selections,

$$^6C_4 = \frac{6 \times 5 \times 4 \times 3}{4!}$$

$$= \frac{6 \times 5 \times 4 \times 3}{1 \times 2 \times 3 \times 4}$$

$$= 15.$$

Example 9

(a) *A committee of 2 men and 3 women is to be chosen from 5 men and 4 women. How many different committees can be formed?*

The 2 men can be selected in $^5C_2 = \dfrac{5 \times 4}{2!} = 10$ ways and

the 3 women can be chosen in $^4C_3 = \dfrac{4 \times 3 \times 2}{3!} = 4$ ways.

Each of the 10 ways of choosing the men can be taken with each of the 4 ways of choosing the women. Therefore there are $10 \times 4 = 40$ possible committees.

(b) *If one of the women refuses to serve on the same committee as a particular man, how many committees are now possible?*

First find the number of committees which will include *both* this particular man and woman. We select therefore one more man from the 4 others (4 choices) and 2 more women from the remaining 3 ($^3C_2$ choices). Hence the number of committees having both the particular man and woman is $4 \times {}^3C_2 = 4 \times \dfrac{3 \times 2}{2!} = 12$.

Out of the 40 possible committees then, 12 will include both these people. Hence $40 - 12 = 28$ committees do *not* include both of them.

Exercise 12.3

1 In how many ways can three equal prizes be given to 5 boys?

2 In how many ways can 4 boys be chosen from 6?

3 In how many ways can 3 men and 2 women be chosen from 6 men and 4 women?

4 In how many ways can a team of 11 players be selected from 13 boys? (Hint: calculation can be made easier if you think of the boys *not* selected).

5 5 points are marked on a piece of paper so that no three of them are collinear (i.e., lie on a straight line). How many straight lines can be drawn to join any two points?

6 A committee of 3 boys and 4 girls is to be formed from 5 boys and 6 girls. How many committees are possible? If one of the girls refuses to serve if a certain other girl is chosen, how many committees are possible?

7 In how many ways can I choose 2 red balls and 3 black balls from a bag containing 5 red and 6 black balls?

8 There are six shoe shops, eight shops selling shirts and 3 book shops near my home. In how many ways can I buy a pair of shoes, 3 shirts and 2 books? (Only one from each shop.)

9 In how many ways can a party of 4 men be selected from 7 men if one at least of two particular men must not be included?

10 A bag contains 2 white and 3 red balls. In how many ways can I choose 3 balls **(a)** if at least one ball must be white, **(b)** if at least one ball must be red?

11 Eight men are to travel in two four-seater cars. In how many ways can the men be allocated to the cars **(a)** if all eight can drive,

(b) if only two can drive? (The same four men in a car count as one way no matter in what positions the men are sitting).   (C)

12 In how many ways can a committee, consisting of a Chairman, Secretary, Treasurer and four ordinary members, be chosen from eight persons? (Committees with different Chairmen, Secretaries or Treasurers count as different committees).   (C)

13 An examination paper has two parts, Part A and Part B. There are 6 questions in Part A and 8 questions in Part B. A candidate is required to do 4 questions from Part A (which must include either Question 1 or Question 2) and any 3 from Part B. In how many ways can he complete the paper?

14 Find the values of $^4C_2$, $^8C_4$, $^9C_6$, $^7C_3$.

15 Prove that $^7C_3 = \dfrac{7!}{3! \times 4!}$ and that $^8C_5 = \dfrac{8!}{5! \times 3!}$.

   Now prove that $^nC_r = \dfrac{n!}{r! \times (n-r)!}$.

16 Explain why $^9C_7 = {}^9C_2$ without calculating either, and then generalise this result.

17 Show that $^7C_3 + {}^7C_2 = {}^8C_3$ and that $^6C_4 + {}^6C_5 = {}^7C_5$.
   (These are examples of a general result: $^nC_r + {}^nC_{r+1} = {}^{n+1}C_{r+1}$).

Probability

Suppose there are 3 red and 4 black balls (all identical in size and material) in a bag. I shut my eyes, put my hand inside the bag and draw out one ball. What is the chance that I have picked a red ball?

We are asking for a numerical estimate of the chance or the **probability** of a certain event (picking a *red* ball) happening. It seems reasonable to express this in terms of the choices available to me.

Since it is impossible for me to tell which ball is which, it is *equally likely* that I shall pick any one of the 7. There are 7 possible choices, only 3 of which will give a red ball. Hence we say the chance of picking a red ball is $\frac{3}{7}$ and write

p (a red ball is chosen) $= \frac{3}{7}$ (often written as a decimal, 0·43).
Similarly p (a black ball is chosen) $= \frac{4}{7}$ (or 0·57).

We can also interpret such probabilities as meaning: if I performed the operation of drawing a ball from such a bag a large number of times (say 700) then I would expect $\frac{4}{7}$ ths of them (i.e., 400) to result in a black ball being picked.

This will give us a working idea of probability or chance. In general, if there are a equally likely possible ways of performing a certain operation and b of them produce a certain outcome or event E, then

$$p(E) = \frac{b}{a} = \frac{\text{number of ways in which event E can happen}}{\text{total number of ways in which the operation can be done}}$$

Then p(the event E will *not* happen) $= \dfrac{a - b}{a} = 1 - \dfrac{b}{a} = 1 - p(E)$.

Hence p(E will not happen) $+ p(E) = 1$, which is the numerical measure of certainty, as either E will happen or it will not. Conversely, if an event cannot possibly happen, its probability $= 0$ (impossibility). So the probability or chance of an event is expressed as a fraction p, where $0 \leqslant p \leqslant 1$. The greater the fraction, the more likely is the chance of the event happening.

It is important to notice the words *equally likely* used above. They are difficult to define exactly, but their meaning will be clear enough. Another phrase with the same meaning is *at random*. In choosing a ball from a bag, we expect the choice to be made at random or all the possible choices to be equally likely.

Returning to the bag of balls, if I pick a ball it must either be red or black. These are **exclusive** events; it must be one *or* the other. Now p(red ball) $= \frac{3}{7}$ and p(black ball) $= \frac{4}{7}$.

Hence p(red ball) $+ p$(black ball) $= 1 = p$(red *or* black ball).

So if A and B are exclusive events, $p(A \text{ } \textbf{or} \text{ } B) = p(A) + p(B)$, i.e., the probabilities are *added*.

Suppose I now pick a ball and then replace it in the bag. I now repeat the operation. What is the probability that I have picked two red balls? p(I pick a red ball first time) $= \frac{3}{7}$. As the ball is replaced, and I start again, the p(I pick a red ball second time) also $= \frac{3}{7}$. The two events are **independent**; one has no effect on the probability of the other. Then p(I pick two red balls) $= \frac{3}{7} \times \frac{3}{7} = \frac{9}{49}$, i.e., the probabilities are *multiplied*.

The diagram illustrates this result. Each cross indicates a possible choice of two balls, 49 pairs in all. The shading covers the crosses showing two red balls, 9 pairs in all. Hence out of a possible 49 choices of two balls, 9 will be both red.

Therefore p(both red) $= \frac{9}{49}$

2nd ball

B × × × × × × ×
B × × × × × × ×
B × × × × × × ×
B × × × × × × ×
R × × × × × × ×
R × × × × × × ×
R × × × × × × ×
　　R R R B B B B
　　　　　　1st ball

So if A and B are independent events,
$p(A \text{ } \textbf{and} \text{ } B) = p(A) \times p(B)$

Example 10

There are 3 red and 4 black balls in a bag. I pick a ball and do not replace it. I then pick a second ball. (a) *What is the probability that both are red?*

$$p(\text{1st ball is red}) = \tfrac{3}{7}.$$

The bag now contains 2 red and 4 black balls.

$$p(\text{2nd ball is red}) = \tfrac{2}{6} = \tfrac{1}{3}$$
$$\text{Hence } p(\text{both are red}) = \tfrac{3}{7} \times \tfrac{1}{3} = \tfrac{1}{7}.$$

(b) *What is the chance that the first is black and the second red?*

$$p(\text{1st ball is black}) = \tfrac{4}{7}.$$

There are now 3 red and 3 black balls left.

$$p(\text{2nd ball is red}) = \tfrac{3}{6} = \tfrac{1}{2}.$$

Therefore $p(\text{1st is black, 2nd is red}) = \tfrac{4}{7} \times \tfrac{1}{2} = \tfrac{2}{7}.$
The result in (b) is twice as likely to happen as the result in (a).

(c) *What is the chance that both will be the same colour?*
Either both will be red or both will be black. These are exclusive events.

$$p(\text{both red}) = \tfrac{1}{7};$$
$$p(\text{both black}) = \tfrac{4}{7} \times \tfrac{3}{6} = \tfrac{2}{7}$$
$$\text{Hence } p(\text{both are same colour}) = \tfrac{1}{7} + \tfrac{2}{7} = \tfrac{3}{7}.$$

(d) *If three balls are picked (without replacements) what is the chance that all three are black?*

$$p(\text{1st is black}) = \tfrac{4}{7};$$
$$p(\text{2nd is black}) = \tfrac{3}{6} = \tfrac{1}{2};$$
$$p(\text{3rd is black}) = \tfrac{2}{5}.$$
$$\text{Hence } p(\text{all three are black}) = \tfrac{4}{7} \times \tfrac{1}{2} \times \tfrac{2}{5} = \tfrac{4}{35}.$$

Example 11

Two dice are thrown together. What is the probability that the total shown is 8?

Number on 2nd dice

It is easier to see all the possibilities in a graphical type of diagram. Each cross marks a possible pair of numbers that might be obtained (36 pairs in all). The shaded area covers all number pairs giving a total of 8: there are 5 such number pairs. Hence $p(\text{total is 8}) = \tfrac{5}{36}.$

```
6  × × × × × ×
5  × × × × × ×
4  × × × × × ×
3  × × × × × ×
2  × × × × × ×
1  × × × × × ×
   1 2 3 4 5 6
```

Number on 1st dice

Exercise 12.4

1 Two children are to be chosen at random from 4 boys and 5 girls. Find the chance that both are girls.

2 A letter is chosen at random from the word LIVELY. What is the chance that it is L?

3 A card is drawn from a pack at random. What is the chance that it is **(a)** an ace, **(b)** a spade, **(c)** any heart except the ace?

4 A bag contains 3 red, 2 white and 4 green balls. If one is chosen at random, what is the chance that a green ball is picked?

5 A thinks he has a chance of $\frac{1}{4}$ of passing a History exam and a chance of $\frac{1}{3}$ in passing a Geography exam. What is the chance he will pass in both (assuming that his effort in History has no effect on his Geography result)? What is the chance he will fail in both?

6 A dice is tossed three times in succession. What is the probability that it will show 2 each time?

7 If I put 6 letters into 6 envelopes at random, what is the probability that one particular letter is in the correct envelope?

8 A number of two digits (less than 30) is written down at random. What is the chance it will be an even number?

9 A bag contains 4 white and 5 yellow balls. Three balls are picked one after the other, without replacing them in the bag. What is the chance that they are all yellow?

10 A bag contains 5 red and 4 green balls. Two are picked at random. What is the chance that they are of different colours?

11 Two dice are thrown at the same time. What is the probability that the total shown on them will be **(a)** exactly 7, **(b)** at least 7?

12 Three men are selected from a group of 7. What is the chance that one particular man and his friend are chosen?

13 In a selected group of 100 people, 60 have brown hair and 45 have blue eyes. Find the chance that a particular member of the group has **(a)** brown hair and blue eyes, **(b)** brown hair and eyes which are not blue, **(c)** neither brown hair nor blue eyes.   (C)

14 Two dice are thrown together. **(a)** What is the chance that they will show the same number? **(b)** What is the probability of the total being 5? 7? 10? **(c)** What is the most probable total?

15 A soldier expects to hit a practice target 3 times out of 5. If he has three shots, what is the chance that he will miss each time?

13 Simple Trigonometrical Problems in Three Dimensions

The Solution of Triangles

The methods of solving triangles, discussed in Chapter 7, can be applied to problems on heights and distances. The terms used in such problems are illustrated in fig 13.1. An observer at O measures

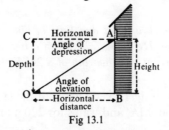

Fig 13.1

an **angle of elevation**, $B\hat{O}A$, at a horizontal distance, OB, to a height, AB. An observer at A measures an **angle of depression**, $C\hat{A}O$, from the horizontal, to a point O, at a horizontal distance, OB. Note the angle of depression is equal to the angle of elevation. If E is the angle of elevation, then

$$\tan E = \frac{AB}{BO} = \frac{\text{height}}{\text{horizontal distance}}$$

$$\text{and} \quad \cot E = \frac{\text{horizontal distance}}{\text{height}}.$$

The compass and bearings

Bearings or directions are given either as points of the compass or as an angle measured clockwise from the North (fig 13.2).

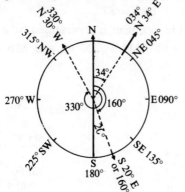

Fig 13.2
Bearings and the compass

The principal points of the compass used are N, E, S, W and the midway points NE, SE, SW, NW. Intermediate directions are given as an angle measured from the North or the South, e.g. N 34°E which is 34° from N towards the E, or S 25° W which is 25° from S towards W.

Bearings are usually now given as a three-figure angles always measured clockwise from the North. For example due E is given as 090°, N 34° E is given as 034°, S 20° E is 160° and N 30° W is 330°.

Example 1

A man wishing to reach a point due East of him, finds a lake between him and his objective. He walks 220 m on a bearing of N 63° E, and then straight to the point on a bearing of S 24° E. How far was he from his objective at first?

Draw a diagram, as in fig 13.3, showing all the facts. The man, at M, starts in the direction, MA, and walks for 220 m; hence MA is 220 m.

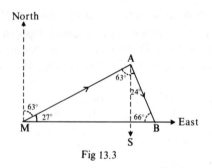

Fig 13.3

He then walks in the direction AB. MB is due East, hence $\widehat{BMA}$ = 90° − 63° = 27°. SA is at 90° to MB; hence $\widehat{SAM}$ = 63°, and $\widehat{SAB}$ = 24° ∴ $\widehat{MAB}$ = (24° + 63°) = 87°.
$\widehat{ABM}$ = (90° − 24°) = 66°. The problem is represented by a triangle with M = 27°, A = 87°, B = 66°, MA = 220 m, and MB is to be found.

$$\frac{MB}{\sin A} = \frac{MA}{\sin B}$$

$$\therefore MB = \frac{MA \sin A}{\sin B}$$

$$= \frac{220 \text{ m} \times \sin 87°}{\sin 66°}$$

$$= 240.5 \approx 241 \text{ m}$$

No.	Log
220	2·3424
sin 87°	$\bar{1}$·9994
	2·3418
sin 66°	$\bar{1}$·9607
240·5	2·3811

The man originally was 241 m from his objective.

(**Note**: The bearing of N 63° E could be written as 063°; the bearing of S 24° E could be written as 156°).

Exercise 13.1

1 A boat sails due South for 2·6 km and then S 47° W for 5·3 km. Find its distance from its starting point.

2 Two towers are 7·3 km apart, and one is due South of the other. A house is due West from one tower and its bearing from the other tower is 233°. Find the distance of the house from both of the towers.

3 A boy walks 4·2 km due North and then 5·3 km 077°. Find his distance and bearing from his starting point.

4 A ship is 5·5 km distant from a lighthouse on a bearing of 040° from the lighthouse. A second ship is 2 km distant from the lighthouse on a bearing of 335° from the lighthouse. What is the distance between the two ships?

5 Two ships leave harbour at the same time. One sails at 10 km/h on a bearing of N 35° E and the other at 9 km/h on a bearing of S 53° E. What is the distance, in km, between the ships after they have been sailing for 3 hours?

6 Point A is 312 m East of point C. Point B is 297 m North of point C. What is the distance between points A and B?

7 A and B are two towers 4·5 km apart. The bearing of A from B is 050°. It is required to locate a point, C, which is 8·5 km from tower B on a bearing of 150° from tower A. What is the distance between tower A and point C?

Planes

A flat surface is called a plane. Examples of planes are: the surface of a wall, the surface of still water. There are two important positions in which a plane can be located – vertical and horizontal. The walls of a house form vertical planes. The floor of a house is in a horizontal plane.

Lines normal to a plane

Take a piece of flat cardboard (i.e. whose surface is a plane) and on it draw several lines, all intersecting at a point O (see fig 13.4). Stand a

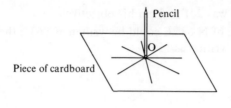

Fig 13.4

pencil on point O so that it is perpendicular to all the lines drawn in the plane. The pencil is now perpendicular to the plane. It is said to be **normal** to the plane.

A line is normal to a plane at a point on the plane when it is at right angles to every line in the plane passing through this point.

Angle between a line and a plane

In fig 13.5, a line AB meets a plane (shaded in the figure) at the point B. A normal to the plane, AC, is drawn from the point A, meeting the

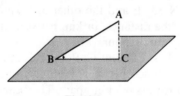

Fig 13.5

plane at point C. BC is then joined; it is a line in the plane. BC is called the **projection** of the line on the plane. If a light were placed vertically above the line AB, then the shadow it would cast on the plane would be BC. CB̂A is the angle between the line and the plane.

The angle between a line and a plane is the angle between the line and its projection on the plane.

The angle between two planes

In fig 13.6, two planes intersect in a common line ADGB. Line DC is drawn in plane X perpendicular to the line ADGB from the point D. Line DE is drawn in plane Y perpendicular to the line ADGB from the point D. The angle CDE is the angle between the two planes.

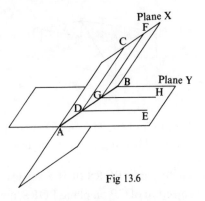

Fig 13.6

> The angle between two planes is the angle between two lines, one in each plane, which intersect on the common line of the two planes and are each at right angles to this line.

Notice that many pairs of lines, each intersecting on the common line, and perpendicular to it, can be drawn. In fig 13.6, GH and GF are such a pair of lines; they intersect on the common line at G, and the angle between the planes is $\widehat{HGF}$, which is equal to $\widehat{CDE}$.

Problems in Three Dimensions

Previously we used trigonometry to solve problems in two dimensions, where all lines lie in one plane. We can use similar methods to solve three-dimensional problems.

In order to do this, a clear diagram must be drawn, and the relevant triangles taken from the diagram. At a later stage, it will be possible to visualise the relevant triangles without having to draw them separately, but in all cases of doubt, the drawing of such triangles is a help. All right-angled triangles, formed by horizontal and vertical lines, should be clearly marked, as this type of triangle is the most useful. The following examples show how these principles are put into practice.

Example 2

A room in a house has its corners labelled ABCD *on the ceiling and* PQRS *on the floor, with corner* A *above corner* P. *A line is drawn from* P *to the diagonally opposite corner* C. *The room is* 10 m *long by* 6 m *wide and* 4 m *high. Find the angle between the line* PC *and the floor.*

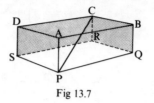

Fig 13.7

First draw a sketch, as in fig 13.7. The angle between CP and the plane PQRS has to be found. This angle is the angle between CP and its projection on the plane.

CR is perpendicular to plane PQRS (it is a vertical side of the wall). Hence RP is the projection of CP on plane PQRS, making the required angle $C\hat{P}R$.

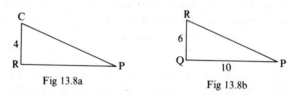

Fig 13.8a

Fig 13.8b

In fig 13.8a, the triangle CPR is drawn. CR = 4 m (height of room). PR is found from the triangle PQR (fig 13.8b), where PR is the diagonal across the floor of the room.

$$PR^2 = PQ^2 + QR^2 = 10^2 + 6^2 = 100 + 36 = 136$$

$$\therefore PR = \sqrt{136} = 11\cdot66 \text{ m}$$

From triangle CRP, $\tan C\hat{P}R = \dfrac{CR}{PR} = \dfrac{4}{11\cdot66}$

$\therefore C\hat{P}R = 18° 56'$.

No.	Log
4	0·6021
11·66	1·0667
tan 18° 56′	$\bar{1}$·5354

Example 3

From a point, A, due West of a wireless mast, the angle of elevation to the top of the mast is 34°, and from a point, B, due South of the mast, the angle of elevation is 28° 30′. If the distance AB is 150 m, find the height of the mast.

First draw a sketch, as in fig 13.9. DC is the wireless mast. The diagram forms three right-angled triangles, shown in fig 13.10. The common factor between the triangles in figs 13.10a and b is h, the height of the mast. In fig 13.10c, $A\hat{C}B$ is a right angle as A is due West and B due South.

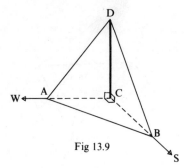

Fig 13.9

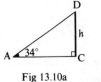

Fig 13.10a

Fig 13.10b

Fig 13.10c

Now AC $= h \cot 34°$ and BC $= h \cot 28° \, 30'$

In the triangle ABC, $AB^2 = AC^2 + BC^2$

i.e. $(150)^2 = (h \cot 34°)^2 + (h \cot 28° \, 30')^2$

$\quad\quad\quad = h^2(1\cdot4826)^2 + h^2(1\cdot8418)^2$

$\quad\quad\quad = h^2(2\cdot197 + 3\cdot393) = h^2(5\cdot590)$

$\therefore h^2 = \dfrac{(150)^2}{5\cdot590}$

$\therefore h = 63\cdot45$ m

$\quad\quad \approx 63\cdot5$ m.

No.	Log
150^2	$2\cdot1761 \times 2$
	$4\cdot3522$
$5\cdot590$	$0\cdot7474$
(h^2)	$3\cdot6048 \div 2$
$63\cdot45$	$1\cdot8024$

Example 4

A regular tetrahedron, OABC, has all its sides 10 cm long. The tetrahedron is cut by a plane PQR such that OP = 5 cm on side OA; OQ = 8 cm on side OB; and OR = 8 cm. Find the angle between the plane PQR and the plane OBC.

A sketch of the solid with plane PQR is shown in fig 13.11i. The common line of the two planes is QR, which is parallel to BC. On the face OBC let OM be the perpendicular bisector of BC and therefore also of QR (mid-point N).

The plane AOM will be a plane of symmetry (as the tetrahedron is regular) and will be perpendicular to QR, cutting it at N. As PN lies in the plane AOM, PN is perpendicular to QR. Hence the angle between the planes PQR and OBC is angle PNO. We concentrate therefore on triangle PNO (fig 13.11ii).

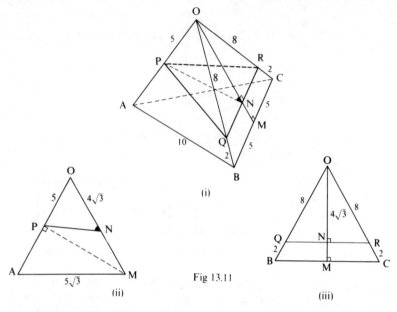

Fig 13.11

From fig 13.11iii $\dfrac{ON}{OM} = \dfrac{OQ}{OB} = \dfrac{8}{10}$

and $OM = 10\sin 60° = 10 \times \dfrac{\sqrt{3}}{2} = 5\sqrt{3}$

Therefore $ON = \dfrac{8}{10} \times 5\sqrt{3} = 4\sqrt{3}$

Since the tetrahedron is regular, $AM = OM = 5\sqrt{3}$.

Triangle OAM is isosceles and P is the mid-point of OA.

Therefore angle $OPM = 90°$ and $\cos P\widehat{O}M = \dfrac{OP}{OM} = \dfrac{5}{5\sqrt{3}} = \dfrac{1}{\sqrt{3}}$.

Also $PN^2 = OP^2 + ON^2 - 2 \times OP \times ON \times \cos P\widehat{O}N$

$\qquad = 5^2 + (4\sqrt{3})^2 - 2 \times 5 \times 4\sqrt{3} \times \dfrac{1}{\sqrt{3}} = 33$

and $\sin P\widehat{O}N = \sqrt{1 - \cos^2 P\widehat{O}M} = \sqrt{\dfrac{2}{3}}$

Then finally in triangle PON, $\dfrac{OP}{\sin P\widehat{N}O} = \dfrac{PN}{\sin P\widehat{O}N}$

therefore $\sin P\widehat{N}O = \dfrac{OP \times \sin P\widehat{O}N}{PN} = \dfrac{5 \times \sqrt{\dfrac{2}{3}}}{\sqrt{33}} = \dfrac{5\sqrt{2}}{\sqrt{99}}$

giving angle $PNO = 45° \ 17'$ which is the angle required.

Volumes of Solids

The basic volumes are those of prisms and pyramids.

Prisms

The volume of a right prism is given by:
Volume = (area of base) × (height) as shown in fig 13.12a.

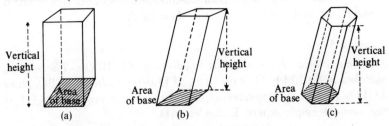

Fig 13.12   Volume of prism = Area of base × Vertical height

The prism can be considered to be made up of a series of thin slices, each slice having a volume given by (area) × (thickness). If the prism is slanted at an angle, as in fig 13.12b, the area of the parallelogram formed by a side A′B′C′D′ is the same as the area of a rectangle ABCD on the same base and of the same vertical height. A series of slices parallel to the side will then have each slice of the same area as the rectangle, and hence the volume will be the same for slices of the same thickness. The volume of the slanted prism, which is the volume of all the slices, will be the same as the volume of the right prism if the area of the base is the same. Hence the volume of any prism (see fig 13.12c) is given by:

Volume of prism = (area of base) × (vertical height)

whether the prism is a right prism or a slanted, i.e. oblique, one. A circular cylinder is a particular form of a prism, where the area of the base is the area of a circle.

Pyramids

The volume of a right pyramid, as proved by advanced mathematics, is Volume = $\frac{1}{3}$ (area of base) × (vertical height), as in fig 13.13a.

(a) Right pyramid   (b) Right cone   (c) Tetrahedron   (d) Oblique pyramid

Fig 13.13   Volume of pyramid = $\frac{1}{3}$ × (Area of base) × (Vertical height)

Special cases of pyramids are **(a)** cones, fig 13.13b, where the area of the base is the area of a circle; **(b)** tetrahedrons, fig 13.13c, where the area of the base is the area of a triangle. In all cases, the same formula is applicable. Oblique cones or pyramids have their volumes given by the same formula, as illustrated in fig 13.13d. An argument based on taking thin slices, can be made as in the case of prisms.

The top vertex of a pyramid is called the **apex**. A sloping side is called a **slant** side and for the cone, the **slant height**.

Example 5

Find the volume of a regular tetrahedron of side 10 cm. The solid is sketched in fig 13.14i. O is the apex. If D is the midpoint of BC, plane AOD is a plane of symmetry as the tetrahedron is regular and OE is the vertical height where E lies on AD.

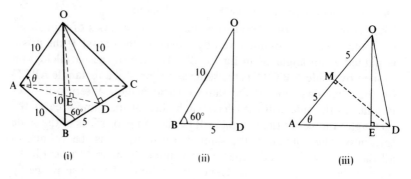

(i)   (ii)   (iii)

Fig 13.14

To find the volume we need the area of the base (ABC) and the vertical height (OE).

To find OE, we use triangle OAE, where angle OAE $= \theta$.

To find θ, we use triangle OAD, first finding the length of OD from triangle BOD (fig 13.14ii).

$$\text{OD} = 10 \sin 60° = 10 \times \frac{\sqrt{3}}{2} = 5\sqrt{3}.$$

Now AD = OD by symmetry and so triangle OAD is isosceles. Let M be the mid-point of AO (fig 13.14iii).

Then
$$\cos \theta = \frac{\text{AM}}{\text{AD}} = \frac{5}{5\sqrt{3}} = \frac{1}{\sqrt{3}}.$$

In triangle OAE, $\text{OE} = 10 \sin \theta = 10\sqrt{1 - \cos^2 \theta}$

$$= 10 \times \sqrt{1 - \frac{1}{3}} = 10\sqrt{\frac{2}{3}}.$$

Hence the volume $= \frac{1}{3}$(area of triangle ABC) $\times$ OE

$$= \frac{1}{3}(\tfrac{1}{2} \times 10 \times 5\sqrt{3}) \times 10 \times \sqrt{\frac{2}{3}}$$

$$= \frac{10 \times 5\sqrt{3} \times 10 \times \sqrt{2}}{3 \times 2 \times \sqrt{3}} \approx 118 \text{ cm}^3.$$

Exercise 13.2

1 A right pyramid, 12 cm high, stands on a rectangular base 6 cm by 10 cm. Calculate (a) the length of an edge of the pyramid; (b) the angles the triangular faces make with the base; (c) the volume of the pyramid.

2 ABC is an equilateral triangle marked out on horizontal ground and each side is 65 m long. At A there is a tree. The angle of elevation of the top of the tree from C is 25°. Find (a) the height of the tree; (b) the angle of elevation of the top of the tree from the midpoint of BC.

3 From a ship at a point A, the base, B, (which is at sea-level) of a lighthouse is 450 m away. The bearing of the lighthouse from the ship is N 39° E. The ship sails East to a point, C, which is due South of the lighthouse; from C the angle of elevation of the top of the lighthouse is 17° 30′. Calculate (a) the distance BC; (b) the height of the lighthouse.

4 A pyramid stands on a square base of side 8 cm. The four triangular faces slope at 70° to the horizontal. Calculate (a) the height of the pyramid; (b) the angle between one of the sloping edges and the base of the pyramid.

5 The angles of depression to two buoys, from a point on a cliff, are found to be 24° 15′ and 30° 23′. The two buoys are 1270 m apart, and are respectively NE and NW of the point on the cliff. Find the height of the cliff in metres, correct to 3 significant figures.

6 The angle of elevation of the top of a tower due North of an observer is 22° 18′. The observer walks 150 m due West and finds the angle of elevation is 16° 47′. What is the height of the tower, and what is the distance between the observer and the tower when he makes the second observation?

7 ABCD is a regular tetrahedron, each edge of which is 10 cm long. Calculate (a) the angle between AB and the plane BCD; (b) the angle between the plane ACD and the plane BCD; (c) the volume of the tetrahedron.   (L)

8 A right pyramid, VABCD, has a square base, ABCD, of side

10 cm. The vertex, V, is 8 cm above the centre of the base. Calculate
(a) the angle between the plane VAB and the base; (b) the angle
between VA and the base. (c) A piece of cotton is stretched tightly
from V to a point X on CA produced where AX = 7 cm. Calculate
the angle AVX.   (C)

9 The square face, ABCD, of a cube is horizontal, and DZ is the
vertical edge through D. The point P on the edge BC is such that
BP = 5 cm and PC = 7 cm. Calculate (a) the length of the perpendicular from D to the line AP; (b) the angle which the plane
APZ makes with the horizontal.   (C)

10 The points P, Q, and R lie on the surface of a sphere of radius
a. The straight lines PQ, QR, RP are each of length a. Find the
length of the shortest path on the surface of the sphere from P
to Q.
Another point S on the surface of the sphere is such that the
straight lines PS, SQ and PQ are equal. Calculate the angle between
the planes PQR and PQS correct to the nearest degree.   (L)

11 The interior dimensions of a rectangular box are 3 m by 4 m by 5 m.
Find the length, to the nearest tenth of a metre, of the longest thin
straight rod which can be put into the box.
When the longest possible rod is in the box, calculate (a) the
angles made by the rod with the edges of the box; (b) the angle
made by the rod with the longest face of the box.   (L)

12 The points A, B and F lie in a horizontal plane. The point A is
due West of a vertical tower FT, and the point B is South-west
of the tower. The angles of elevation from A and B are each 42°.
Find the bearing of B from A.   (C)

13 ABC is a triangle in a horizontal plane such that AB = 8 m,
AC = 9 m and the angle BAC is 120°. V is the point vertically
above A, such that the angle BVC is a right angle Calculate the
height of V above A. Calculate the volume of the pyramid with
ABC as base and V as vertex.   (C)

14 OA, OB, OC are three straight lines such that each is perpendicular
to the other two. OA = 6 cm, OB = 8 cm, OC = 15 cm. Calculate (a) the angle ABC; (b) the area of the triangle ABC. By
considering the volume of the tetrahedron OABC, calculate the
length of the perpendicular from O to the plane ABC.   (C)

15 Three points P, F, and Q lie on horizontal ground. The point
P is due East and the point Q due South of F. The angle of elevation
of the top of a vertical tower FT from P is 32° and from Q is 48°.
Given that FT = 80 m, calculate the distance PQ and the bearing
of Q from P.   (C)

14 Compound and Multiple Angles: Trigonometrical Identities

Sum of Two Angles

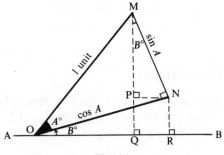

Fig 14.1

A triangle MNO is right-angled at N and has $\hat{NOM} = A$ (fig 14.1). The hypotenuse, OM, is of unit length, so $MN = \sin A$ and $ON = \cos A$. The triangle is placed with its base, ON, on a horizontal line, AB. The triangle is then rotated anticlockwise about point O, through an angle of B. Angle $B = \hat{BON}$. Now $\hat{BOM} = (A + B)$ and it is required to find $\sin (A + B)$ and $\cos (A + B)$ in terms of $\sin A$, $\cos A$, $\sin B$ and $\cos B$.

(1) Drop a perpendicular MQ to AB.

(2) Draw NP perpendicular to MQ, meeting it at P.

(3) Drop a perpendicular NR to AB.

Now $\sin (A + B) = \dfrac{MQ}{OM} = MQ$ (since OM is 1 unit long)

In triangle MPN, $\hat{PMN} = B$ ($B = \hat{PNO} = 90° - \hat{MNP} = \hat{PMN}$) and $\hat{MPN}$ is a right angle (PN ⊥ PM)

$$\therefore \qquad \cos B = \frac{MP}{MN}$$

i.e. $\qquad MP = MN \cos B = \sin A \cos B.$

In triangle ONR, $\hat{NRO}$ is a right angle (NR ⊥ AB)

$$\therefore \qquad \frac{NR}{ON} = \sin B$$

i.e. $\qquad NR = ON \sin B = \cos A \sin B$

but PNRQ is a rectangle (PQ and NR ⊥ QR; NP ⊥ MQ)

$$\therefore \qquad NR = PQ$$

i.e. $PQ = \cos A \sin B$

Now $\sin (A + B) = MQ = MP + PQ$

$= \sin A \cos B + \cos A \sin B.$

Now $\cos (A + B) = \dfrac{OQ}{OM} = OQ$

$= OR - RQ = OR - PN$

$= ON \cos B - MN \sin B$

$= \cos A \cos B - \sin A \sin B.$

Difference of Two Angles

Start with the triangle MNO again, but this time rotate it clockwise about point O, through an angle B, as in fig 14.2.

Now $\widehat{BOM} = (A - B)$ and it is required to find $\sin (A - B)$ and $\cos (A - B)$.

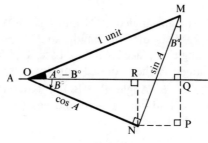

Fig 14.2

(1) Draw MQ perpendicular to AB.
(2) Produce MQ to P and draw NP perpendicular to MP.
(3) Draw NR perpendicular to AB.

Now $\sin (A - B) = \dfrac{MQ}{OM} = MQ$ (OM is 1 unit long)

In triangle MNP, $\widehat{NMP} = B$.

Hence $MP = MN \cos B = \sin A \cos B$

and $NP = MN \sin B = \sin A \sin B.$

In triangle ONR, $\widehat{ORN}$ is a right angle.

Hence $RN = ON \sin B = \cos A \sin B$

and $OR = ON \cos B = \cos A \cos B.$

Now $\sin (A - B) = MQ = MP - QP$

$= MP - RN$ (RN = QP: opp. sides of

rectangle)

$= \sin A \cos B - \cos A \sin B.$

$$\cos (A - B) = \frac{OQ}{OM} = OQ$$
$$= OR + RQ = OR + NP \qquad (RQ = NP: \text{opp.}$$
$$= \cos A \cos B + \sin A \sin B \qquad \text{sides of rectangle).}$$

These results are very important and must be remembered; they are summarised below:

$$\boxed{\begin{array}{l} \sin (A + B) = \sin A \cos B + \cos A \sin B \\ \sin (A - B) = \sin A \cos B - \cos A \sin B \\ \cos (A + B) = \cos A \cos B - \sin A \sin B \\ \cos (A - B) = \cos A \cos B + \sin A \sin B \end{array}}$$

These formulae are **identities**, as they are true for any value of angles A and B. Although a proof has only been given for acute angles, the identities can be proved true for angles in any quadrant, and of any magnitude.

The formulae can be used as an alternative way of determining the ratios of angles of any magnitude. For example:

$$\sin 330° = \sin (360° - 30°)$$
$$= \sin 360° \cos 30° - \cos 360° \sin 30°$$
$$= 0 \times \cos 30° - (1) \times \sin 30°$$
$$= -\sin 30° = -\tfrac{1}{2}.$$

$$\cos 240° = \cos (180° + 60°)$$
$$= \cos 180° \cos 60° - \sin 180° \sin 60°$$
$$= (-1) \times \cos 60° - 0 \times \sin 60°$$
$$= -\cos 60° = -\tfrac{1}{2}.$$

They can also be used to find the value of negative angles. For example:

$$\sin(-A) = \sin (0° - A)$$
$$= \sin 0° \cos A - \cos 0° \sin A$$
$$= 0 \times \cos A - (1) \times \sin A$$
$$= -\sin A.$$

$$\cos(-A) = \cos (0° - A)$$
$$= \cos 0° \cos A + \sin 0° \sin A$$
$$= (1) \times \cos A + 0 \times \sin A$$
$$= \cos A.$$

The formulae can also be used to find the ratios of compound angles in a simple way. Care must be taken over the *signs* of the ratios involved.

Example 1

If $\sin A = \dfrac{4}{5}$ *and* $\cos B = \dfrac{12}{13}$, *find the values of*

$\qquad$ $\sin (A + B)$ *and* $\cos (A + B)$ *without using tables,*

(i) *if A and B are both acute angles,* **(ii)** *if A is obtuse and B is acute.*

If $\sin A = \dfrac{4}{5}$ then $\cos A = \sqrt{1 - \sin^2 A}$

$\qquad\qquad\qquad = \sqrt{1 - \dfrac{16}{25}} = \pm\dfrac{3}{5}$ $\qquad$ (+ if A is acute, − if A is obtuse).

If $\cos B = \dfrac{12}{13}$, then $\sin B = \sqrt{1 - \cos^2 B}$

$\qquad\qquad\qquad = \sqrt{1 - \dfrac{144}{169}} = \dfrac{5}{13}$ $\qquad$ (+ as B is acute).

(i) $\sin (A + B) = \sin A \cos B + \cos A \sin B$

$\qquad\qquad = \dfrac{4}{5} \times \dfrac{12}{13} + \dfrac{3}{5} \times \dfrac{5}{13} = \dfrac{63}{65}.$

$\cos (A + B) = \cos A \cos B - \sin A \sin B$

$\qquad\qquad = \dfrac{3}{5} \times \dfrac{12}{13} - \dfrac{4}{5} \times \dfrac{5}{13} = \dfrac{16}{65}.$

(ii) $\sin (A + B) = \dfrac{4}{5} \times \dfrac{12}{13} + \left(-\dfrac{3}{5}\right) \times \dfrac{5}{13} = \dfrac{33}{65}$

$\cos (A + B) = \left(-\dfrac{3}{5}\right) \times \dfrac{12}{13} - \dfrac{4}{5} \times \dfrac{5}{13} = -\dfrac{56}{65}.$

If special angles are used, the answer can usually be expressed in surd form.

Example 2

Find the value of $\sin 15°$, *leaving the answer in surd form.*

$\sin 15° = \sin (45° - 30°) = \sin 45° \cos 30° - \cos 45° \sin 30°$

$\qquad\qquad\qquad = \dfrac{1}{\sqrt{2}} \times \dfrac{\sqrt{3}}{2} - \dfrac{1}{\sqrt{2}} \times \dfrac{1}{2}$

$\qquad\qquad\qquad = \dfrac{\sqrt{3} - 1}{2\sqrt{2}} = \dfrac{\sqrt{6} - \sqrt{2}}{4}.$

The formulae can also be used to reduce certain trigonometrical expressions to a single ratio.

Example 3

Express as a single ratio: **(a)** $\cos 32° \cos 48° - \sin 32° \sin 48°$;

(b) $\dfrac{1}{\sin 46° \cos 44° + \cos 46° \sin 44°}$.

(a) $\cos (A + B) = \cos A \cos B - \sin A \sin B$
and $A = 32°$; $B = 48°$
∴ $\cos 32° \cos 48° - \sin 32° \sin 48° = \cos (32° + 48°) = \cos 80°$.

(b) $\sin (A + B) = \sin A \cos B + \cos A \sin B$
and $A = 46°$; $B = 44°$

$$\therefore \frac{1}{\sin 46° \cos 44° + \cos 46° \sin 44°} = \frac{1}{\sin (46° + 44°)}$$

$$= \frac{1}{\sin 90°} = 1.$$

Exercise 14.1

1 If $\cos A = \frac{4}{5}$ and $\cos B = \frac{12}{13}$, both angles being acute, find the value of $\sin (A + B)$ without using tables.

2 Given $\cos A = 0.8$ and $\cos B = 0.6$, find, without using tables, the value of **(a)** $\sin (A - B)$; **(b)** $\cos (A - B)$; **(c)** the value of $(A + B)$ in degrees.

3 $\sin A = \frac{5}{13}$ and A is obtuse; $\cos B = \frac{4}{5}$ and B is acute. Find the value of **(a)** $\sin (A + B)$; **(b)** $\cos (A + B)$; **(c)** $\cos (A - B)$.

4 Express as single trigonometrical ratios:
   (a) $\sin 36° \cos 29° + \cos 36° \sin 29°$;
   (b) $\cos 63° \cos 27° + \sin 63° \sin 27°$;
   (c) $\sin 120° \cos 54° - \cos 120° \sin 54°$;
   (d) $\cos 23° \cos 58° - \sin 23° \sin 58°$.

5 Find the value of $\cos 75°$ in surd form, using the ratios of special angles.

6 Find the value of $\cos 15°$ in surd form.

7 Using the identity $\cos 90° = \cos (45° + 45°)$, prove that
$$\cos 90° = 0.$$

8 Express the following as single trigonometrical ratios:
   (a) $\frac{1}{2} \cos \theta + \dfrac{\sqrt{3}}{2} \sin \theta$;   **(b)** $\dfrac{1}{\sqrt{2}} \cos \theta - \dfrac{1}{\sqrt{2}} \sin \theta$.

9 By using the formulae for compound angles, prove that:
   (a) $\sin (180° + \theta) = - \sin \theta$;   **(b)** $\cos (270° + \theta) = \sin \theta$.

10 Find, in surd form, the value of the following ratios:
   (a) $\cos 105°$;   **(b)** $\sin 165°$;   **(c)** $\sin 345°$.

Tangents of Compound Angles

The tangents of compound angles can be deduced from the formulae for the sines and cosines of compound angles.

$$\tan (A + B) = \frac{\sin (A + B)}{\cos (A + B)}$$

$$= \frac{\sin A \cos B + \cos A \sin B}{\cos A \cos B - \sin A \sin B}$$

$$= \frac{\dfrac{\sin A \cos B}{\cos A \cos B} + \dfrac{\cos A \sin B}{\cos A \cos B}}{\dfrac{\cos A \cos B}{\cos A \cos B} - \dfrac{\sin A \sin B}{\cos A \cos B}} \qquad \text{(divide top and bottom by } \cos A \cos B)$$

$$= \frac{\tan A + \tan B}{1 - \tan A \tan B}$$

Similarly

$$\tan (A - B) = \frac{\sin (A - B)}{\cos (A - B)}$$

$$= \frac{\sin A \cos B - \cos A \sin B}{\cos A \cos B + \sin A \sin B} \qquad \text{(divide top and bottom by } \cos A \cos B)$$

$$= \frac{\tan A - \tan B}{1 + \tan A \tan B}$$

These formulae are important and must be remembered; they are summarised below.

$$\boxed{\begin{aligned} \tan (A + B) &= \frac{\tan A + \tan B}{1 - \tan A \tan B} \\[2mm] \tan (A - B) &= \frac{\tan A - \tan B}{1 + \tan A \tan B} \end{aligned}}$$

The tangent formulae are used in the same way as those for sine and cosine.

Example 4

If $\tan A = \frac{1}{3}$ *and* $\tan B = \frac{3}{4}$, *both A and B being acute angles, find the value of* $\tan (A + B)$ *without using tables.*

$$\tan (A + B) = \frac{\tan A + \tan B}{1 - \tan A \tan B}$$

$$= \frac{\frac{1}{3} + \frac{3}{4}}{1 - \frac{1}{3} \times \frac{3}{4}} = \frac{\frac{4+9}{12}}{1 - \frac{1}{4}}$$

$$= \frac{\frac{13}{12}}{\frac{3}{4}} = \frac{13}{12} \times \frac{4}{3} = \frac{13}{9}.$$

Example 5

If $\tan (A - B) = \frac{1}{5}$ *and* $\tan A = 2$, *find the value of* $\tan B$.

$$\tan (A - B) = \frac{\tan A - \tan B}{1 + \tan A \tan B} = \frac{1}{5}$$

$$\frac{2 - \tan B}{1 + 2 \tan B} = \frac{1}{5}$$

i.e. $\qquad 1 + 2 \tan B = 10 - 5 \tan B$

$\therefore \qquad 7 \tan B = 9$ i.e. $\tan B = \frac{9}{7}$.

Example 6

Express the following as a single trigonometrical ratio:

$$\frac{\sqrt{3} + \tan \theta}{1 - \sqrt{3} \tan \theta}.$$

$$\tan (A + B) = \frac{\tan A + \tan B}{1 - \tan A \tan B}$$

Hence $\qquad \tan A = \sqrt{3}$ i.e. $A = 60°$ (or $240°$) and $B = \theta$

$$\therefore \qquad \frac{\sqrt{3} + \tan \theta}{1 - \sqrt{3} \tan \theta} = \tan (60° + \theta) \text{ or } \tan (240° + \theta).$$

Exercise 14.2

1 Find the value of the following expressions without using trigono-metrical tables, leaving your answers in surd form:
(a) $\tan (45° + 30°)$;   **(b)** $\tan 15°$;   **(c)** $\tan 105°$;
(d) $\cot 165°$.

2 If $\tan A = \frac{1}{3}$ and $\tan B = \frac{1}{2}$, both A and B being acute, find the value of $\tan (A - B)$ without using tables.

3 If $\tan A = \frac{1}{2}$ and $\tan B = -\frac{3}{4}$, A being acute and B being obtuse, find the value of $\tan (A + B)$ without using trigonometrical tables.

4 Express the following as single trigonometrical ratios:

(i) $\dfrac{\tan \theta - \sqrt{3}}{1 + \sqrt{3} \tan \theta}$;   **(ii)** $\dfrac{1 + \tan \theta}{1 - \tan \theta}$.

5 Find, without using trigonometrical tables, the value of:

(i) $\dfrac{\tan 20° + \tan 25°}{1 - \tan 20° \tan 25°}$;

(ii) $\dfrac{1 + \tan 15°}{1 - \tan 15°}$;

(iii) $\dfrac{\tan 75° + \sqrt{3}}{1 - \sqrt{3}\tan 75°}$.

6 Find, without using trigonometrical tables, the value of tan B if tan $A = \frac{1}{4}$ and tan $(A + B) = 2$.

7 Find, without using trigonometrical tables, the value of cot A if cot $B = \frac{1}{2}$ and cot $(A - B) = 4$.

(Hint: $\tan (A - B) = \dfrac{1}{\cot (A - B)}$.)

8 If sin $A = \frac{3}{5}$ and cos $B = -\frac{12}{13}$, where A and B are both obtuse, find, without using trigonometrical tables, the values of

(i) tan $(A + B)$; (ii) tan $(A - B)$.

Multiple Angles

The formulae for compound angles can be used to find values of multiple angles. In the first instance, angle B is made equal to angle A.

Hence $\sin 2A = \sin (A + A) = \sin A \cos A + \cos A \sin A$

$\qquad\qquad\qquad\qquad = 2 \sin A \cos A$

$\cos 2A = \cos (A + A) = \cos A \cos A - \sin A \sin A$

$\qquad\qquad\qquad\qquad = \cos^2 A - \sin^2 A$

But $\qquad\qquad\qquad \cos^2 A + \sin^2 A = 1$

$\qquad\qquad\qquad\qquad \therefore \sin^2 A = 1 - \cos^2 A$

$\therefore \cos 2A = \cos^2 A - (1 - \cos^2 A)$

$\qquad\qquad\quad = 2 \cos^2 A - 1$

Also $\qquad\qquad \cos^2 A = 1 - \sin^2 A$

$\therefore \qquad\qquad \cos 2A = (1 - \sin^2 A) - \sin^2 A$

$\qquad\qquad\qquad\quad = 1 - 2 \sin^2 A$

$\tan 2A = \tan (A + A) = \dfrac{\tan A + \tan A}{1 - \tan A \tan A}$

$\qquad\qquad\qquad\qquad = \dfrac{2 \tan A}{1 - \tan^2 A}$

The formulae for multiple angles are important, and should be memorised; they are summarised next page.

$$\sin 2A = 2 \sin A \cos A$$
$$\cos 2A = \cos^2 A - \sin^2 A$$
$$= 2 \cos^2 A - 1$$
$$= 1 - 2 \sin^2 A$$
$$\tan 2A = \frac{2 \tan A}{1 - \tan^2 A}$$

Higher multiple angles can be found by building up on the results already found, e.g. $\sin 3A = \sin (2A + A)$; the right hand expression is expanded, using the compound angle formula, then $\sin 2A$ is substituted from the results above. Similarly, $\cos 4A = \cos (2A + 2A)$. In this way multiples of A can be found in terms of ratios of single angles.

Half Angles

Half angles can be substituted in the compound angle formulae:

$$\sin A = \sin \left(\frac{A}{2} + \frac{A}{2} \right) = 2 \sin \frac{A}{2} \cos \frac{A}{2}$$

Similarly, $\cos A = \cos^2 \frac{A}{2} - \sin^2 \frac{A}{2}$

$$= 2 \cos^2 \frac{A}{2} - 1$$

$$= 1 - 2 \sin^2 \frac{A}{2}$$

and $\tan A = \dfrac{2 \tan \dfrac{A}{2}}{1 - \tan^2 \dfrac{A}{2}}$.

A useful substitution in later trigonometrical work can be derived by expressing the trigonometrical ratios in terms of the tangent of the half angle.

$$\sin A = 2 \sin \frac{A}{2} \cos \frac{A}{2}$$

$$= \frac{2 \sin \dfrac{A}{2} \cos \dfrac{A}{2}}{\sin^2 \dfrac{A}{2} + \cos^2 \dfrac{A}{2}}$$

$(\sin^2 \frac{A}{2} + \cos^2 \frac{A}{2} = 1$

$\therefore$ original expression is unaltered)

$$= \frac{\dfrac{2 \sin A/2 \cos A/2}{\cos^2 A/2}}{\dfrac{\sin^2 A/2}{\cos^2 A/2} + \dfrac{\cos^2 A/2}{\cos^2 A/2}} \qquad \text{(dividing top and bottom}$$
by $\cos^2 \dfrac{A}{2}$)

$$= \frac{2 \tan \dfrac{A}{2}}{\tan^2 \dfrac{A}{2} + 1} \qquad \text{Now put } \tan \dfrac{A}{2} = t$$

We have $\sin A = \dfrac{2t}{1 + t^2}$

$$\cos A = \cos^2 \frac{A}{2} - \sin^2 \frac{A}{2}$$

$$= \frac{\cos^2 \dfrac{A}{2} - \sin^2 \dfrac{A}{2}}{\cos^2 \dfrac{A}{2} + \sin^2 \dfrac{A}{2}}$$

$$= \frac{1 - t^2}{1 + t^2} \qquad \text{(after dividing top and bottom}$$
by $\cos^2 \dfrac{A}{2}$)

$$\tan A = \frac{2 \tan \dfrac{A}{2}}{1 - \tan^2 \dfrac{A}{2}}$$

$$= \frac{2t}{1 - t^2}.$$

To summarise the results:

$$\boxed{\begin{aligned} \sin A &= \frac{2t}{1 + t^2} \\[2mm] \cos A &= \frac{1 - t^2}{1 + t^2} \\[2mm] \tan A &= \frac{2t}{1 - t^2} \end{aligned}}$$

These are called the t formulae, where $t = \tan \dfrac{A}{2}$.

Example 7

Express sin 3*A* *in terms of* sin *A*.

$$\begin{aligned}
\sin 3A &= \sin (2A + A) \\
&= \sin 2A \cos A + \cos 2A \sin A \\
&= (2 \sin A \cos A) \cos A + (1 - 2 \sin^2 A) \sin A \\
&= 2 \sin A \cos^2 A + (1 - 2 \sin^2 A) \sin A \\
&= 2 \sin A (1 - \sin^2 A) + (1 - 2 \sin^2 A) \sin A \\
&= 2 \sin A - 2 \sin^3 A + \sin A - 2 \sin^3 A \\
&= 3 \sin A - 4 \sin^3 A.
\end{aligned}$$

Example 8

Express simply $1 - 2 \sin^2 42°$.

Now $\cos 2A = 1 - 2 \sin^2 A$;   put $A = 42°$.

$\therefore$ $\cos 84° = 1 - 2 \sin^2 42°$.

Example 9

Evaluate, without using tables, $\dfrac{2 \tan 30°}{1 - \tan^2 30°}$.

Now $\tan 2A = \dfrac{2 \tan A}{1 - \tan^2 A}$;   put $A = 30°$.

$$\tan 60° = \frac{2 \tan 30°}{1 - \tan^2 30°} = \sqrt{3}.$$

Example 10

If $\tan \theta = \frac{3}{4}$, *find* **(i)** sin 2$\theta$; **(ii)** tan 2$\theta$ *without using tables.*

Now $\sin^2 \theta + \cos^2 \theta = 1$

i.e. $\tan^2 \theta + 1 = \sec^2 \theta$ (dividing throughout by $\cos^2 \theta$)

$\therefore$ $\cos^2 \theta = \dfrac{1}{1 + \tan^2 \theta}$

$\therefore$ $\cos \theta = \dfrac{1}{\sqrt{1 + \tan^2 \theta}} = \dfrac{1}{\pm\sqrt{1 + \frac{9}{16}}}$

$\qquad\quad = \dfrac{\pm\sqrt{16}}{\sqrt{25}} = \pm \dfrac{4}{5}$

Note: If θ is in the 1st quadrant, $\cos \theta = +\frac{4}{5}$; if θ is in the 3rd quadrant, $\cos \theta = -\frac{4}{5}$.

$$\sin \theta = \pm \sqrt{1 - \cos^2 \theta} = \pm \sqrt{1 - \frac{16}{25}} = \pm \sqrt{\frac{9}{25}} = \pm \frac{3}{5}$$

($\sin \theta$, similarly, can be $+\frac{3}{5}$ in the 1st quadrant, and $-\frac{3}{5}$ in the 3rd quadrant).

$$\sin 2\theta = 2 \sin \theta \cos \theta = 2 \times \tfrac{3}{5} \times \tfrac{4}{5}$$
$$= \tfrac{24}{25}$$

(Only a positive result as, if θ is in the 1st quadrant, both $\sin \theta$ and $\cos \theta$ are positive; if θ is in the 3rd quadrant, both ratios are negative).

$$\tan 2\theta = \frac{2 \tan \theta}{1 - \tan^2 \theta} = \frac{2 \times \frac{3}{4}}{1 - \frac{9}{16}} = \frac{\frac{6}{4}}{\frac{7}{16}}$$

$$= \tfrac{24}{7} \; (+ \text{ as } \tan \theta \text{ is given as } +\tfrac{3}{4}).$$

Example 11

θ is an obtuse angle and $\tan 2\theta = \frac{5}{12}$. *Without using tables, find the value of* (**a**) $\tan \theta$; (**b**) $\cos 2\theta$; (**c**) $\cos 4\theta$.

$$\tan 2\theta = \frac{2 \tan \theta}{1 - \tan^2 \theta} \qquad \text{Let } \tan \theta = t$$

Then $\tan 2\theta = \dfrac{2t}{1 - t^2} = \dfrac{5}{12}$

i.e. $24t = 5 - 5t^2$ which is $5t^2 + 24t - 5 = 0$
i.e. $(5t - 1)(t + 5) = 0$ whence $t = \frac{1}{5}$ or -5.
Since θ is obtuse, $\tan \theta = -5$
If θ is obtuse, and $\tan 2\theta$ is positive, 2θ must lie in the 3rd quadrant.

$$\cos 2\theta = \frac{1}{-\sqrt{1 + \tan^2 2\theta}} = \frac{1}{-\sqrt{1 + \frac{25}{144}}}$$

$$= -\sqrt{\frac{144}{169}} = -\frac{12}{13}$$

$$\cos 4\theta = 2 \cos^2 2\theta - 1 = 2 \times \left(\frac{-12}{13}\right)^2 - 1$$

$$= 2 \times \frac{144}{169} - 1 = \frac{119}{169}.$$

(Hence 4θ lies in the 1st or 4th quadrant).

Exercise 14.3

1 Express more simply:
   (**a**) $2 \sin 40° \cos 40°$;
                                    (**b**) $2 \cos^2 32° - 1$;

(c) $1 - 2 \sin^2 65°$;   (d) $\dfrac{2 \tan 55°}{1 - \tan^2 55°}$;

(e) $2 \cos^2 2\theta - 1$.

2 Evaluate, without using trigonometrical tables:

(a) $\sin 75° \cos 75°$;   (b) $\dfrac{2 \tan 30°}{1 - \tan^2 30°}$;

(c) $1 - 2 \sin^2 15°$;   (d) $\cos^2 30° - \sin^2 30°$;

(e) $2 \cos^2 30° - 1$.

3 Find, without using tables, the value of $\sin 2\theta$ when

(a) $\sin \theta = \frac{12}{13}$;   (b) $\sin \theta = \dfrac{\sqrt{3}}{2}$.

4 If $\cos 2\theta = \frac{119}{169}$, find the value of (a) $\sin \theta$; (b) $\cos \theta$ without using tables.

5 If $\tan 2\theta = \frac{120}{119}$, find, without using tables, the value of (a) $\tan \theta$; (b) $\cos \theta$; (c) $\sin \theta$.

6 If $\tan \theta = \frac{1}{2}$, find the values of (a) $\tan 2\theta$; (b) $\tan 4\theta$; (c) $\tan (4\theta + 45°)$.

7 Given $\tan 2A = -\frac{8}{15}$, find, without using tables, the following ratios, given that angle A is acute:
(a) $\tan A$; (b) $\sin A$; (c) $\cos 2A$.

8 If $\tan A$ and $\tan B$ are the roots of the equation $x^2 + bx + c = 0$, find in terms of b and c the value of (a) $\tan (A + B)$; (b) $\cos (A + B)$.

9 θ is an acute angle and $\cos 2\theta = -\frac{3}{5}$. Without using trigonometrical tables, find (a) $\tan \theta$; (b) $\tan 4\theta$.

10 Express $\cos 3A$ in terms of $\cos A$.

The Factor Formulae

Using the compound angle formulae, the sum of two sines can be written as:

$\sin (A + B) + \sin (A - B)$
$$= \sin A \cos B + \cos A \sin B + \sin A \cos B - \cos A \sin B$$
$$= 2 \sin A \cos B$$

Now let $A + B = X$ and $A - B = Y$

then $X + Y = 2A$ $\therefore A = \dfrac{X + Y}{2}$

and $X - Y = 2B$ $\therefore B = \dfrac{X - Y}{2}$

Substituting the values of X and Y for A and B, we have:

$$\sin X + \sin Y = 2 \sin \frac{X + Y}{2} \cos \frac{X - Y}{2}$$

The formula for the sum of two cosines can be written as:

$\cos (A + B) + \cos (A - B)$

$\qquad = \cos A \cos B - \sin A \sin B + \cos A \cos B + \sin A \sin B$

$\qquad = 2 \cos A \cos B$

Substituting the values of X and Y for A and B, we have:

$$\cos X + \cos Y = 2 \cos \frac{X + Y}{2} \cos \frac{X - Y}{2}$$

Example 12

Express in factors: $\sin 5\theta + \sin \theta$.

The factors are obtained from the formula for the sum of two sines.

$$\sin X + \sin Y = 2 \sin \frac{X + Y}{2} \cos \frac{X - Y}{2} \qquad \text{(here } X = 5\theta$$
$$\text{and } Y = \theta)$$

$$\therefore \sin 5\theta + \sin \theta = 2 \sin \frac{5\theta + \theta}{2} \cos \frac{5\theta - \theta}{2}$$

$$= 2 \sin 3\theta \cos 2\theta$$

The difference of two sines can be written as:

$\sin (A + B) - \sin (A - B)$

$\qquad = \sin A \cos B + \cos A \sin B - (\sin A \cos B - \cos A \sin B)$

$\qquad = 2 \cos A \sin B$

Substituting the values of X and Y for A and B, we have

$$\sin X - \sin Y = 2 \cos \frac{X + Y}{2} \sin \frac{X - Y}{2}$$

The difference of two cosines can be written as:

$\cos (A + B) - \cos (A - B)$

$\qquad = \cos A \cos B - \sin A \sin B - (\cos A \cos B + \sin A \sin B)$

$\qquad = -2 \sin A \sin B$

Substituting X and Y for A and B, we have:

$$\cos X - \cos Y = -2 \sin \frac{X + Y}{2} \cos \frac{X - Y}{2}$$

Note: Remember the negative sign in this formula.

Example 13

Express in factors: $\cos 5\theta - \cos 3\theta$.

$$\cos X - \cos Y = - 2 \sin \frac{X + Y}{2} \sin \frac{X - Y}{2}$$

$$\cos 5\theta - \cos 3\theta = -2 \sin \frac{5\theta + 3\theta}{2} \sin \frac{5\theta - 3\theta}{2}$$
$$= -2 \sin 4\theta \sin \theta.$$

The factor formulae are important and should be memorised; they are summarised below:

$$\sin X + \sin Y = 2 \sin \frac{X + Y}{2} \cos \frac{X - Y}{2}$$

$$\sin X - \sin Y = 2 \cos \frac{X + Y}{2} \sin \frac{X - Y}{2}$$

$$\cos X + \cos Y = 2 \cos \frac{X + Y}{2} \cos \frac{X - Y}{2}$$

$$\cos X - \cos Y = -2 \sin \frac{X + Y}{2} \sin \frac{X - Y}{2}$$

These formulae can also be remembered verbally, as under:

sine + sine $= 2 \sin (\frac{1}{2}$ sum$) \cos (\frac{1}{2}$ difference$)$
sine − sine $= 2 \cos ($ ” $) \sin ($ ” $)$
cos + cos $= 2 \cos ($ ” $) \cos ($ ” $)$
cos − cos $= -2 \sin ($ ” $) \sin ($ ” $)$

Example 14

Express $2 \sin 2\theta \cos 5\theta$ *as a difference between two trigonometrical ratios.*

$$\sin X - \sin Y = 2 \cos \frac{X + Y}{2} \sin \frac{X - Y}{2}$$

i.e. $\dfrac{X + Y}{2} = 5\theta \qquad \therefore X + Y = 10\theta$ . . . . . .(1)

and $\dfrac{X - Y}{2} = 2\theta \qquad \therefore X - Y = 4\theta$ . . . . . .(2)

Adding (1) and (2), $2X = 14\theta \qquad \therefore X = 7\theta$
Subtracting (2) from (1), $2Y = 6\theta \qquad \therefore Y = 3\theta$
$\therefore \qquad\qquad 2 \sin 2\theta \cos 5\theta = 2 \cos 5\theta \sin 2\theta$
$$= \sin 7\theta - \sin 3\theta.$$

Exercise 14.4

1 Express the following in factors:

    (a) $\sin 5\theta + \sin 3\theta$;   **(b)** $\cos 5\theta + \cos 3\theta$;

(c) $\sin 2\theta - \sin \theta$;   (d) $\cos 5\theta - \cos 3\theta$;
(e) $\sin 3\theta - \sin 7\theta$;   (f) $\cos 2\theta - \cos 8\theta$.

2 Express the following as the sum or difference of two ratios:
   (a) $2 \sin 3\theta \cos \theta$;   (b) $2 \cos 3A \sin A$;
   (c) $2 \cos 5x \cos 3x$;   (d) $-2 \sin 5\theta \cos 3\theta$;
   (e) $-2 \cos 5A \sin A$;   (f) $2 \sin 7\theta \sin \theta$.

3 Express in factors:
   (a) $\cos 2\theta + \cos 60°$;   (b) $\sin 4\theta - \sin 120°$;
   (c) $\cos 2\theta - \cos 60°$.

4 Express the following as the sum or difference of two ratios:
   (a) $2 \sin (2\theta + 30°) \cos (2\theta - 30°)$
   (b) $2 \cos (\theta + 40°) \sin (\theta - 40°)$
   (c) $-2 \sin (A + 62°) \sin (A - 62°)$

5 Express as factors:
   (a) $\sin (\theta + 20°) + \sin \theta$;   (b) $\cos 2\theta + \cos (90° - 3\theta)$;
   (c) $\cos (\theta + 60°) - \cos \theta$;   (d) $\cos (2\theta + 10°) + \sin (100° - 2\theta)$.
                                                      (Hint: $\cos \theta = \sin (90° - \theta)$)

Simple Identities

A trigonometrical identity is an expression that is valid for all values of the angles contained in the expression,
e.g. $\sin^2 \theta + \cos^2 \theta = 1$ is true for all values of θ.

The three basic identities, already found, are:

$$\sin^2\theta + \cos^2\theta = 1$$
$$1 + \tan^2\theta = \sec^2\theta$$
$$1 + \cot^2\theta = \mathrm{cosec}^2\theta$$

These three basic identities, together with the identities for compound and multiple angles and the identities of the factor formulae, can be used to manipulate trigonometrical expressions into different forms.

Exercises on trigonometrical identities give two expressions which have to be proved equal; the proof depends on substituting the trigonometrical formulae (themselves identities) given so far. There is no general method of attacking such problems, but the following suggestions are useful.

(1) Multiplication, or division, by $(\sin^2 \theta + \cos^2 \theta)$ can always be carried out, as the expression is equal to unity, and hence does

not alter the value of the original expression. Any of the alternative identities from Pythagoras' theorem can also be used.

(2) Always look for $(\sin^2 \theta + \cos^2 \theta)$, or its alternative forms, when simplifying an expression, and replace it by unity.

(3) Expressions involving $\tan \theta$ and $\cot \theta$ are often simplified by replacing $\tan \theta$ by $\dfrac{\sin \theta}{\cos \theta}$ and $\cot \theta$ by $\dfrac{\cos \theta}{\sin \theta}$.

(4) Do not be worried by the size of a trigonometrical expression; the larger expressions usually are simplified more easily.

(5) You can either prove that the left hand side (LHS) of an expression is equal to the right hand side (RHS) or RHS = LHS. Start with the more complicated expression and simplify it. Sometimes it is easier to rearrange the identity first and then prove the rearrangement (see Example 16 below).

(6) Try to get the simplest or most elegant proof. The following examples will illustrate the methods.

Example 15

Prove that $\cot \theta + \tan \theta = \operatorname{cosec} \theta \sec \theta$.

$$\text{LHS} = \frac{\cos \theta}{\sin \theta} + \frac{\sin \theta}{\cos \theta} = \frac{\cos^2 \theta + \sin^2 \theta}{\sin \theta \cos \theta}$$

$$= \frac{1}{\sin \theta \cos \theta} \qquad (\sin^2 \theta + \cos^2 \theta = 1)$$

$$= \operatorname{cosec} \theta \sec \theta \qquad (\operatorname{cosec} \theta = \frac{1}{\sin \theta}, \ \sec \theta = \frac{1}{\cos \theta}).$$

Example 16

Prove that $\tan^2 \theta + \sin^2 \theta = \sec^2 \theta - \cos^2 \theta$.

$\text{LHS} = \sec^2 \theta - 1 + 1 - \cos^2 \theta = \sec^2 \theta - \cos^2 \theta$

Alternatively, if the original identity is rearranged as

$$\cos^2 \theta + \sin^2 \theta = \sec^2 \theta - \tan^2 \theta$$

This is seen to be true, as both sides equal 1.

Hence the original identity must be true.

Identities involving Compound Angles

Example 17

Prove that $\cot (A - B) = \dfrac{1 + \cot A \cot B}{\cot B - \cot A}$.

$$\text{LHS} = \frac{\cos (A - B)}{\sin (A - B)} = \frac{\cos A \cos B + \sin A \sin B}{\sin A \cos B - \cos A \sin B}$$

As the RHS begins with the figure 1, dividing by sin A sin B, top and bottom, should produce an expression of the correct form.

$$\text{LHS} = \frac{\dfrac{\cos A \cos B}{\sin A \sin B} + \dfrac{\sin A \sin B}{\sin A \sin B}}{\dfrac{\sin A \cos B}{\sin A \sin B} - \dfrac{\cos A \sin B}{\sin A \sin B}}$$

$$= \frac{\cot A \cot B + 1}{\cot B - \cot A}$$

$$= \text{RHS}.$$

Example 18

Prove that $2 \sin A \cos A \cos B$
$$= \sin B + \sin A \cos (A + B) + \cos A \sin (A - B).$$

Start with the RHS, which is more complicated.

$$\text{RHS} = \sin B + \sin A (\cos A \cos B - \sin A \sin B)$$
$$+ \cos A(\sin A \cos B - \cos A \sin B)$$
$$= \sin B + \sin A \cos A \cos B - \sin^2 A \sin B$$
$$+ \cos A \sin A \cos B - \cos^2 A \sin B$$
$$= \sin B (1 - \sin^2 A - \cos^2 A) + 2 \sin A \cos A \cos B$$

but $\sin^2 A + \cos^2 A = 1 \qquad \therefore (1 - \sin^2 A - \cos^2 A) = 0$
$\therefore \text{RHS} = 2 \sin A \cos A \cos B = \text{LHS}.$

Identities involving Multiple Angles

Example 19

Prove that $\tan 2\theta \operatorname{cosec} \theta = \dfrac{1}{\cos \theta + \sin \theta} + \dfrac{1}{\cos \theta - \sin \theta}.$

(1) Simplify RHS, which is more complicated.
(2) Look for sin 2θ and cos 2θ to give tan 2θ.

$$\text{RHS} = \frac{(\cos \theta - \sin \theta) + (\cos \theta + \sin \theta)}{(\cos \theta + \sin \theta) (\cos \theta - \sin \theta)}$$

$$= \frac{2 \cos \theta}{\cos^2 \theta - \sin^2 \theta}$$

$$= \frac{2 \cos \theta}{\cos 2\theta} \qquad (2 \cos \theta \text{ has to be changed to sin } 2\theta)$$

$$= \frac{2 \cos \theta}{\cos 2\theta} \times \frac{\sin \theta}{\sin \theta}$$

$$= \frac{2 \sin \theta \cos \theta}{\cos 2\theta \sin \theta}$$

$$= \frac{\sin 2\theta}{\cos 2\theta \sin \theta}$$

$$= \tan 2\theta \csc \theta = \text{LHS}.$$

Example 20

Prove that $2 \cos \theta \sin 3\theta = \sin 2\theta (2 \cos 2\theta + 1)$.

$$\text{RHS} = 2 \sin 2\theta \cos 2\theta + \sin 2\theta$$
$$= \sin 4\theta + \sin 2\theta$$
$$= 2 \sin 3\theta \cos \theta \quad \text{(using formula for the sum of two sines).}$$

Identities involving the Factor Formulae

Example 21

Prove that $\dfrac{\sin \theta - \sin 2\theta + \sin 3\theta}{\cos \theta - \cos 2\theta + \cos 3\theta} = \tan 2\theta$.

(1) Simplify LHS which is more complicated.
(2) There are no half angles in RHS, so the ratios must be paired to avoid half angles.

$$\text{LHS} = \frac{(\sin 3\theta + \sin \theta) - \sin 2\theta}{(\cos 3\theta + \cos \theta) - \cos 2\theta}$$

$$= \frac{2 \sin 2\theta \cos \theta - \sin 2\theta}{2 \cos 2\theta \cos \theta - \cos 2\theta}$$

$$= \frac{\sin 2\theta (2 \cos \theta - 1)}{\cos 2\theta (2 \cos \theta - 1)}$$

$$= \tan 2\theta = \text{RHS}.$$

Example 22

Prove $\dfrac{\cos 2(X + Y) + \cos 2X + \cos 2Y + 1}{\cos 2(X + Y) - \cos 2X - \cos 2Y + 1} = - \cot X \cot Y$.

(1) Simplify LHS, which is more complicated.
(2) Simplify double compound angles and double angles.
(3) Expand compound angles.

$$\text{LHS} = \frac{[2 \cos^2 (X + Y) - 1] + 2 \cos (X + Y) \cos (X - Y) + 1}{[2 \cos^2 (X + Y) - 1] - 2 \cos (X + Y) \cos (X - Y) + 1}$$

$$= \frac{2 \cos (X + Y) [\cos (X + Y) + \cos (X - Y)]}{2 \cos (X + Y) [\cos (X + Y) - \cos (X - Y)]}$$

$$= \frac{\cos (X + Y) + \cos (X - Y)}{\cos (X + Y) - \cos (X - Y)} \quad \left(\frac{\text{sum of two cosines}}{\text{difference of two cosines}} \right)$$

$$= \frac{2 \cos X \cos Y}{-2 \sin X \sin Y}$$

$$= - \cot X \cot Y = \text{RHS}.$$

Exercise 14.5

Prove the following identities:

1 $\sin^4 \theta - \cos^4 \theta = \sin^2 \theta - \cos^2 \theta$.

2 $(\operatorname{cosec} A + \cot A)(\operatorname{cosec} A - \cot A) = 1$.

3 $\dfrac{\operatorname{cosec} \theta + \sec \theta}{\cot \theta + \tan \theta} = \dfrac{\cot \theta - \tan \theta}{\operatorname{cosec} \theta - \sec \theta}$.

4 $\dfrac{1 - \tan^2 x}{1 + \tan^2 x} = 2 \cos^2 x - 1$.

5 $(\tan \theta + \sec \theta)^2 = \dfrac{1 + \sin \theta}{1 - \sin \theta}$.

6 If $m = \dfrac{1 - \cos A}{\sin A}$, then $\dfrac{1}{m} = \dfrac{1 + \cos A}{\sin A}$.

7 $\tan A - \tan B = \dfrac{\sin (A - B)}{\cos A \cos B}$.

8 $\sin A \sin (A - B) = \cos B - \cos A \cos (A - B)$.

9 $\cos (A + B) = \dfrac{\operatorname{cosec} A \operatorname{cosec} B - \sec A \sec B}{\sec A \sec B \operatorname{cosec} A \operatorname{cosec} B}$.

10 $\sin 2\theta (\tan \theta + \cot \theta) = 2$.

11 $\dfrac{\sin (X + Y) + \sin (X - Y)}{\cos (X - Y) - \cos (X + Y)} = \cot Y$.

12 $\dfrac{2 \sin \theta}{\cos 3\theta} = \tan 3\theta - \tan \theta$.

13 $\tan 2A = \dfrac{2}{\cot A - \tan A}$.

14 $\cos 3\theta + \sin 3\theta = (1 + 2 \sin 2\theta)(\cos \theta - \sin \theta)$.

15 $\dfrac{\sin 3\theta + \sin \theta}{\cos 3\theta + \cos \theta} = \tan 2\theta$.

16 $\cos B + \sin 2B - \cos 3B = \sin 2B (2 \sin B + 1)$.

17 $\dfrac{\sin 2(A + B) - \sin 2A - \sin 2B}{\sin 2(A + B) + \sin 2A + \sin 2B} = - \tan A \tan B$.

18 $\dfrac{\cos 2 (C + D) + \cos 2C - \cos 2D - 1}{\sin 2 (C + D) - \sin 2C - \sin 2D} = \cot D$.

Exercise 14.6   Miscellaneous

1 Write down the expansion of cos $(A + B)$.

Write $90° - C$ for A in this expansion and thus find the corresponding expansion for sin $(B - C)$.

If A is an acute angle such that sin $A = \frac{3}{5}$ and if B is an obtuse angle such that cos $B = -\frac{5}{13}$, calculate, without using tables, the values of sin B + sin A cos $(A + B)$ + cos A sin $(A - B)$. (C)

2 Tables may NOT be used in this question.

(i) Given that tan $A = \frac{1}{5}$, find the values of tan $2A$, tan $4A$, and tan $(45° - 4A)$.

(ii) Find a formula for cos 3α in terms of cos α. (Formulae for cos 2α and sin 2α may be used without proof). Use your result to find the exact value of $8 \cos^3 20° - 6 \cos 20°$. (C)

3 (i) If sin $\theta = -\frac{3}{5}$, find, without the use of tables, the possible values of cos θ and tan θ.

(ii) Use tables to find the values, correct to four significant figures, of tan $215°$, sin $312°$, and cos $210°$.

(iii) If A and B are acute angles such that sin $A = 0.28$ and cos $B = 0.8$, find, without the use of tables, the values of:

(a) sin $(A + B)$; (b) cos $(A - B)$; and (c) tan $(A + B)$.

(L)

4 Given that tan $2A = \frac{3}{4}$ and that the angle A is acute, calculate, without using tables, the values of (a) cos $2A$; (b) sin A; (c) tan A; (d) tan $3A$. (C)

5 Write down the formula for cos $(A + B)$ in terms of sin A, cos A, sin B and cos B.

Without using tables, find, in surd form, the values of: sin $75°$, cos $105°$, tan $285°$ Show that sin x + sin $3x$ + sin $5x$ + sin $7x$ is equal to K cos x cos $2x$ sin $4x$, where K is a constant, and find the value of K. (L)

6 (i) Given that $A = B + C$, prove that

tan A − tan B − tan C = tan A tan B tan C.

(ii) Prove that sin $(\alpha + 30°) = \cos \alpha + \sin (\alpha - 30°)$. (C)

7 It is given that tan A and tan B are the roots of the equation $t^2 - pt + q = 0$. Find, in terms of p and q, expressions for:

(a) tan $(A + B)$; (b) sin^2 $(A + B)$; (c) cos $2 (A + B)$ (C)

8 Given that tan $(A + B) = 1$ and that tan $(A - B) = \frac{1}{7}$, find, without using tables, the values of tan A and tan B. (C)

15 Angle between Straight Lines: Normals

Angle between Two Straight Lines

When two straight lines intersect (fig 15.1) the angle between them is either ϕ or $180° - \phi$, one of which will be acute. This acute angle

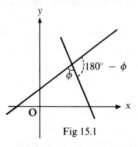

Fig 15.1

is normally taken as the angle between the lines. If the lines are perpendicular, both ϕ and $180° - \phi$ will equal $90°$.

In fig 15.2 let the two intersecting lines have equations:
$$y = m_1 x + c_1 \quad (1)$$
$$\text{and} \quad y = m_2 x + c_2 \quad (2)$$

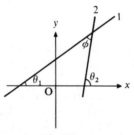

Fig 15.2

The angle made by line (1) with the x-axis is θ_1, where $\tan \theta_1 = m_1$. Similarly, line (2) makes an angle θ_2 with the x-axis, where $\tan \theta_2 = m_2$. Let ϕ be the angle between the two lines.

Then $\theta_2 = \theta_1 + \phi$ or $\phi = \theta_2 - \theta_1$

Hence $\tan \phi = \tan (\theta_2 - \theta_1) = \dfrac{\tan \theta_2 - \tan \theta_1}{1 + \tan \theta_2 \tan \theta_1} = \dfrac{m_2 - m_1}{1 + m_2 m_1}$.

Thus the angle ϕ between two lines of gradients m_1 and m_2 is given by:

$$\tan \phi = \frac{m_2 - m_1}{1 + m_2 m_1}.$$

In practice only the numerical value of this fraction is taken, to obtain the acute angle ϕ. It is then irrelevant which gradient is taken as m_1 or m_2.

Example 1
Find the acute angle between the lines $y = 2x - 6$ and $x + 4y = 12$.
The gradient of the first line is 2 and so $m_1 = 2$.

The second line is $y = -\dfrac{x}{4} + 3$ and so $m_2 = -\dfrac{1}{4}$.

Hence $\tan \phi = \dfrac{-\frac{1}{4} - 2}{1 + (-\frac{1}{4}) \times 2} = \dfrac{-\frac{9}{4}}{\frac{1}{2}} = -\dfrac{9}{2} = -4 \cdot 5$

Taking the numerical value $4 \cdot 5$, the acute angle $\phi = 77° 28'$.

Perpendicular Lines

If two lines, with gradients m_1 and m_2 respectively, are perpendicular, the angle between them is 90°.

Then $\tan 90° = \dfrac{m_2 - m_1}{1 + m_2 m_1}$

As $\tan 90°$ is infinite, and m_1 and m_2 are both finite, then
$$1 + m_2 m_1 = 0 \text{ or } m_1 = -\frac{1}{m_2}.$$

Hence if a line of gradient m_1 is perpendicular to another line of gradient m_2:

$$m_1 = -\frac{1}{m_2}.$$

Example 2
Find the equation of the line passing through $(3, -1)$ and perpendicular to $y = 4x - 6$.

The gradient of the given line $= 4$. The gradient of a perpendicular line $= -\frac{1}{4}$.

Hence the required line has a gradient of $-\frac{1}{4}$ and passes through the point $(3, -1)$; its equation is therefore $y = -\dfrac{x}{4} + c$ where $-1 = -\frac{3}{4} + c$ and $c = -\frac{1}{4}$.

The required line is $y = -\dfrac{x}{4} - \dfrac{1}{4}$

$$\text{i.e., } 4y + x + 1 = 0.$$

Example 3

Find the equation of the perpendicular bisector of the line joining the points $(3, -2)$ and $(7, -6)$.

The midpoint of the two given points is $(5, -4)$.

The gradient of the line joining the given points is

$$\frac{y_2 - y_1}{x_2 - x_1} = \frac{-6 - (-2)}{7 - 3} = \frac{-4}{4} = -1.$$

Hence the gradient of the perpendicular bisector $= \dfrac{-1}{-1} = 1$ and it passes through the midpoint $(5, -4)$.

Its equation is therefore $\dfrac{y - (-4)}{x - 5} = 1,$

$$\text{i.e., } y = x - 9.$$

Normals

A line perpendicular to a tangent at a given point is called the **normal** at that point. Hence if the gradient of the tangent is m, the gradient of the normal is $-\dfrac{1}{m}$.

If a curve has equation $y = f(x)$ the gradient of the tangent is given by $\dfrac{dy}{dx}$. The gradient of the normal is therefore $-\dfrac{1}{dy/dx}$.

Example 4

Find the gradient of the normal to the curve $y = 4x - x^2$ at the point where $x = 1$ and find the coordinates of the point where this normal cuts the curve again.

If $y = 4x - x^2$ then $\dfrac{dy}{dx} = 4 - 2x$ and when $x = 1$ this has the value 2. Hence the gradient of the tangent where $x = 1$ is 2. The gradient of the normal is therefore $-\frac{1}{2}$ at the point $(1, 3)$.

The equation of the normal is $\dfrac{y - 3}{x - 1} = -\frac{1}{2}$, i.e., $2y + x = 7$.

To find where this meets the curve, solve the equations $2y + x = 7$ and $y = 4x - x^2$ simultaneously.

Then $4x - x^2 = \frac{1}{2}(7 - x)$ giving $2x^2 - 9x + 7 = 0$
or $(2x - 7)(x - 1) = 0$ and this gives $x = 3\frac{1}{2}$ or 1.

Hence the normal at $(1, 3)$ meets the curve again where $x = 3\frac{1}{2}$, $y = 1\frac{3}{4}$.

Exercise 15

Find the acute angle between the following pairs of lines:

1 $y = \frac{3}{2}x - 4$ and $y = \frac{1}{5}x + 7$.

2 $y = x - 12$ and $y = \frac{1}{3}x + 5$.

3 $2x + y - 4 = 0$ and $y - 3x + 7 = 0$.

4 $y - \sqrt{2}x + 3\sqrt{2} = 0$ and $y - x - 5\sqrt{2} = 0$.

5 Find the equations of the normals to these curves at the points given: **(a)** $y = 3x^2 - 4$ at $x = 1$; **(b)** $y = \sqrt{x}$ at $x = 4$; **(c)** $y = 1/x$ at $x = -2$.

6 Find the equation of the normal to the curve $y = x^3 - x^2 - 2$ where $x = 1$.

7 Find the equations of the tangent and normal to the curve
$$y = (x + 1)/(x - 2) \text{ where } x = 3.$$

8 Find the equation of the normal to the curve $4y = x^2$ at the point $(2, 1)$ and find where it cuts the curve again.

9 Find the equation of a straight line which is perpendicular to the line $2y + 5x - 6 = 0$, and passes through the point $(3, -1)$.

10 Find the equation of a straight line which is perpendicular to the line $y - 3x + 7 = 0$ and has an intercept of 2 on the y-axis.

11 A line, AB, joins the points $(3, 4)$ and $(7, -2)$. Find the equation of the perpendicular bisector of AB.

12 Calculate the distance between the points A $(-1, -2)$ and B $(5, 6)$. Obtain the equations of the line AB and of the perpendicular bisector of AB. Obtain also the equation of the line parallel to AB through the point $(3, -5)$, and find the distance between this line and AB.

(L)

13 The coordinates of the vertices A, B, and C of the triangle ABC are (3, 1), (8, 2) and (−1, 11) respectively, and D is the foot of the perpendicular from A to BC. Find **(a)** the equation of BC; **(b)** the equation of AD; **(c)** the length of AD.   (L)

14 Two diagonally opposite vertices of a square are (0, 8) and (6, 4). Prove that (1, 3) is also a vertex of the square and find the coordinates of the fourth vertex. Calculate the length of the side of the square.   (C)

15 The line through A (4, 7) with gradient m meets the x-axis at P and the y-axis at R. The line through B (8, 3) with gradient $-\dfrac{1}{m}$ meets the x-axis at Q and the y-axis at S. Find, in terms of m, the coordinates of P, Q, R, and S. Obtain expressions for OP.OQ and OR.OS, where O is the point (0, 0).   (C)

16 Calculus (4): Differentiation of Trigonometrical Functions

In this Chapter we extend our technique of differentiation to include the trigonometrical functions. We begin by finding a very important limit which is required later:

$$\lim_{x \to 0} \frac{\sin x}{x}$$

Look at the sine table in your book of tables, where the angle is small and compare the value of sin x with x (expressed in radians).

Angle x		sin x
in degrees	in radians	
10	0·1745	0·1736
5	0·0873	0·0872
3	0·0524	0·0523
1	0·0175	0·0175
30′	0·008 73	0·0087
6′	0·001 75	0·0017
3′	0·000 87	0·0009

Clearly when x is small, say less than $5°$, $\sin x \approx x$ or $\dfrac{\sin x}{x} \approx 1$. It is impossible to make any more precise comparison using 4 figure tables, but it would seem reasonable to predict that $\lim\limits_{x \to 0} \dfrac{\sin x}{x} = 1$, provided we work in radian measure.

Here is a simple proof of this result.

In fig 16.1 OAB is a sector of a circle radius r, $A\hat{O}B$ (acute) $= \theta$ radians, AB is a chord and AC is the tangent at A.
$AC = r \tan \theta$.

Then $\triangle AOB <$ area of sector OAB $< \triangle OAC$
i.e., $\frac{1}{2}r^2 \sin \theta < \frac{1}{2}r^2 \theta < \frac{1}{2}r^2 \tan \theta$
or $\sin \theta < \theta < \tan \theta$

Hence $1 < \dfrac{\theta}{\sin \theta} < \dfrac{1}{\cos \theta}$

Now as $\theta \to 0$, $\cos \theta \to 1$ and $\dfrac{1}{\cos \theta}$ also $\to 1$.

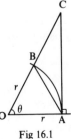

Fig 16.1

The left hand term of the inequality is fixed at 1 and the right hand term decreases and tends to 1, as $\theta \to 0$, and the middle term must also tend to 1.

Hence $\lim\limits_{\theta \to 0} \dfrac{\theta}{\sin \theta} = 1$ and reverting to x as the variable

$$\lim_{x \to 0} \frac{\sin x}{x} = 1 \qquad \text{where } x \text{ is in rad.}$$

This result plays a crucial part in finding the derivative of sin x. To take advantage of it, we must work in radians throughout.

Derivative of Sin x

Working as before from first principles (Chapter 9) let $y = \sin x$. Take an increment δx in x to produce a corresponding increment δy in y.

Then
$$y + \delta y = \sin (x + \delta x)$$
Hence
$$\delta y = \sin (x + \delta x) - \sin x$$
$$= 2 \cos (x + \tfrac{1}{2} \delta x) \sin (\tfrac{1}{2} \delta x)$$
using one of the factor formulae.

Therefore
$$\frac{\delta y}{\delta x} = 2 \cos (x + \frac{\delta x}{2}) \frac{\sin (\frac{\delta x}{2})}{\delta x}$$

$$= \cos (x + \frac{\delta x}{2}) \frac{\sin (\frac{\delta x}{2})}{(\frac{\delta x}{2})}$$

Now as $\delta x \to 0$,
$$\frac{\delta y}{\delta x} \to \frac{dy}{dx},$$

$$\cos (x + \frac{\delta x}{2}) \to \cos x$$

and using the above limit,

$$\frac{\sin (\frac{\delta x}{2})}{(\frac{\delta x}{2})} \to 1.$$

Hence
$$\frac{dy}{dx} = \cos x.$$

$$\boxed{\frac{d(\sin x)}{dx} = \cos x}$$ where x is in rad.

The gradient at any point on the sine curve ($y = \sin x$) is the value of $\cos x$ for that value of x. A comparison of the two curves (fig 16.2) illustrates this.

Composite functions involving the sine can now be differentiated.

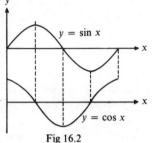

Fig 16.2

Example 1

$$\frac{d(\sin 3x)}{dx} = \cos 3x \times 3.$$

First differentiate sin ... wrt (3x) and then differentiate 3x wrt x.

So, if $y = \sin (2x + 5)$,

$$\frac{dy}{dx} = [\cos (2x + 5)] \times 2$$

$$= 2 \cos (2x + 5).$$

Again, if $y = \sin^2 3x$, $\dfrac{dy}{dx} = $ $\underbrace{2 \sin 3x}$ $\underbrace{\times \cos 3x}$ $\underbrace{\times 3}$

                          differentiate   differentiate   differentiate
                          $\sin^2 3x$ as a $\sin 3x$ wrt $3x$ $3x$ wrt x
                          power wrt
                          $\sin 3x$

$$= 6 \sin 3x \cos 3x$$

$$= 3 \sin 6x.$$

Example 2

$y = \sin x°$. *Find* $\dfrac{dy}{dx}$.

We must first convert the angle to radians.

$$x° = \frac{\pi}{180} x \text{ radians.}$$

Hence $y = \sin x° = \sin \dfrac{\pi}{180} x.$

Then $\dfrac{dy}{dx} = \dfrac{\pi}{180} \cos \dfrac{\pi}{180} x = \dfrac{\pi}{180} \cos x°.$

Derivative of Cos x

Since $\qquad \cos x = \sin \left(\frac{\pi}{2} - x\right),$

$$\frac{d(\cos x)}{dx} = \cos \left(\frac{\pi}{2} - x\right) \times (-1)$$

$$= - \cos \left(\frac{\pi}{2} - x\right)$$

$$= - \sin x.$$

Alternatively we can work again from first principles.
Let $y = \cos x$, then $y + \delta y = \cos (x + \delta x)$ taking as before increments δx and δy.

Then $\qquad \delta y = \cos (x + \delta x) - \cos x$

$$= - 2 \sin (x + \tfrac{1}{2} \delta x) \sin (\tfrac{1}{2} \delta x)$$

Therefore $\dfrac{\delta y}{\delta x} = - 2 \sin (x + \tfrac{1}{2} \delta x) \dfrac{\sin \left(\frac{\delta x}{2}\right)}{\delta x}$

$$= - \sin (x + \tfrac{1}{2} \delta x) \dfrac{\sin \left(\frac{\delta x}{2}\right)}{\dfrac{\delta x}{2}}$$

and hence in the limit, when $\delta x \to 0$, $\dfrac{dy}{dx} = - \sin x$

$$\boxed{\dfrac{d(\cos x)}{dx} = - \sin x} \qquad \text{where } x \text{ is in radians.}$$

Example 3

If $y = \cos 5x$, $\dfrac{dy}{dx} = - \sin 5x \times 5 = - 5 \sin 5x.$

If $y = \cos^2 (2x - 3)$, $\dfrac{dy}{dx} = 2 \cos (2x - 3) \times [- \sin (2x - 3)] \times 2$

$$= - 4 \cos (2x - 3) \sin (2x - 3)$$

$$= - 2 \sin 2(2x - 3).$$

The methods previously used for differentiating products, quotients, implicit functions and for finding maximum and minimum values can also be applied where necessary.

Example 4

Find the derivative of tan x.

If $y = \tan x = \dfrac{\sin x}{\cos x}$, then by the quotient rule,

$$\frac{dy}{dx} = \frac{\cos x(\cos x) - \sin x(-\sin x)}{\cos^2 x}$$

$$= \frac{\cos^2 x + \sin^2 x}{\cos^2 x}$$

$$= \frac{1}{\cos^2 x}$$

Hence

$$\boxed{\dfrac{d(\tan x)}{dx} = \sec^2 x.}$$

Example 5

If $y = (1 + x^2) \sin 2x$, *find* $\dfrac{dy}{dx}$ *and its value when* $x = \dfrac{\pi}{2}$.

By the product rule,

$$\frac{dy}{dx} = (1 + x^2) 2 \cos 2x + (2x) \sin 2x$$

When $x = \dfrac{\pi}{2}$, $\dfrac{dy}{dx} = (1 + \dfrac{\pi^2}{4}) 2 \cos (2 \times \dfrac{\pi}{2}) + (2 \times \dfrac{\pi}{2}) \sin (2 \times \dfrac{\pi}{2})$

$$= - 2 \left(1 + \frac{\pi^2}{4}\right) \quad \text{as } \cos \pi = -1 \text{ and } \sin \pi = 0.$$

Example 6

If $xy = \sin 2x$, *prove that* $x\dfrac{d^2y}{dx^2} + 2\dfrac{dy}{dx} + 4xy = 0$.

Differentiate wrt x, treating xy as a product:

$$x\frac{dy}{dx} + y = 2 \cos 2x.$$

Now differentiate again wrt x, treating $x\dfrac{dy}{dx}$ as a product:

$$x\frac{d^2y}{dx^2} + \frac{dy}{dx} + \frac{dy}{dx} = - 2 \sin 2x \times 2$$

$$= - 4 \sin 2x$$

$$= - 4xy$$

Hence

$$x\frac{d^2y}{dx^2} + 2\frac{dy}{dx} + 4xy = 0.$$

Example 7

Differentiate $\tan^3 2x$.

$$\text{If } y = \tan^3 2x,$$
$$\frac{dy}{dx} = 3 \tan^2 2x \times \sec^2 2x \times 2$$
$$= 6 \tan^2 2x \sec^2 2x.$$

($\tan^3 \ldots$ is first differentiated as a power, then $\tan \ldots$ is differentiated, and finally $2x$ is differentiated).

Example 8

Find the maximum and minimum values of $3 \sin x + 4 \cos x$ *and the values of x at which they occur.*

Take $\qquad\qquad\qquad y = 3 \sin x + 4 \cos x.$

$$\frac{dy}{dx} = 3 \cos x - 4 \sin x$$

and $\qquad\qquad\qquad \dfrac{dy}{dx} = 0$ at the turning points.

If $\dfrac{dy}{dx} = 0$, then $3 \cos x - 4 \sin x = 0$ or $\tan x = \frac{3}{4}$.

In the range $0° - 360°$ this gives $x = 36° \, 52'$ or $216° \, 52'$.
Our usual test to distinguish between the maximum and minimum values is a little awkward in this case, but we can ease the working if we modify the expression for $\dfrac{dy}{dx}$.

$$\frac{dy}{dx} = 3 \cos x - 4 \sin x$$
$$= \cos x(3 - 4 \tan x)$$
$$= 4 \cos x(\tfrac{3}{4} - \tan x).$$

x	$< 36° \, 52'$	$36° \, 52'$	$> 36° \, 52'$	$< 216° \, 52'$	$216° \, 52'$	$> 216° \, 52'$
	$\cos x \, +$ $\tan x < 0{\cdot}75$		$\cos x \, +$ $\tan x > 0{\cdot}75$	$\cos x \, -$ $\tan x < 0{\cdot}75$		$\cos x \, -$ $\tan x > 0{\cdot}75$
$\dfrac{dy}{dx}$	$+$	0	$-$	$-$	0	$+$
	/	___	\	\	___	/
		maximum			minimum	

When $\tan x = \frac{3}{4}$, $\sin x = \frac{3}{5}$ and $\cos x = \frac{4}{5}$ (x in the first quadrant)

Fig 16.3

(fig 16.3) or $\sin x = -\frac{3}{5}$, $\cos x = -\frac{4}{5}$ (x in the third quadrant).
Therefore the maximum value of y when $x = 36° 52'$ is $3(\frac{3}{5}) + 4(\frac{4}{5}) = 5$
and the minimum value of y when $x = 216° 52'$ is -5. (These values
will recur every 360°).

Exercise 16

Differentiate wrt x (x in radians):

1 $\sec x$ [take as $(\cos x)^{-1}$]   **2** $\csc x$ **3** $\cot x$
4 $\sin 4x$ **5** $\cos 6x$ **6** $\tan 2x$ **7** $\cot 3x$

8 $\cos (2x - 8)$ **9** $\sin (3x + \frac{\pi}{3})$ **10** $x \tan x$ **11** $x \cos 2x$

12 $x \sin 3x$ **13** $\cos 2x + \sin x$ **14** $\sin x - \cos x$
15 $\cos^2 x$ **16** $\sin^2 5x$ **17** $\tan^2 x$

18 $\sin^2 (x - \frac{\pi}{4})$ **19** $4x^2 + \sin 4x$ **20** $\dfrac{\sin x}{1 + \cos x}$

21 $\dfrac{\cos x + \sin x}{\cos x - \sin x}$ **22** $\dfrac{2}{1 + \cos 2x}$ **23** $\sin x \tan x$

24 $\sec^2 x$ **25** $\tan \dfrac{x}{2}$ **26** $x^3 \tan x$

27 $(1 + x^2) \tan x$ **28** $\dfrac{1 + x}{\sin x}$ **29** $2 \cos^2 x - \sin^2 x$

30 $(1 + \sin x)(1 - \cos x)$

31 If $y = \sin 2x$, find $\dfrac{dy}{dx}, \dfrac{d^2y}{dx^2}$ and $\dfrac{d^3y}{dx^3}$.

32 Find the maximum and minimum values of $2 \cos x + \sin x$ and
the values of x at which they occur (in the range 0° to 360°).

33 If $x \cos y = \sin x$, prove that $\dfrac{dy}{dx} = \dfrac{\cos y(\cos y - \cos x)}{\sin x \sin y}$.   (C)

34 Given that $y = x \sin 2x$, prove that

$$x^2 \frac{d^2y}{dx^2} - 2x\frac{dy}{dx} + 2y + 4x^2y = 0.$$ (C)

35 Given that $y = \dfrac{x}{2 + \cos x}$, find the values of $\dfrac{dy}{dx}$ when $x = 0$,
$\dfrac{\pi}{2}$ and π.   (C)

36 If $y = 3x \sin 3x + \cos 3x$, show that $x\dfrac{d^2y}{dx^2} + 9xy = 2\dfrac{dy}{dx}$.   (C)

37 A particle is moving in a straight line and its distance s from a fixed point of the line after t s is given by $s = \sin 2t$.

Find its velocity and acceleration at this time and prove that its acceleration is always numerically 4 times its distance from the fixed point and directed towards the point. What is its velocity and acceleration at times $0, \dfrac{\pi}{4}, \dfrac{\pi}{2}, \pi, \dfrac{3\pi}{2}, 2\pi$ s? Make a diagrammatic sketch of the motion. (Such a motion is called Simple Harmonic Motion).

38 If $y = P \cos 2x + Q \sin 2x$, where P and Q are constants, prove that $\dfrac{d^2y}{dx^2} + 4y = 0$. If $y = 1$ when $x = 0$ and $\dfrac{dy}{dx} = 2$ when $x = \dfrac{\pi}{2}$, find the values of P and Q.

39 If $y = 2 \cos x + 3 \sin x - \cos 2x$, prove that $\dfrac{d^2y}{dx^2} + y = 3 \cos 2x$.

40 If $y = \sin x + 3 \cos 2x$, solve the equation $\dfrac{dy}{dx} = 0$ in the range 0 to 2π and hence find the maximum and minimum values of y in that range.

41 If $y = \tan 2x$, prove that $\dfrac{dy}{dx} = 2(1 + y^2)$.

42 If $y = \dfrac{3 \sin 2x}{x}$, prove that $x\dfrac{d^2y}{dx^2} + 2\dfrac{dy}{dx} + 3xy = 0$.

43 Prove that the maximum value of $\cos x - \sin x$ is $\sqrt{2}$.

44 If $y = \sin x$, prove that $\dfrac{\dfrac{d^2y}{dx^2}}{\left[1 + \left(\dfrac{dy}{dx}\right)^2\right]^{\frac{3}{2}}} = -\dfrac{2}{3\sqrt{3}}$ when $x = \dfrac{\pi}{4}$.

45 If $r = 1 - \cos \theta$, use the result of No 33 on page 176 to prove that $r\dfrac{d\theta}{dr} = \cot\dfrac{\theta}{2}$.

17 The Binomial Theorem

You will already be familiar with the identity $(A + B)^2 = A^2 + 2AB + B^2$ and so the square of a **binomial** (expression of two terms) such as $(x + a)^2 = x^2 + 2xa + a^2$. We now seek similar expansions for higher powers of such a binomial, the general result being the **Binomial Theorem**.

By long multiplication, find expansions for $(x + a)^3$, $(x + a)^4$ and $(x + a)^5$.

(For the first, multiply $x^2 + 2ax + a^2$ by $x + a$, then multiply this result by $x + a$ and so on.)

You should find that $(x + a)^3 = x^3 + 3x^2a + 3xa^2 + a^3$
$$(x + a)^4 = x^4 + 4x^3a + 6x^2a^2 + 4xa^3 + a^4$$
and $(x + a)^5 = x^5 + 5x^4a + 10x^3a^2 + 10x^2a^3 + 5xa^4 + a^5$

Note the following points in these expansions:

(1) Each expansion is **homogeneous** in x and a, i.e., the sum of the powers of x and a is constant and equal to the power of the binomial.

(2) The powers of x are in descending order and consequently the powers of a are in ascending order. With the powers arranged in this way,

(3) The coefficients form a pattern as shown below:

Binomial	Coefficients
$(x + a)^1$	1 1
$(x + a)^2$	1 2 1
$(x + a)^3$	1 3 3 1
$(x + a)^4$	1 4 6 4 1
$(x + a)^5$	1 5 10 10 5 1

There are many features of this pattern which can be picked out. Each line is symmetrical, beginning and ending with 1 and having one more term than the degree of the binomial. The second and second last terms equal the degree of the binomial. The most important feature is how each line is generated from the preceding line. Can you see how this is done?

$$(0) \underbrace{\quad}_{1} 1 \underbrace{\quad}_{5} 4 \underbrace{\quad}_{10} 6 \underbrace{\quad}_{10} 4 \underbrace{\quad}_{5} 1 \underbrace{\quad}_{1} (0)$$

Now write the line for a power of 6, giving the coefficients for $(x + a)^6$ and check by long multiplication.

(The coefficients are 1, 6, 15, 20, 15, 6, 1).

Now write the succeeding line (for a power of 7).

This pattern of coefficients is called **Pascal's Triangle**. (Pascal was a French mathematician and philosopher of the 17th century). It can be used to find the coefficients for particular expansions, using $(x + a)$ as a standard binomial.

Example 1

Expand $(a - b)^6$, i.e., $[a + (-b)]^6$.

The coefficients will be 1, 6, 15, 20, 15, 6, 1. The powers of a will decrease and those of b increase.

$$\text{Then } (a - b)^6 = 1a^6 + 6a^5(-b)^1 + 15a^4(-b)^2 + 20a^3(-b)^3$$
$$+ 15a^2(-b)^4 + 6a(-b)^5 + (-b)^6$$
$$= a^6 - 6a^5b + 15a^4b^2 - 20a^3b^3 + 15a^2b^4 - 6ab^5 + b^6.$$

Example 2

Expand $(3x - 2)^5$, i.e., $[3x + (-2)]^5$.

The coefficients are 1, 5, 10, 10, 5, 1.

$$\text{Then } (3x - 2)^5 = 1(3x)^5 + 5(3x)^4(-2) + 10(3x)^3(-2)^2$$
$$+ 10(3x)^2(-2)^3 + 5(3x)(-2)^4 + 1(-2)^5$$
$$= 243x^5 - 810x^4 + 1080x^3 - 720x^2 + 240x - 32 \text{ on multiplying}$$

out the powers of the terms in the brackets.

Example 3

Find the exact value of $(1 \cdot 02)^5$ *using the binomial theorem.*

$$(1 \cdot 02)^5 = (1 + 0 \cdot 02)^5 = 1 + 5(1)^4(0 \cdot 02) + 10(1)^3(0 \cdot 02)^2$$
$$+ 10(1)^2(0 \cdot 02)^3 + 5(1)(0 \cdot 02)^4 + (0 \cdot 02)^5.$$
$$= 1 + 5 \times 0 \cdot 02 + 10 \times 0 \cdot 0004 + 10 \times 0 \cdot 000\,008$$
$$+ 5 \times 0 \cdot 000\,000\,16 + 0 \cdot 000\,000\,003\,2$$
$$= 1 \cdot 104\,080\,803\,2.$$

Check this by ordinary multiplication. Discuss any advantages in using the binomial expansion.

Pascal's Triangle is useful in quickly finding the required coefficients for low powers of a binomial but a general formula for the expansion of a binomial can be found by a second approach, using the idea of combinations.

The Binomial Theorem

Consider $(x + a)^n = (x + a)(x + a)(x + a)$ ......$(x + a)$, n brackets.

We find the expansion of the right hand side, placing the powers of x in descending order.

The highest power of x is clearly n, obtained by multiplying all the x terms together from each bracket. There is only one way of doing this, and so the first term is x^n.

The second highest power of x is now $n - 1$. This term will be obtained by taking a from any bracket and x from each of the remaining brackets and multiplying them all together. So we can choose one a from any of n brackets and the number of ways of doing this is $^nC_1$. Hence the second term of the expansion will be $^nC_1 x^{n-1} a$.

Similarly the third term will be $^nC_2 x^{n-2} a^2$ as we should choose any two a's from the n brackets and multiply by the remaining x's.

This method can continue until we reach the final term, which is a^n. In this case *all* the a's are chosen and there are $^nC_n = 1$ way of doing this.

A general term $x^{n-r} a^r$ will involve choosing any r a's from the n brackets, and the number of choices will be ^nC_r.

So
$$\boxed{\begin{aligned}(x + a)^n &= x^n + {}^nC_1 x^{n-1} a + {}^nC_2 x^{n-2} a^2 + {}^nC_3 x^{n-3} a^3 + \ldots\ldots \\ &\quad + {}^nC_r x^{n-r} a^r \ldots\ldots + {}^nC_n a^n\end{aligned}}$$

which is the Binomial Theorem for all values of x and a provided n is a positive integer.

Notes

(1) There are $(n + 1)$ terms in the expansion.

(2) The terms are homogeneous, of degree n.

(3) The coefficients are symmetrical. The first coefficient is 1 (i.e. ^nC_o) and the last $^nC_n = 1$. Again, choosing 3 items from n gives $^nC_3$ ways which is equal to leaving $(n - 3)$ behind in $^nC_{n-3}$ ways. Hence $^nC_3 = {}^nC_{n-3}$ and in general $^nC_r = {}^nC_{n-r}$.

(4) We can see how the pattern in Pascal's triangle develops. For example,

$(x + a)^6$ 1 $^6C_1$ $^6C_2$ $^6C_3$ (coefficients only)

$(x + a)^7$ $^7C_3$

and this will follow as $^6C_2 + {}^6C_3 = {}^7C_3$ (No. 17 Exercise 12.3).

(5) $^nC_1 = n, \quad ^nC_2 = \dfrac{n(n-1)}{2!}, \quad ^nC_3 = \dfrac{n(n-1)(n-2)}{3!}$ and so on.

The following examples show ways in which the theorem can be used.

Example 4

Find the first five terms of the expansion of $(1 - x)^{10}$ *in ascending powers of* x.

This is the expansion of $[1 + (-x)]^{10}$, $n = 10$.

The first 5 terms are: $(1)^{10} + {}^{10}C_1(1)^9(-x) + {}^{10}C_2(1)^8(-x)^2$
$$+ {}^{10}C_3(1)^7(-x)^3 + {}^{10}C_4(1)^6(-x)^4.$$

$$= 1 - 10x + \frac{10 \times 9}{2}x^2 - \frac{10 \times 9 \times 8}{1 \times 2 \times 3}x^3 + \frac{10 \times 9 \times 8 \times 7}{1 \times 2 \times 3 \times 4}x^4$$

$$= 1 - 10x + 45x^2 - 120x^3 + 210x^4.$$

Example 5

Using the expansion of Example 4, find the value of $(0·98)^{10}$ *correct to 5 places of decimals.*

Write $0·98$ as $1 - 0·02$ and substitute $x = 0·02$ (2×10^{-2}) in the expansion.

Then $(0·98)^{10} \approx 1 - 10(2 \times 10^{-2}) + 45(2 \times 10^{-2})^2 - 120(2 \times 10^{-2})^3$
$$+ 210(2 \times 10^{-2})^4$$

$$= 1 - 2 \times 10^{-1} + 180 \times 10^{-4} - 960 \times 10^{-6} + 3360 \times 10^{-8}.$$

$$= 1 - 0·2 + 0·018 - 0·000\,96 + 0·000\,033\,6$$

$$= 0·817\,07 \text{ (to 5 places).}$$

It is wise in such approximations to check if sufficient terms have been taken to give the accuracy required. The next term would have been ${}^{10}C_5(1)^5(-x)^5$ i.e. with $x = 0·02$,
$$-252 \times 32 \times 10^{-10} = -8064 \times 10^{-10},$$
which would not affect the result already given.

Example 6

Expand $(1 + x + x^2)^8$ *in ascending powers of* x, *as far as the term in* x^3.

The first four terms in the expansion will be of the form $1 + ax + bx^2 + cx^3$.

Treat $1 + x + x^2$ as the binomial $[1 + (x + x^2)]$.

Then $(1 + x + x^2)^8 = 1 + {}^8C_1(x + x^2) + {}^8C_2(x + x^2)^2$
$$+ {}^8C_3(x + x^2)^3 + {}^8C_4(x + x^2)^4 + \ldots\ldots$$

$$= 1 + 8(x + x^2) + 28(x^2 + 2x^3 + x^4) + 56(x^3 + \ldots)$$

all further powers being higher than 3.

$$= 1 + 8x + 8x^2 + 28x^2 + 56x^3 + 56x^3 \ldots$$

$$= 1 + 8x + 36x^2 + 112x^3.$$

Example 7

Calculate, without using tables, the value of $(1 + \sqrt{2})^4 - (1 - \sqrt{2})^4$.

Expand both, taking the coefficients from Pascal's Triangle for simplicity.

$$(1 + \sqrt{2})^4 = 1 + 4(\sqrt{2}) + 6(\sqrt{2})^2 + 4(\sqrt{2})^3 + (\sqrt{2})^4$$
$$(1 - \sqrt{2})^4 = 1 + 4(-\sqrt{2}) + 6(-\sqrt{2})^2 + 4(-\sqrt{2})^3 + (-\sqrt{2})^4$$

On subtraction the 1st, 3rd and 5th terms cancel out, leaving

$$(1 + \sqrt{2})^4 - (1 - \sqrt{2})^4 = 8(\sqrt{2}) + 8(\sqrt{2})^3$$
$$= 8\sqrt{2} + 16\sqrt{2}$$
$$= 24\sqrt{2}.$$

Example 8

Find the first four terms in the expansion of $(1 + x)^4 (1 - 2x)^5$ *in ascending powers of x.*

The expansion will take the form $1 + ax + bx^2 + cx^3$. So we expand each binomial as far as x^3 and then take the products of the terms.

$$(1 + x)^4 = 1 + 4x + 6x^2 + 4x^3 \ldots$$
$$(1 - 2x)^5 = 1 + 5(-2x) + 10(-2x)^2 + 10(-2x)^3 \ldots$$
$$= 1 - 10x + 40x^2 - 80x^3 \ldots$$

Hence $(1 + x)^4 (1 - 2x)^5$

$$= (1 + 4x + 6x^2 + 4x^3)(1 - 10x + 40x^2 - 80x^3)$$

$=1 - 10x + 40x^2 -$	$80x^3$	multiplying the second bracket by 1 of the first		
$+ 4x - 40x^2 + 160x^3$	$\ldots$	$\ldots$	$4x$	$\ldots$
$6x^2 - 60x^3$	$\ldots$	$\ldots$	$6x^2$	$\ldots$
$+ 4x^3$	$\ldots$	$\ldots$	$4x^3$	$\ldots$

$$\overline{1 - 6x + 6x^2 + 24x^3} \ldots\ldots \text{ retaining only terms up to } x^3.$$

Example 9

Find the term in x^2 *and the term independent of x in the expansion of* $(x - \dfrac{1}{2x})^{12}$.

The general term will be $^{12}C_r(x)^{12-r}(-\dfrac{1}{2x})^r$

$$= {}^{12}C_r x^{12-r}(\dfrac{1}{x^r})(-\tfrac{1}{2})^r$$

Now $x^{12-r}(\dfrac{1}{x^r}) = x^{12-2r}$.

When this is x^2, then $12 - 2r = 2$ or $r = 5$ (i.e., the 6th term).

Hence the term in x^2 is $^{12}C_5 x^7 (-\dfrac{1}{2x})^5$

$$= - \frac{12 \times 11 \times 10 \times 9 \times 8}{1 \times 2 \times 3 \times 4 \times 5} \times \frac{1}{32} x^2.$$

$$= - \frac{99x^2}{4}$$

If the term is to be independent of x, then $12 - 2r = 0$ or $r = 6$ and this term (the 7th term) is then $^{12}C_6 x^6 \left(-\frac{1}{2x}\right)^6$

$$= \frac{12 \times 11 \times 10 \times 9 \times 8 \times 7}{1 \times 2 \times 3 \times 4 \times 5 \times 6} \times \frac{1}{64}$$

$$= \frac{231}{16} \text{ which is independent of } x.$$

Exercise 17.1

Write down the expansions of the following, simplifying the coefficients:

1 $(x + 2)^4$ **2** $(x - 3)^5$ **3** $(2x - 1)^3$ **4** $(3x - 2)^4$

5 $(1 + x)^6$ **6** $\left(1 + \frac{x}{2}\right)^4$ **7** $\left(x - \frac{1}{x}\right)^5$ **8** $\left(x + \frac{1}{2x}\right)^4$

9 Find the first five terms in the expansion of $(1 - 2x)^{10}$, simplifying them.

10 Find the exact value of $(1 \cdot 2)^5$ using a binomial expansion. Compare with the value obtained using logarithm tables.

11 Expand $(1 - x + x^2)^6$ in ascending powers of x as far as the term in x^3.

12 Find the first three terms of the expansion of $(1 + x - 2x^2)^5$ in ascending powers of x.

13 By putting $x = 0 \cdot 01$ in the expansion of $(1 - x)^{10}$, find the value of $(0 \cdot 99)^{10}$ correct to four decimal places.

14 Without using tables, find the value of $(3 + \sqrt{2})^5 - (3 - \sqrt{2})^5$.

15 Find the term in x^2 and the term independent of x in the expansion of $(2x - \frac{1}{x})^8$.

16 Find the coefficient of x in the expansion of $\left(x - \frac{2}{3x}\right)^7$.

17 Write down the first five terms in the binomial expansion of $(x + y)^8$, simplifying the coefficients.
Find the value of $(1 \cdot 004)^8$ correct to five decimal places.   (C)

18 (i) Expand $(2 + x)^5 - (2 - x)^5$ in ascending powers of x and simplify your result.
(ii) If the first three terms of the expansion of $(1 + ax)^n$ in ascending powers of x are $1 + 12x + 64x^2$, find n and a.   (C)

19 Write down the first five terms of the series $(1 - 2x)^{10}$. Use your expansion to find the value, correct to seven decimal places, of $(0.998)^{10}$.

20 (i) Expand $(1 - 2x)^4$ and $(1 - x)^8$, simplifying the coefficients. For what values of x is the sum of the first four terms of the first expansion equal to the sum of the first three terms of the second? (ii) Use the binomial theorem to express $(16.32)^3$ in the form $a(2^n)$. (Hint: Remove the highest power of 2 from 16.32).   (C)

21 In the expansion of $(x + y)^n$, the second term is 1620, the third term is 4320 and the fourth term is 5760. Find the values of x, y and n.

22 Show that a first approximation for $(1 + x)^n$, if x is small, is $1 + nx$ and that a better approximation is $1 + nx + \frac{1}{2}n(n - 1)x^2$. Using this second approximation show that $(1.03)^{10} \approx 1.34$. Taking the first four terms of the expansion of $(1 + x)^{10}$ find an approximate value for $(1.03)^{10}$ correct to 3 decimal places and compare your answer with that obtained using logarithms.

The General Binomial Theorem

The binomial expansion of $(x + a)^n$ above was restricted to positive integral powers. With some modification however, the theorem can be made general and used with any rational index. The general form of the binomial theorem (which we shall not prove) states that

$$(1 + x)^n = 1 + nx + \frac{n(n - 1)}{2!} x^2 + \frac{n(n - 1)(n - 2)}{3!} x^3 + \cdots\cdots$$

where n is any rational number (positive or negative) and x is numerically less than 1. This theorem is due to Newton.

Note:

(1) If n is not a positive integer, the expansion is no longer finite but is an infinite series of ascending powers of x, as none of the numbers $(n - 1)$, $(n - 2)$, $(n - 3)$, ... in the coefficients will ever be zero.

(2) We can no longer write the coefficients in the form $^nC_1$, $^nC_2$, etc as these symbols would be meaningless. The coefficients are written in the expanded form as shown above.

(3) The expansion is only valid **if the numerical value of x is less than 1, i.e.** $-1 < x < 1$.

(4) The binomial is written in the form $(1 + x)$. Other binomials must be adapted to this form.

Subject to these points, expansions can be used as in the Examples already given.

Example 10

Give, as far as the x^4 term, the expansions of **(a)** $\dfrac{1}{(1 + x)^3}$; **(b)** $\sqrt{1 - 2x}$;

(c) $\sqrt{1 - 2x} - \dfrac{1}{(1 + x)^3}$, *in ascending powers of x and state the ranges of values of x for which the expansions are valid.*

(a) $\dfrac{1}{(1 + x)^3} = (1 + x)^{-3}$ so $n = -3$ in the general theorem.

$$(1 + x)^{-3} = 1 + (-3)x + \frac{(-3)(-4)}{2!}(x)^2 + \frac{(-3)(-4)(-5)}{3!}(x)^3$$
$$+ \frac{(-3)(-4)(-5)(-6)}{4!}(x)^4$$

$$= 1 - 3x + 6x^2 - 10x^3 + 15x^4.$$

and the expansion is valid if $-1 < x < 1$.

(b) $\sqrt{1 - 2x} = (1 - 2x)^{\frac{1}{2}}$ and $n = \frac{1}{2}$.

$$(1 - 2x)^{\frac{1}{2}} = 1 + \tfrac{1}{2}(-2x) + \frac{\frac{1}{2}(\frac{1}{2} - 1)}{2!}(-2x)^2 + \frac{\frac{1}{2}(\frac{1}{2} - 1)(\frac{1}{2} - 2)}{3!}(-2x)^3$$
$$+ \frac{\frac{1}{2}(\frac{1}{2} - 1)(\frac{1}{2} - 2)(\frac{1}{2} - 3)}{4!}(-2x)^4$$

$$= 1 - x + \tfrac{1}{2}(-\tfrac{1}{2})(\tfrac{1}{2})(4x^2) + (\tfrac{1}{2})(-\tfrac{1}{2})(-\tfrac{3}{2})(\tfrac{1}{6})(-8x^3)$$
$$+ (\tfrac{1}{2})(-\tfrac{1}{2})(-\tfrac{3}{2})(-\tfrac{5}{2})(\tfrac{1}{24})(-2x)^4$$

$$= 1 - x - \frac{x^2}{2} - \frac{x^3}{2} - \frac{5x^4}{8}.$$

This expansion is only valid if $2x$ is numerically less than 1, i.e. $-1 < 2x < 1$ or $-\tfrac{1}{2} < x < \tfrac{1}{2}$.

(c) $\sqrt{1 - 2x} - \dfrac{1}{(1 + x)^3} = (\, 1 - x - \dfrac{x^2}{2} - \dfrac{x^3}{2} - \dfrac{5x^4}{8})$
$$- (1 - 3x + 6x^2 - 10x^3 + 15x^4)$$
$$= 2x - \frac{13x^2}{2} + \frac{19x^3}{2} - \frac{125x^4}{8}.$$

This combined expansion will only be valid if x satisfies *both* conditions: $-1 < x < 1$ and $-\tfrac{1}{2} < x < \tfrac{1}{2}$ which means that x must lie in the range $-\tfrac{1}{2}$ to $+\tfrac{1}{2}$.

Example 11

Find the value of $\sqrt{0.9}$, *correct to four significant figures.*

$\sqrt{0.9} = \sqrt{1 - 0.1}$ and so we substitute $x = 0.1$ in the expansion of $\sqrt{1 - x}$.

This expansion will be valid if x is numerically less than 1 which is true if $x = 0.1$.

$$\sqrt{1 - x} = (1 - x)^{\frac{1}{2}} = 1 - \frac{x}{2} - \frac{x^2}{8} - \frac{x^3}{16} - \frac{5x^4}{128} \quad \text{(Check this)}$$

Now put $x = 0.1$

Then $\sqrt{0.9} \approx 1 - \frac{1}{2}(0.1) - \frac{1}{8}(0.1)^2 - \frac{1}{16}(0.1)^3 - \frac{5}{128}(0.1)^4$

$$= 1 - 0.05 - \frac{0.01}{8} - \frac{0.001}{16} - \frac{5}{128}(0.1)^4$$
$$= 1 - 0.05 - 0.001\ 25 - 0.000\ 062\ 5 - 0.000\ 003\ 9$$
$$= 1 - 0.051\ 316\ 4 = 0.948\ 7 \text{ to 4 significant figures.}$$

Example 12

Expand $(2 - 3x)^{-2}$ *as far as the term in* x^3. *For what range of values of* x *will the expansion be valid?*

Convert to the standard form: $(2 - 3x)^{-2} = \left[2\left(1 - \frac{3x}{2}\right)\right]^{-2}$

$$= 2^{-2}\left(1 - \frac{3x}{2}\right)^{-2} \text{ and now expand.}$$

$(2 - 3x)^{-2} = 2^{-2}\left[1 + (-2)\left(-\frac{3x}{2}\right) + \frac{(-2)(-3)}{2}\left(-\frac{3x}{2}\right)^2\right.$

$$\left. + \frac{(-2)(-3)(-4)}{3!}\left(-\frac{3x}{2}\right)^3\right]$$

$$= 2^{-2}\left[1 + 3x + \frac{27x^2}{4} + \frac{27x^3}{2}\right]$$

This expansion is only valid if $-1 < \frac{3x}{2} < 1$ i.e. $-\frac{2}{3} < x < \frac{2}{3}$.

Example 13

Find the first five terms, in their simplest form, in the expansion of
$$\frac{1 - x^2}{\sqrt{1 + x}}.$$

$$\frac{1 - x^2}{\sqrt{1 + x}} = (1 - x^2)(1 + x)^{-\frac{1}{2}}.$$

Expand the second binomial as far as x^4.

$(1 - x^2)(1 + x)^{-\frac{1}{2}}$

$$= (1 - x^2)\left[1 + (-\tfrac{1}{2})x + \frac{(-\tfrac{1}{2})(-\tfrac{3}{2})}{2!} x^2 + \frac{(-\tfrac{1}{2})(-\tfrac{3}{2})(-\tfrac{5}{2})}{3!} x^3\right.$$

$$\left. + \frac{(-\tfrac{1}{2})(-\tfrac{3}{2})(-\tfrac{5}{2})(-\tfrac{7}{2})}{4!} x^4\right]$$

$$= (1 - x^2)(1 - \frac{x}{2} + \frac{3x^2}{8} - \frac{5x^3}{16} + \frac{35x^4}{128} \ldots)$$

$$= 1 - \frac{x}{2} + \frac{3x^2}{8} - \frac{5x^3}{16} + \frac{35x^4}{128}$$

$$\qquad - x^2 + \frac{x^3}{2} - \frac{3x^4}{8}$$

$$\cdots\cdots\cdots\cdots\cdots\cdots\cdots\cdots\cdots$$

$$= 1 - \frac{x}{2} - \frac{5x^2}{8} + \frac{3x^3}{16} - \frac{13x^4}{128}.$$

Proof of $\dfrac{\mathrm{d}x^n}{\mathrm{d}x} = nx^{n-1}$ (Chapter 9, page 164)

Assuming the general binomial theorem we can now show that the derivative of x^n is nx^{n-1} where n is any rational number.

Let x take an increment δx, the corresponding increment in y being δy.

Then $\qquad y + \delta y = (x + \delta x)^n$

and $\qquad \delta y = (x + \delta x)^n - x^n$

$$= x^n\left(1 + \frac{\delta x}{x}\right)^n - x^n$$

$$= x^n\left[1 + n\frac{\delta x}{x} + \frac{n(n-1)}{2}\left(\frac{\delta x}{x}\right)^2 + \ldots\ldots\right] - x^n$$

$$= nx^n\frac{\delta x}{x} + \frac{n(n-1)}{2}x^n\left(\frac{\delta x}{x}\right)^2 + \ldots\ldots$$

Therefore $\dfrac{\delta y}{\delta x} = nx^{n-1} + \dfrac{n(n-1)}{2}x^{n-2}(\delta x) +$ terms each containing

a power of (δx).

Let $\qquad \delta x \to 0$ and $\dfrac{\delta y}{\delta x} \to \dfrac{\mathrm{d}y}{\mathrm{d}x}$.

Then $\dfrac{dy}{dx} = nx^{n-1}.$

The expansion is valid if $\dfrac{\delta x}{x} < 1$ which must be true as $\delta x \to 0$.

Exercise 17.2

Give the first four terms in the expansions of each of the following, simplifying the coefficients. Also state the range of values of x for which your expansions are valid.

1 $(1 - x)^{-2}$ **2** $(1 + x)^{-2}$ **3** $(1 + x)^{-4}$ **4** $(1 + x)^{\frac{1}{2}}$

5 $\sqrt{1 + 2x}$ **6** $\dfrac{1}{1 - x}$ **7** $(1 + 2x)^{-2}$ **8** $(1 - 8x)^{\frac{1}{8}}$

9 $(1 + 3x)^{\frac{1}{3}}$ **10** $\dfrac{1}{\sqrt{1 + x}}$ **11** $\dfrac{1}{(1 - 2x)^2}$ **12** $(3 - x)^{-1}$

13 $(2 + x)^{\frac{1}{2}}$ **14** $\left(1 - \dfrac{x}{2}\right)^{\frac{1}{2}}$ **15** $(4 - 3x)^{-1}$ **16** $(1 + x)^{-1}$

17 Using the expansion in Question 4, find the value of $\sqrt{1·01}$ correct to four decimal places.

18 Use the expansion in Question 9 to find $\sqrt[3]{1·03}$ correct to four decimal places.

19 By substituting $x = 0·01$ in the expansion of Question 14, calculate an approximate value for $\sqrt{0·995}$ correct to four decimal places.

20 Find the first three terms in the expansion, in ascending powers of x, of $\dfrac{1 + x}{\sqrt{1 - x}}$.

21 Find the expansion in ascending powers of x of
$$(1 + x)^{-1} - (1 - x)^{\frac{1}{2}}$$
as far as the term in x^3.

22 Write down and simplify the first four terms of the binomial expansion of $(1 + x)^{\frac{1}{2}}$ when $x = 0·08$. Show that $(1·08)^{\frac{1}{2}} = \frac{3}{5}\sqrt{3}$ and hence, using your expansion, find an approximation for $\sqrt{3}$.
(C)

23 Expand $(1 + x)^{-1} + (1 + 2x)^{-\frac{1}{2}}$ as far as the term in x^3. State the range of values of x for which the expansion is valid. (C)

18 Calculus (5): Integration

Integration or Anti-differentiation

You have probably wondered if it is possible to find a function given its derivative. For example, if $\dfrac{dy}{dx} = x^2 + 2x - 3$, is it possible to find y?

We can think of this as a process of anti-differentiation, the inverse operation to differentiation, though it is actually called **integration**, and the result an **integral**.

The principle is easily found for single terms. We remember that if we differentiate x^n we get nx^{n-1}. So if we differentiate x^{n+1} we get $(n + 1)x^n$ and if we differentiate $\dfrac{1}{n+1} x^{n+1}$ we get x^n.

Hence if $\dfrac{dy}{dx} = x^n$, $y = \dfrac{x^{n+1}}{n+1}$ i.e., the integral of x^n (wrt x) is $\dfrac{x^{n+1}}{n+1}$.

For example, if $\dfrac{dy}{dx} = x^5$, $y = \dfrac{x^6}{6}$;

if $\dfrac{dy}{dx} = 3x^2$, $y = \dfrac{3x^3}{3} = x^3$;

if $\dfrac{dy}{dx} = 6$, $y = 6x$;

if $\dfrac{dy}{dx} = \dfrac{1}{x^2} = x^{-2}$, $y = \dfrac{x^{-1}}{-1} = -\dfrac{1}{x}$.

Check each of these results by differentiating y wrt x.

However, there is one important point to notice before proceeding further. If we differentiate $x^3 - x + 5$, $x^3 - x - 5$, $x^3 - x$ wrt x we obtain $3x^2 - 1$ in each case. On integrating $3x^2 - 1$ the constant term cannot be recovered, without further information. To show that there is a constant term in the integral, we add an **arbitrary constant c** (which may be zero).

Thus if $\dfrac{dy}{dx} = ax^n$, $y = \dfrac{ax^{n+1}}{n+1} + c$. This is known as the **indefinite** integral of ax^n, and the constant c should always be added. We shall discuss this matter further below.

Our working rule is: increase the index of the term by 1 and divide by the new index, leaving coefficients as they are, and add an arbitrary constant. The result can always be tested by differentiation.

Example 1

Integrate wrt x **(a)** 5; **(b)** $x^{\frac{3}{2}}$; **(c)** $4\sqrt{x}$; **(d)** $\dfrac{1}{\sqrt{x}}$.

(a) If $\dfrac{dy}{dx} = 5$, then $y = 5x + c$ (In using the rule, 5 can be thought of as $5x^0$.)

(b) If $\dfrac{dy}{dx} = x^{\frac{3}{2}}$, then $y = \dfrac{x^{\frac{5}{2}}}{\frac{5}{2}} = \dfrac{2}{5}x^{\frac{5}{2}} + c$.

(c) If $\dfrac{dy}{dx} = 4\sqrt{x} = 4x^{\frac{1}{2}}$, then $y = \dfrac{4x^{\frac{3}{2}}}{\frac{3}{2}} = \dfrac{8}{3}x^{\frac{3}{2}} + c$.

(d) If $\dfrac{dy}{dx} = \dfrac{1}{\sqrt{x}} = x^{-\frac{1}{2}}$, then $y = \dfrac{x^{\frac{1}{2}}}{\frac{1}{2}} = 2x^{\frac{1}{2}} + c$ or $2\sqrt{x} + c$.

If $\dfrac{dy}{dx}$ is given as a polynomial, integrate term by term. In this way it is possible to integrate $\dfrac{x^4 + x - 3}{x^3}$ as $x + \dfrac{1}{x^2} - \dfrac{3}{x^3}$ but not $\dfrac{x^4 + x - 3}{x + 1}$; $(2x + 3)^2$ can be integrated if expanded first.

Example 2

If $\dfrac{dy}{dx} = 3x^3 - 4x^2 + 5x - 1 + \dfrac{1}{x^2}$, *find* y.

$$y = \dfrac{3x^4}{4} - \dfrac{4x^3}{3} + \dfrac{5x^2}{2} - x + \dfrac{x^{-1}}{-1}$$

$$= \dfrac{3x^4}{4} - \dfrac{4x^3}{3} + \dfrac{5x^2}{2} - x - \dfrac{1}{x} + c.$$

Note: There is one important exception. If $\dfrac{dy}{dx} = \dfrac{1}{x} = x^{-1}$ what is y?

If we use the rule, then $y = \dfrac{x^0}{0}$ which is meaningless. Hence $\dfrac{1}{x}$ cannot be integrated (at least by this method). There is an integral, a surprising one. The integral is a logarithm function but the work is too advanced for this stage.

Summarizing,

$$\boxed{\text{If } \dfrac{dy}{dx} = ax^n,\ y = \dfrac{ax^{n+1}}{n+1} + c}$$

provided $n \neq -1$.

Exercise 18.1

Integrate wrt x, simplifying your results where appropriate. Check the first ten by differentiation.

1 x^2 **2** x^3 **3** $2x^4$ **4** $3x$

5 $4x^5$ **6** 8 **7** x^7 **8** $\dfrac{1}{x^3}$

9 $2\sqrt{x}$ **10** $4x^{\frac{1}{2}}$ **11** $x^{\frac{2}{3}}$ **12** $x^{-\frac{1}{2}}$

13 1 **14** $x^{\frac{3}{4}}$ **15** $\dfrac{x^6}{2}$ **16** $\dfrac{1}{3\sqrt{x}}$

17 $x^3 + x^2 + x + 1$ **18** $3x^4 - x + 2$

19 $x - \sqrt{x}$ **20** $\dfrac{x^4 + x + 2}{x^3}$ **21** $\dfrac{3x^3 + x - 4}{2x^3}$

22 $(x + 3)^2$ **23** $(x - 1)^2$ **24** $\left(x - \dfrac{1}{x} \right)^2$ **25** $(1 + x^2)^2$

26 $(x - 1)^3$ **27** $\dfrac{(x + 3)^2}{2x^4}$

The Arbitrary Constant

If we differentiate $y = 3x^2 + 2x + 5$ we obtain $\dfrac{dy}{dx} = 6x + 2$. On integrating $\dfrac{dy}{dx}$ we must write $y = 3x^2 + 2x + c$. Without further information the actual solution could be for example, $y = 3x^2 + 2x + 5$ or $y = 3x^2 + 2x$ or $y = 3x^2 + 2x - 3$ (fig 18.1). Each of the curves shown is the graph of a solution of the **differential equation** $\dfrac{dy}{dx} = 6x + 1$. They are identical in shape (all parabolas) and differ

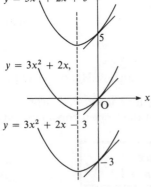

$y = 3x^2 + 2x + 5$

$y = 3x^2 + 2x$,

$y = 3x^2 + 2x - 3$

Fig 18.1

only in position, which depends on the value of the constant term. For any particular value of x, $\dfrac{dy}{dx}$ is the same for all the curves, so the tangents at the corresponding points are parallel, i.e. the curves are parallel.

Summarizing, if $\dfrac{dy}{dx} = 6x + 2$, the indefinite integral or general solution is
$$y = 3x^2 + 2x + c$$
which represents a family of parallel curves. c is an arbitrary constant. To find the actual value of c, thus identifying

a particular member of the family, a pair of values of x and y (i.e., one point on the curve) must be given. See Example 4 below.

Notation for Integration

If $\frac{dy}{dx} = 6x + 2$, then we write $y = \int (6x + 2)dx$ (read 'integral $(6x + 2)dx$'). $\int$ is the sign of integration or the integral sign and $\int$ and dx must both be written. The function to be integrated, called the **integrand**, is placed between them. dx is written to show that the integrand is to be integrated wrt x.

So if $\boxed{\dfrac{dy}{dx} = f(x), \ y = \int f(x)\,dx + c}$ where c is any constant.

Example 3

$$\int (x^2 + 3x + 4)\,dx = \frac{x^3}{3} + \frac{3x^2}{2} + 4x + c;$$

$$\int (s^2 + 4s)\,ds = \frac{s^3}{3} + 2s^2 + c;$$

$$\int \frac{t^3 + 3t^2 - 1}{t^2}\,dt = \int \left(t + 3 - \frac{1}{t^2}\right) dt = \frac{t^2}{2} + 3t + \frac{1}{t} + c.$$

Example 4

If $\frac{dy}{dx} = 6x + 2$, find y given that $y = 3$ when $x = 1$.

$$y = \int (6x + 2)dx = 3x^2 + 2x + c.$$

Substitute the given information.

Then $\qquad\qquad 3 = 3 + 2 + c$ giving $c = -2$

Hence $\qquad\qquad y = 3x^2 + 2x - 2.$

Example 5

A particle moves in a straight line such that its acceleration after time t s is a m s^{-2} where $a = 2t^2 + t$. If its initial velocity was 3 m s^{-1} find an expression for s, the distance (in m) travelled from the start in t s.

$$a = \frac{dv}{dt} = 2t^2 + t.$$

Hence $\qquad\qquad v = \int (2t^2 + t)dt$

$$= \frac{2t^3}{3} + \frac{t^2}{2} + c$$

When $t = 0$, $v = 3$ which gives $c = 3$.

Hence
$$v = \frac{2t^3}{3} + \frac{t^2}{2} + 3$$

Now $v = \dfrac{ds}{dt}$ and hence $s = \displaystyle\int\left(\frac{2t^3}{3} + \frac{t^2}{2} + 3\right)dt$

$$= \frac{2t^4}{12} + \frac{t^3}{6} + 3t + c$$

(A different arbitrary constant, though the same letter is used). When $t = 0$, $s = 0$ which gives $c = 0$.

Hence $s = \dfrac{t^4}{6} + \dfrac{t^3}{6} + 3t$ which is the expression required.

Integration of Trigonometrical Functions

If $y = \sin x$, $\dfrac{dy}{dx} = \cos x$: hence $\displaystyle\int \cos x\, dx = \sin x + c$.

If $y = \cos x$, $\dfrac{dy}{dx} = -\sin x$: hence $\displaystyle\int \sin x\, dx = -\cos x + c$.

If $y = \tan x$, $\dfrac{dy}{dx} = \sec^2 x$: hence $\displaystyle\int \sec^2 x\, dx = \tan x + c$.

Further, if $y = \sin ax$, $\dfrac{dy}{dx} = a\cos ax$, where a is a constant.

Hence
$$\int \cos ax\, dx = \frac{1}{a}\sin ax + c.$$

Similarly
$$\int \sin ax\, dx = -\frac{1}{a}\cos ax + c$$

and
$$\int \sec^2 ax\, dx = \frac{1}{a}\tan ax + c.$$

Example 6

$$\int \sin 3x\, dx = -\frac{1}{3}\cos 3x + c;$$

$$\int\left(\cos 2x - \sin \frac{x}{2}\right)dx = \tfrac{1}{2}\sin 2x + 2\cos \frac{x}{2} + c;$$

$$\int \sec^2 4\theta\, d\theta = \frac{1}{4}\tan 4\theta + c.$$

Exercise 18.2

Find

1 $\int x \, dx$ **2** $\int 3 \, dx$ **3** $\int (x^2 + 1) \, dx$

4 $\int \sin 2x \, dx$ **5** $\int (x + 3)^2 \, dx$

6 $\int \frac{dx}{x^2}$ (abbreviated form of $\int \frac{1}{x^2} \, dx$) **7** $\int \cos 5x \, dx$

8 $\int \sqrt{x} \, dx$ **9** $\int (x - \frac{1}{x})^2 \, dx$ **10** $\int \sec^2 \frac{x}{2} \, dx$

11 $\int (\sqrt{x} - \frac{1}{\sqrt{x}}) \, dx$ **12** $\int \frac{x^2 + 1}{x^2} \, dx$

13 $\int (\cos 2x - \sin 4x) \, dx$ **14** $\int \frac{x^3 - x^2 + 1}{x^2} \, dx$

15 $\int x(x - 3) \, dx$ **16** $\int (x + 1)(x - 2) \, dx$ **17** $\int (3t + 4t^2) \, dt$

18 $\int (x + \cos \frac{x}{3}) \, dx$ **19** $\int (t^3 - t) \, dt$ **20** $\int \sec^2 (\frac{2\theta}{3}) \, d\theta$

21 $\int (\sqrt{x} - \frac{1}{x})^2 \, dx$

22 $\int (\cos x + \cos 2x + \cos 3x) \, dx$

23 $\int \frac{2(x + 3)}{x^3} \, dx$

24 A curve is given by the differential equation $\frac{dy}{dx} = x + 2$, and

it passes through the point (2, 0). Find its equation and sketch the curve.

25 If a curve is given by $\frac{dy}{dx} = 2x + 1$ and passes through the point (1, 2), find its equation and sketch the curve.

26 The rate of change of a quantity A is given by $\frac{dA}{dt} = t^2 - 1$.

If $A = \frac{4}{3}$ when $t = 1$ find A in terms of t.

27 The velocity of a particle moving in a straight line at time t s is given by $v = 2t^2 - 3t$. Find an expression for the distance (s m) travelled, if $s = 0$ when $t = 0$.

28 A particle starts from rest at a point O and moves in a straight line in such a way that its velocity, $v \, m \, s^{-1}$, after time t s, is given by $v = 12t - 3t^2$, until it comes to rest again at A after 4 s. Calculate (a) the distance OA; (b) the greatest velocity of the particle.   (C)

29 If $\dfrac{dy}{d\theta} = \dfrac{1}{\theta^2} + \dfrac{1}{2} \cos 2\theta$, find y if $y = 0$ when $\theta = \dfrac{\pi}{2}$.

30 A particle is moving in a straight line and at time t s its acceleration is $(6 - kt) \,\mathrm{m\,s^{-2}}$ where k is a constant. When $t = 9$, the acceleration of the particle is zero and its velocity is $30 \,\mathrm{m\,s^{-1}}$. Find the numerical value of the velocity when $t = 0$ and the distance between its positions when $t = 0$ and $t = 9$.   (C)

31 (i) A curve passes through the point $(0, 1)$ and is such that at every point of the curve $\dfrac{dy}{dx} = x^2$. Sketch the curve.

(ii) A particle is given an initial velocity of $12 \,\mathrm{m\,s^{-1}}$ and travels in a straight line so that its retardation after t s is equal to $6t \,\mathrm{m\,s^{-2}}$ until it comes to rest. If the particle then remains stationary, calculate the distance travelled.   (C)

32 If $\dfrac{d^2y}{dx^2} = 6x - 4$ find $\dfrac{dy}{dx}$ given that $\dfrac{dy}{dx} = 3$ when $x = 0$. If also $y = 0$ when $x = 0$ find y.

33 Integrate wrt x:
(a) $\sin^2 x$ (Use the double angle formula $\cos 2x = 1 - 2 \sin^2 x$ in the form $\sin^2 x = \dfrac{1 - \cos 2x}{2}$ and now integrate.)

(b) $\cos^2 2x$ (Use the double angle formula for $\cos 4x$.)

Applications of Integration (1): Areas

Suppose $y = f(x)$ is the equation of a curve. We assume for the moment that the portion of the curve between the ordinates $x = a$ and $x = b$ ($b > a$) lies entirely above the x-axis, i.e. $y > 0$ (fig 18.2). We also

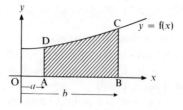

Fig 18.2

assume that the curve is 'continuous', i.e. that there are no breaks or gaps in it. We now find a method of calculating the area enclosed by the curve, the x-axis and the ordinates at A and B, i.e. the area ABCD. You will appreciate that up to now only areas which could be dissected into triangles or trapezia could be found by *calculation*, other shapes being found by approximate methods. Our new method is therefore very important, as it will apply to areas such as ABCD, bounded partly by a *curve*.

Let P be a variable point on the x-axis between A and B where OP = x (fig 18.3). Draw the ordinate PQ (length y) and let the shaded area APQD = A. A is thus a function of x and when x = a, A = 0. Now take an increment δx in x and the area A is increased by an amount δA i.e. the portion PRSQ. RS is y + δy.

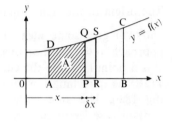

Fig 18.3

Now from fig 18.4 it is seen that
    area PRTQ < δA < area PRSU
where QT, US are parallel to the x-axis.
∴   y × δx < δA < (y + δy) × δx

or $y < \dfrac{\delta A}{\delta x} < (y + \delta y)$

If now δx → 0, δy → 0

and $\dfrac{\delta A}{\delta x} \to \dfrac{dA}{dx}$

and hence $\dfrac{dA}{dx} = y,$

as the right hand term of the above inequality tends to y.

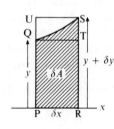

Fig 18.4

Therefore $A = \int y \, dx + c = \int f(x) dx + c.$

The value of c can be found from the fact that when x = 0, A = 0. We then have A expressed as a function of x and can substitute x = b to obtain the area ABCD of fig 18.2.

Note: If the curve has a negative gradient in the range considered (fig 18.5) the above must be modified as follows.

The inequality will now be
        area PRTQ > δA > area PRSU
or $y\,\delta x > \delta A > (y + \delta y)\,\delta x$

i.e. $y > \dfrac{\delta A}{\delta x} > y + \delta y$ and

in the limit the same result is obtained. If the curve contains a turning point in the range, further modification could easily be devised.

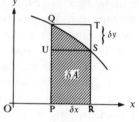

Fig 18.5

The origin of the sign $\displaystyle\int$ may be of interest

and will give some idea of an alternative approach to the question of area calculation. P is a point (x, y) on the curve $y = f(x)$ and PQRS is a rectangle whose side SR is δx (fig 18.6). The area under the curve will contain a series of such rectangles, of area $y.\delta x$.

Then the area under the curve will be approximately the sum of the areas of these rectangles, i.e. sum $(y\delta x)$ for the range considered. As $\delta x \to 0$, the limit of this sum (assuming it exists and can be found) will be the actual area under the curve. The initial S of sum is then written in the form

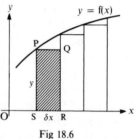

Fig 18.6

$\displaystyle\int$ and δx is written as dx, to show that the limit has been taken.

Example 7
Find the area bounded by the curve $y = x^2 + 3$, the x-axis and the ordinates $x = 1$, $x = 3$.   (fig 18.7)

By the above, $A = \displaystyle\int y\,dx$

$\qquad\qquad = \displaystyle\int (x^2 + 3)\,dx$

$\qquad\qquad = \dfrac{x^3}{3} + 3x + c$

When $x = 1$, $A = 0$.

Hence $0 = \frac{1}{3} + 3 + c$, giving $c = -3\frac{1}{3}$.

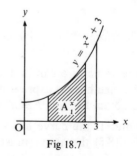

Fig 18.7

Therefore $A_1^x = \dfrac{x^3}{3} + 3x - \dfrac{10}{3}$, A_1^x meaning the area from 1 to x.

Now put $x = 3$.

Then $A_1^3 =$ the required area $= \dfrac{27}{3} + 9 - \dfrac{10}{3} = \dfrac{44}{3}$ square units.

The Definite Integral

We can now generalize the above process, introducing a very important technique.

Consider the curve $y = f(x)$, $y > 0$ in the range of $x = a$ to $x = b$ (fig 18.8). Then the area A between the curve and the x-axis is given by

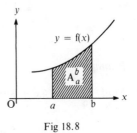

Fig 18.8

$$A = \int f(x)\,dx + c = g(x) + c \quad \text{say.}$$

When $x = a$, the area $= 0$.

Thus $0 = g(a) + c$ or $c = -g(a)$

Then $A_a^x = g(x) - g(a)$ and so

$\quad A_a^b = g(b) - g(a)$

$\quad\quad = $ (value of integral when $x = b$)

$\quad\quad - $ (value of integral when $x = a$)

which is written $\displaystyle\int_a^b f(x)\,dx$. This is called the

definite integral of $f(x)$ wrt x between the **limits** a (the **lower** limit) and b (the **upper** limit). It is a function of a and b. The arbitrary constant c disappears in the subtraction.

Hence if $y = f(x)$ is the equation of a curve, the area between the curve, the x-axis and the ordinates $x = a$, $x = b$ ($b > a$) it given by $\displaystyle\int_a^b y\,dx = \int_a^b f(x)\,dx$. In the next section we examine some complications which may arise in finding areas.

The actual technique in evaluating definite integrals is shown in the following examples.

Example 8

Evaluate $\displaystyle\int_1^3 (x^2 - 1)\,dx$

$\displaystyle\int_1^3 (x^2 - 1)\,dx = \left[\dfrac{x^3}{3} - x\right]_1^3$ Square brackets round the indefinite integral but omit the arbitrary constant.

$$= (\tfrac{27}{3} - 3) - (\tfrac{1}{3} - 1)$$

(value of integral	(value of integral
for upper limit,	for lower limit,
$x = 3$)	$x = 1$)

$$= 6 + \tfrac{2}{3} = 6\tfrac{2}{3}.$$

Example 9

Find the value of $\displaystyle\int_{-2}^{1} (3t - 2)\,dt$

$$\int_{-2}^{1} (3t - 2)\,dt = \left[\frac{3t^2}{2} - 2t\right]_{-2}^{1}$$

$$= \left(\frac{3 \times (1)^2}{2} - 2\right) - \left(\frac{3 \times (-2)^2}{2} - 2(-2)\right)$$

$$= -\tfrac{1}{2} - 10 = -10\tfrac{1}{2}.$$

Example 10

Find $\displaystyle\int_{0}^{\pi/4} (\cos 4x + \sin 2x)\,dx$

$$\text{Integral} = \left[\frac{\sin 4x}{4} - \frac{\cos 2x}{2}\right]_{0}^{\pi/4}$$

$$= \left(\frac{\sin \dfrac{4\pi}{4}}{4} - \frac{\cos \dfrac{2\pi}{4}}{2}\right) - \left(\frac{\sin 0}{4} - \frac{\cos 0}{2}\right)$$

$$= (0 - 0) - \left(0 - \frac{1}{2}\right) = \tfrac{1}{2}.$$

Exercise 18.3

Evaluate

1 $\displaystyle\int_{0}^{1} dx$ **2** $\displaystyle\int_{0}^{3} x\,dx$ **3** $\displaystyle\int_{0}^{2} x^3\,dx$

4 $\displaystyle\int_{1}^{2} (2x - 1)\,dx$ **5** $\displaystyle\int_{0}^{\pi} \cos x\,dx$ **6** $\displaystyle\int_{0}^{\pi/2} \sin 2x\,dx$

7 $\displaystyle\int_{-1}^{1} 2x^3\,dx$ **8** $\displaystyle\int_{2}^{3} \frac{1}{x^2}\,dx$ **9** $\displaystyle\int_{0}^{2} (x + 1)^2\,dx$

10 $\displaystyle\int_{-2}^{-1} (x^2 + x - 1)\,dx$ **11** $\displaystyle\int_{0}^{\pi/4} \sec^2 x\,dx$

12 $\displaystyle\int_{-1}^{0} x(x-1)\,dx$ **13** $\displaystyle\int_{0}^{t} (x^2 - 3)\,dx$

14 $\displaystyle\int_{0}^{1} (s^2 + 3s - 2)\,ds$ **15** $\displaystyle\int_{0}^{\pi} (\sin 2\theta - \cos \theta)\,d\theta$

Further notes on areas

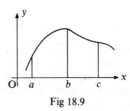

(1) From fig 18.9 it is clear that
$$\int_{a}^{c} f(x)\,dx + \int_{c}^{b} f(x)\,dx = \int_{a}^{b} f(x)\,dx$$

Fig 18.9

(2) If y is negative in the range a to b, then the value obtained from the integral $\displaystyle\int_{a}^{b} f(x)\,dx$ will also be negative (fig 18.10) as dx is essentially positive. Thus the *numerical* value of the area shown shaded will be $- \displaystyle\int_{a}^{b} f(x)\,dx$.

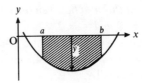

Fig 18.10

(3) If the range includes both positive and negative values of y (fig 18.11) the total area must be found in two parts and will be
$$\int_{a}^{c} f(x)\,dx - \int_{c}^{b} f(x)\,dx.$$

The integral $\displaystyle\int_{a}^{b} f(x)\,dx$ in this case would give the algebraic sum of the two portions. It is wise to sketch a graph before integrating.

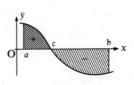

Fig 18.11

(4) The area between a curve and the y-axis and the lines $y = a$, $y = b$ (fig 18.12) will be $\displaystyle\int_{a}^{b} x\,dy$. This can be proved in a manner as before.

Fig 18.12

(5) The area between any two curves $y = f(x)$ and $y = g(x)$ is easily found if the points of intersection or the limits are known (fig 18.13).

The area below $y = f(x)$ is $\int_a^b f(x)\,dx$ and

the area below $y = g(x)$ is $\int_a^b g(x)\,dx$.

Hence the enclosed area (shown shaded) is the difference between the two areas above, i.e.

$$\int_a^b f(x)\,dx - \int_a^b g(x)\,dx = \int_a^b [f(x) - g(x)]\,dx$$

assuming $f(x) > g(x)$. A sketch should always be made to show the relative positions of the curves.

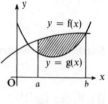

Fig 18.13

Example 11

Find the area between the curve $y = x^2 - x$, *the x-axis and the ordinates* $x = 0$ *and* $x = 2$.

The curve crosses the x-axis where $x = 0$ and $x = 1$ (fig 18.14). Hence the total area numerically

$$= -\int_0^1 y\,dx + \int_1^2 y\,dx$$

$$= -\int_0^1 (x^2 - x)\,dx + \int_1^2 (x^2 - x)\,dx$$

$$= -\left[\frac{x^3}{3} - \frac{x^2}{2}\right]_0^1 + \left[\frac{x^3}{3} - \frac{x^2}{2}\right]_1^2$$

$$= -[(\tfrac{1}{3} - \tfrac{1}{2}) - 0] + [(\tfrac{8}{3} - \tfrac{4}{2}) - (\tfrac{1}{3} - \tfrac{1}{2})]$$

$$= +\tfrac{1}{6} + \tfrac{2}{3} + \tfrac{1}{6} = 1 \text{ square unit.}$$

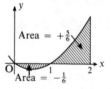

Area $= +\tfrac{5}{6}$

Area $= -\tfrac{1}{6}$

Fig 18.14

Example 12

Find the areas between the curve $y = 2x^2$ *and* **(i)** *the x-axis,* **(ii)** *the y-axis, cut off by lines parallel to the axes through the points on the curve where* $x = 1$ *and* $x = 3$. (fig 18.15)

(i) A_x is required area $= \int_1^3 y\,dx$

$$= \int_1^3 2x^2\,dx$$

$$= \left[\frac{2x^3}{3}\right]_1^3 = 17\tfrac{1}{3}.$$

(ii) $A_y = \displaystyle\int_2^{18} x\,\mathrm{d}y$

$\quad = \displaystyle\int_2^{18} \sqrt{\frac{y}{2}}\,\mathrm{d}y$

$\quad = \dfrac{1}{\sqrt{2}} \displaystyle\int_2^{18} \sqrt{y}\,\mathrm{d}y$

$\quad = \dfrac{1}{\sqrt{2}} \left[\dfrac{2}{3} y^{\frac{3}{2}} \right]_2^{18}$

$\quad = \dfrac{1}{\sqrt{2}} \left[\left(\dfrac{2}{3} \times 18^{\frac{3}{2}} \right) - \left(\dfrac{2}{3} \times 2^{\frac{3}{2}} \right) \right]$

$\quad = \dfrac{1}{\sqrt{2}} \left[\dfrac{2}{3} \times 27 \times 2\sqrt{2} - \dfrac{2}{3} \times 2\sqrt{2} \right]$

$\qquad\qquad$ as $18^{\frac{3}{2}} = (\sqrt{18})^3 = (3\sqrt{2})^3 = 27 \times 2\sqrt{2}$

$\quad = 36 - \frac{4}{3} = 34\frac{1}{3}.$

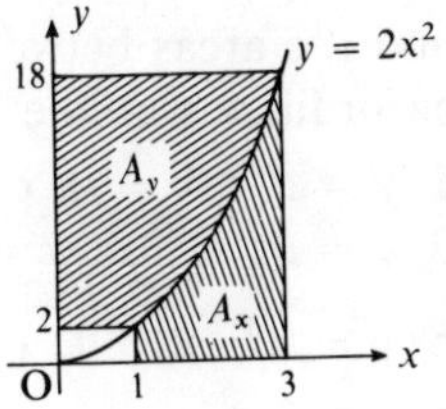

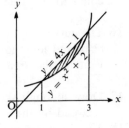

Fig 18.15

Note: In this part the limits must be the limits which y takes to cover the required range, and the equation of the curve must be written

in the form $x = +\sqrt{\dfrac{y}{2}}.$

Example 13

Find the area enclosed between the curve $y = x^2 + 2$ *and the line* $y = 4x - 1$.

The intersections are given by

$x^2 + 2 = 4x - 1$ or $x^2 - 4x + 3 = 0$

which gives $x = 1$ or 3.

The graphs are sketched in fig 18.16. Between $x = 1$ and $x = 3$ the line is above the curve.

Area enclosed

$\quad = \displaystyle\int_1^3 (4x - 1)\,\mathrm{d}x - \int_1^3 (x^2 + 2)\,\mathrm{d}x$

$\quad = \displaystyle\int_1^3 (4x - 1 - x^2 - 2)\,\mathrm{d}x$

$\quad = \displaystyle\int_1^3 (4x - x^2 - 3)\,\mathrm{d}x$

$\quad = \left[2x^2 - \dfrac{x^3}{3} - 3x \right]_1^3$

$\quad = (18 - 9 - 9) - (2 - \frac{1}{3} - 3) = 1\frac{1}{3}.$

Fig 18.16

Exercise 18.4

Find the areas between the following curves and the x-axis between the ordinates at the values given. In each case sketch the curve.

1 $y = x^2$; $x = 0$, $x = 3$ **2** $y = 3x - 2$; $x = 3$, $x = 4$
3 $y = x^2 - 3x$; $x = 0$, $x = 3$ **4** $y = x^2 - 5x + 6$; $x = 2$, $x = 4$

5 $y = x^3$; $x = 1$, $x = 3$ **6** $y = \dfrac{1}{x^2}$; $x = 1$, $x = 3$

7 $y = (x - 1)(x - 3)$; $x = 1$, $x = 3$
8 $y = (1 - x)(x - 2)$; $x = 0$, $x = 3$.
9 Sketch the curve $y = 3x - x^2$ and calculate the area between the curve and the x-axis.
10 Find the area between the curve $y = \sin x$ and the x-axis from $x = 0$ to $x = \pi$.

11 Find the area between the curve $y = \dfrac{1}{x^2}$, the y-axis and the lines $y = 4$, $y = 9$.
12 Find the area enclosed between the curves $y = 2x^2$ and $y^2 = 4x$.
13 Find the area between the x-axis and the part of the curve $y = (x - 3)(2 - x)$ which is above the x-axis.
14 Make a rough sketch of the curve $y = x(x + 1)(x + 3)$ from $x = -4$ to $x = +1$. What are the slopes of the graph at the points where it crosses the x-axis?
Calculate the area enclosed by the x-axis and the curve between $x = -3$ and $x = -1$.   (C)
15 Draw a rough sketch of the curve $y = x(x - 1)(x - 2)$.
If this curve crosses the axis of x at O, A and B (in that order), show that the area included between the arc OA and the x-axis is equal to the area included between the arc AB and the x-axis.
                                                                          (C)
16 The curve $y = ax^2 + bx + c$ passes through the points $(1, 0)$ and $(2, 0)$ and its gradient at the point $(2, 0)$ is 2. Find the numerical value of the area included between the curve and the axis of x.
                                                                          (C)
17 Draw a rough sketch of the curve $y^2 = 16x$. Calculate the area enclosed by the curve and the line $x = 4$.
18 The curve $y = ax^2 + b$ passes through the points $(0, k)$ and $(h, 2k)$. Express a and b in terms of h and k. Show that the area bounded by the curve, the x-axis, the y-axis and the line $x = h$ is $\dfrac{4}{3}hk$.   (C)

19 Calculate the coordinates of the points of intersection of the line $x - y - 1 = 0$ and the curve $y = 5x - x^2 - 4$. In the same diagram sketch the line and the curve for values of x from 0 to

+5. Calculate the area contained between the line and the curve.

(C)

20 Sketch the curve whose equation is $y = (x - 1)^3$. Find the equation of the tangent to this curve at the point where $x = 3$. Calculate the area enclosed by the curve, the tangent at the point where $x = 3$ and the x-axis.

(L)

(2): Volumes of Revolution

A solid which has a central axis of symmetry is a **solid of revolution** for example, a cone, a cylinder, a flower vase, etc. Imagine the area under a portion AB of the curve $y = f(x)$ revolved about the x-axis through four right angles or 360°, the x-axis acting as a kind of hinge (fig 18.17). Each point of the curve describes a circle centred on the x-axis. A solid of revolution can be thought of as created in this way, with two circular plane ends, cutting the x-axis at $x = a$ and $x = b$.

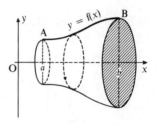

Fig 18.17

Let V be the volume of the solid from $x = a$ up to an arbitrary value of x between a and b (fig 18.18). Given an increment δx in x, y takes an increment δy and V an increment δV.

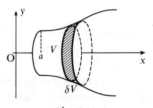

Fig 18.18

Fig 18.19 shows a section through the x-axis and from this it is seen that the slice δV of thickness δx is enclosed between two cylinders of outer radius $y + \delta y$, and inner radius y.

Then $\pi y^2 \, \delta x < \delta V < \pi(y + \delta y)^2 \, \delta x$, with appropriate modification if the curve is falling at this point.

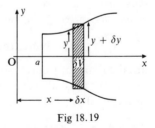

Fig 18.19

Then $\pi y^2 < \dfrac{\delta V}{\delta x} < \pi(y + \delta y)^2$

Now let $\delta x \to 0$ and $\delta y \to 0$ and $\dfrac{\delta V}{\delta x} \to \dfrac{dV}{dx}$. Hence from the above

inequality, $\dfrac{dV}{dx} = \pi y^2$ or

$$V = \int_a^b \pi y^2 \, dx$$

where $y = f(x)$ and V is the volume of solid generated when the curve $y = f(x)$ between limits $x = a$ and $x = b$ is rotated completely around the x-axis.

Example 14

The portion of the curve $y = x^2$ between $x = 0$ and $x = 2$ is rotated completely round the x-axis. Find the volume of the solid created.

$$V = \int_0^2 \pi y^2 \, dx$$

$$= \int_0^2 \pi x^4 \, dx$$

$$= \pi \left[\frac{x^5}{5} \right]_0^2$$

$$= \frac{32\pi}{5} \, (\approx 20 \cdot 1) \text{ units of volume.}$$

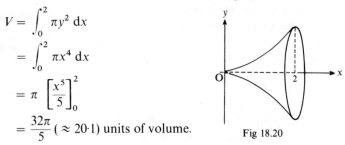

Fig 18.20

Similarly if a portion of the curve $y = f(x)$ between the limits $y = a$ and $y = b$ is rotated completely round the y-axis, the volume of the solid generated will be given by $\int_a^b \pi x^2 \, dy$ which can be proved in the same way.

Example 15

The part of the curve $y = x^3$ from $x = 1$ to $x = 2$ is rotated completely round the y-axis. Find the volume of the solid generated (fig 18.21).

$V = \int_1^8 \pi x^2 \, dy$ Note the limits: these are the limits of y corresponding to $x = 1$, $x = 2$. We must also express the integrand in terms of y, as we are integrating wrt y.

Then $V = \int_1^8 \pi y^{\frac{2}{3}} \, dy = \pi \left[\frac{3}{5} y^{\frac{5}{3}} \right]_1^8$

$$= \pi \left(\frac{3}{5} \times 32 \right) - \pi \left(\frac{3}{5} \times 1 \right)$$

$$= \frac{93\pi}{5}.$$

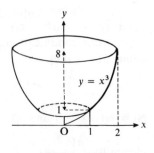

Fig 18.21

Exercise 18.5

Leave your answers in terms of π, as in Example 15.

1 Find the volume generated by rotating the curve $y = x + 1$ from $x = 1$ to $x = 2$ completely round the x-axis.

2 The portion of the curve $y = \frac{1}{2}x^2$ from $x = 0$ to $x = 2$ is rotated about the x-axis through four right angles. Find the volume generated.

3 Sketch the curve $y = x^2 - x$. The part below the x-axis is rotated about the x-axis to form a solid of revolution. Find its volume.

4 If the part of the curve $y = x^2$ from $x = 1$ to $x = 2$ is rotated completely about the y-axis, find the volume of the solid so formed.

5 The part of the line $y = mx$ from $x = 0$ to $x = h$ is rotated about the x-axis through four right angles. Find the volume generated and hence show that the volume of a right circular cone of base radius r and height h is $\frac{1}{3}\pi r^2 h$.

6 If the area enclosed between the curves $y = x^2$ and the line $y = 2x$ is rotated around the x-axis through four right angles, find the volume of the solid generated.

7 Calculate (a) the area bounded by the x-axis and the curve
$$y = x - 3\sqrt{x};$$
(b) the volume generated by revolving this area through four right angles about the x-axis. Leave this result in terms of π.
                                                                      (C)

8 Find the area included between the curves $y^2 = x^3$ and $y^3 = x^2$. Find also the volume obtained by rotating this area through four right angles about the axis of x.   (C)

9 An area is bounded by the curve $y = x + \dfrac{3}{x}$, the x-axis and the ordinates at $x = 1$ and $x = 3$. Calculate the volume of the solid obtained by rotating this area through four right angles about the x-axis.   (C)

10 The equation $x^2 + y^2 = r^2$ represents a circle radius r, centre the origin. The quarter circle in the first quadrant is rotated completely round the x-axis to form a hemisphere. Find its volume and deduce a formula for the volume of a sphere of radius r.

11 The area contained between the curve $y^2 = x - 2$, the x-axis, the y-axis and the line $y = 1$ is rotated about the y-axis through four right angles. Find the volume of the solid generated.

12 Sketch the curve $y^2 = x - 1$. The area contained by this curve, the y-axis and the lines $y = \pm 2$ is completely rotated about the y-axis. Find the volume of the solid so formed.

Exercise 18.6   *Miscellaneous*

1 Find the area enclosed between the curve $y = x^2 - x - 2$ and the x-axis.

2 Find the area enclosed by the curve $y = x^2 + 1$ and the line $y = x + 7$.

3 Find the volume generated when the area (above the x-axis) between the curve $y = \sqrt{3(x - 2)}$, the x-axis and the line $x = 3$ is rotated about the x-axis through 360°.

4 Find the area between the curves $y = x^2 - 2x + 3$ and $y = 6 - x - x^2$.

5 Find the area (in the first quadrant) enclosed by the curves
$$y = \sin x, \ y = \cos x \text{ and the y-axis.}$$

6 Sketch the curve $y = x(x - 1)(x - 2)$. Find the equation of the tangent to the curve at the point where $x = 1$. Calculate the numerical area between the curve and the part of the x-axis from $x = 0$ to $x = 2$.

7 Find the area enclosed between the curves $y^2 = x$ and $x^2 = 8y$.

8 A particle moves along a straight line so that its distance (s) from a fixed point O on the line after t s is given by $s = t^3 - 12t^2 + 45t$. Find the distances from O when the particle is momentarily at rest and the accelerations at these times.

9 Find the area between the curve $y = 2 \cos x + \cos 2x$, the x-axis and the lines $x = 0$ and $x = \dfrac{\pi}{4}$.

10 Sketch the curve $y = 6x - x^2$. Calculate the area contained by this curve, the tangent to the curve at the point where $x = 2$ and the y-axis.

19 The Circle: Parametric Equations

The Equation of a Circle

A circle is defined as the locus of points (in a plane) equidistant from a fixed point, i.e. the centre.

Fig 19.1 shows a circle, with centre C, at (h, k), and a radius of a units. To find the equation of the circle let P be any point, (x, y), on the circle:

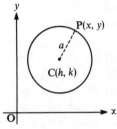

then $CP = a$

i.e. $\sqrt{(x - h)^2 + (y - k)^2} = a$

i.e. $(x - h)^2 + (y - k)^2 = a^2$

which, since P is *any* point, is the equation of the circle.

Fig 19.1

Example 1

A circle has its centre at (5, 4) *and its radius is* 5 *units. Find the equation of the circle.*

By the above, the equation is:

$$(x - 5)^2 + (y - 4)^2 = 5^2$$

i.e. $x^2 - 10x + 25 + y^2 - 8y + 16 = 25$

i.e. $x^2 + y^2 - 10x - 8y + 16 = 0.$

Centre at the origin

In this case, $h = k = 0$, and the equation is:

$(x - 0)^2 + (y - 0)^2 = a^2$ i.e. $x^2 + y^2 = a^2$

Centre and one point given

This information is sufficient to define a circle.

Example 2

Find the equation of the circle which has its centre at $(-1, 3)$ *and passes through the point* (1, 2).

Equation of the circle is:

$$(x + 1)^2 + (y - 3)^2 = a^2, \text{ where } a \text{ is the radius.}$$

Now $a = $ distance from $(-1, 3)$ to $(1, 2)$.

∴ $a^2 = (-1-1)^2 + (3 - 2)^2 = 5$

Hence $(x + 1)^2 + (y - 3)^2 = 5$

∴ $x^2 + 2x + 1 + y^2 - 6y + 9 = 5$

i.e. $x^2 + y^2 + 2x - 6y + 5 = 0.$

Exercise 19.1

1 Find the equations of the following circles with centres at the origin
and **(i)** a radius of 4 units;   **(ii)** a radius of 6 units;
  (iii) a radius of $\sqrt{3}$ units;   **(iv)** a radius of $\sqrt{m^2 + n^2}$;
   (v) which passes through the point (2, 3);
  (vi) which passes through the point (4, 3);
 (vii) which passes through the point (−5, 12);
(viii) which passes through the point $(-2, -\sqrt{3})$.

2 Find the radius of the following circles:
  (i) $x^2 + y^2 = 121$;   **(ii)** $x^2 + y^2 = 49$;
 (iii) $x^2 + y^2 = 12$;   **(iv)** $x^2 + y^2 = (a + b)^2$;
  (v) $x^2 + y^2 = pq$.

3 Show that the point (2, 1) lies on the circle $x^2 + y^2 = 5$.

4 By considering the distance of point $P(3, \frac{1}{4})$ from the origin, determine whether point P is inside or outside the circle $x^2 + y^2 = 10$.

5 Find the equations of the following circles:
  (i) Centre (3, 5), radius 6;
 (ii) Centre (−4, 1), radius 2;
(iii) Centre $(-2, -\frac{3}{2})$, radius $\sqrt{5}$.

6 A circle has its centre at (−1, 6) and passes through the point (2, 3).
Find the equation of the circle, and its radius.

7 A circle has its centre at (−3, −4) and passes through the origin.
Find its equation and its radius.

8 A circle has its centre at (p, q) and passes through the point (−p, 0).
Find its equation and its radius.

The General Form of the Equation of a Circle

The equation of a circle, centre (h, k) and radius a, is
$$(x - h)^2 + (y - k)^2 = a^2$$
i.e. $x^2 - 2hx + h^2 + y^2 - 2ky + k^2 = a^2$
or, rearranging the terms
$$x^2 + y^2 - 2hx - 2ky + (h^2 + k^2 - a^2) = 0.$$
This can be written in a general form as
$$x^2 + y^2 + 2gx + 2fy + c = 0$$
where any of g, f, c can be zero.

This is the standard general form of the equation of a circle.

Note the following points on the general equation of the circle:
(1) There is no term higher than the second power.
(2) The coefficients of the terms in x^2 and y^2 are numerically equal
and of the same sign.

(3) There is no term in xy.

The following equations represent circles:

(a) $x^2 + y^2 + 4x - 5y + 6 = 0$ **(b)** $x^2 + y^2 + 4x + 2y = 0$

(c) $x^2 + y^2 + 3y - 5 = 0$ **(d)** $x^2 + y^2 - 5x = 0.$

The following equations do not represent circles:

(a) $x^2 + 2y^2 + 5x - 3y + 2 = 0$ (coefficients of x^2 and y^2 not equal)

(b) $x^2 - y^2 + 4x - 6y + 1 = 0$ (coefficients of x^2 and y^2 have opposite signs)

(c) $x^2 + y^2 + 3x + 5xy + 4y + 3 = 0$ (a term in xy present)

(d) $x^3 + y^2 + 3x + 4y - 2 = 0$ (x^3 – higher than a second power)

To find the centre and radius, given the equation of a circle.

Example 3

Show that the equation $3x^2 + 3y^2 - 24x + 12y + 11 = 0$ represents a circle and find its centre and radius.

The equation is of the 2nd degree, the coefficients of x^2 and y^2 are equal and there is no term in xy. Hence the equation is that of a circle.

To find the centre and radius, first reduce the coefficients of x^2 and y^2 to unity.

Then $x^2 + y^2 - 8x + 4y + \frac{11}{3} = 0$. Now complete the squares in x and y.

We obtain $(x - 4)^2 + (y + 2)^2 = -\frac{11}{3} + 16 + 4 = \frac{49}{3}$.

i.e. $\sqrt{(x - 4)^2 + (y + 2)^2} = \frac{7}{\sqrt{3}} = \frac{7\sqrt{3}}{3}$

This form also shows that the locus of (x, y) is a circle.

Now compare with the original form of the equation of a circle, $(x - h)^2 + (y - k)^2 = a^2$ and we see that the centre is $(4, -2)$ and

the radius is $\sqrt{\dfrac{49}{3}} = \dfrac{7\sqrt{3}}{3}$ units.

Exercise 19.2

1 Find the centres and radii of the following circles:

    (i) $x^2 + y^2 - 2x - 6y - 15 = 0$;

    (ii) $x^2 + y^2 - 6x + 14y + 49 = 0$;

    (iii) $x^2 + y^2 + x + 4y - 2 = 0$;

    (iv) $x^2 + y^2 - x + 4y = 2$;

    (v) $x^2 + y^2 = 6x.$

2 Do the following equations represent circles? If they do, find the centres and radii. If not, give reasons.

(a) $x^2 + y^2 - 3x + 4y - 18 = 0$;

(b) $x^2 + y^2 - 2x + 2xy + 6 = 0$;

(c) $3x^2 + 3y^2 + 2x = 0$;

(d) $x^2 + y^2 - 4x + 5y = 0$;

(e) $x^2 - y^2 + 2x + 3y + 7 = 0$;

(f) $x^2 + y^2 - 16 = 0$;

(g) $2x^2 + 3y^2 + 3x - 4y + 10 = 0$;

(h) $2x + 4y - 10 - x^2 - y^2 = 0$.

3 Find the centres and radii of the circles given by:

(i) $x^2 + y^2 + 2by = 4 - b^2$;

(ii) $x^2 + y^2 - 2cx + 2cy = 0$;

(iii) $t^2x^2 + t^2y^2 - 2at^3x - 2aty + a^2t^4 - a^2t^2 + a^2 = 0$.

Other methods of defining a circle

The methods of defining a circle, used so far, are:

(1) Given the centre and the radius.

(2) Given the centre and a point on the circle.

Method (2) defines the radius as it is the distance between the point on the circle and the centre.

If one point only is given, an infinite number of circles can be drawn through it.

If two points on the circle are given, e.g., points P and Q in fig 19.2, an infinite number of circles can be drawn through the two points. As PQ is a chord of any one of these circles, then the centre lies on the perpendicular bisector of the chord. The perpendicular bisector, AB, is shown in the diagram, and the centres, C_1 to C_4, of the four circles, F_1 to F_4, are shown lying on the perpendicular bisector.

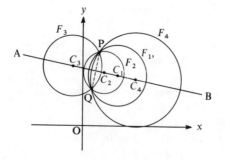

Fig 19.2

To have only one circle, a third point, R, must be taken (see fig 19.3) and this determines the position of the circle. Hence three points define one, and only one, circle.

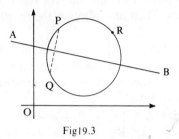

Fig19.3

If *three points* are given, a unique circle is defined. This conclusion can also be reached by considering the general equation of the circle :
$$x^2 + y^2 + 2gx + 2fy + c = 0$$
There are three unknowns in this equation, g, f, and c, and three simultaneous equations are needed to determine their values. This requires three sets of coordinates for x and y, i.e., three points.

If *four or more points* are given, they have to be **concyclic** for a circle to be drawn through them. A circle can always be drawn through any three given points, but more than three points do not all necessarily lie on a circle.

Example 4

Find the equation of the circle passing through the points $(2, 1)$, $(0, -3)$ *and* $(4, 3)$, *and deduce the coordinates of the centre and the radius of the circle.*

Substituting the coordinates of the 3 points in turn in the general equation $x^2 + y^2 + 2gx + 2fy + c = 0$, we have:

$4 + 1 + 4g + 2f + c = 0$ i.e. $4g + 2f + c = -5$ . . . .(1)

$0 + 9 + 0g - 6f + c = 0$ i.e. $6f - c = 9$ . . . .(2)

$16 + 9 + 8g + 6f + c = 0$ i.e. $8g + 6f + c = -25$ . . . .(3)

We have 3 simultaneous equations, to be solved by the methods given in chapter 3.

Check that $g = -11; f = +6; c = +27$.

Therefore the equation of the circle is : $x^2 + y^2 - 22x + 12y + 27 = 0$.

Rearranging and completing the squares :
$$x^2 - 22x + 121 + y^2 + 12y + 36 + 27 = 121 + 36$$
i.e. $(x - 11)^2 + (y + 6)^2 = 130$

∴ centre is $(11, -6)$ and the radius is $\sqrt{130}$.

Example 5

Are the four points (5, 2), (2, 3), (−3, −2) *and* (6, −5) *concyclic?*

A circle can always be drawn through any three points. Take any three of the four points and find the equation of the circle which passes through them. Then find whether the fourth point lies on this circle.

Take the points (5, 2), (2, 3), (−3, −2) and find the equation of the circle passing through them. Using the general form of the equation of the circle, $x^2 + y^2 + 2gx + 2fy + c = 0$, substitute the coordinates of the three points in turn.

Point (5, 2) gives $25 + 4 + 10g + 4f + c = 0$

i.e. $10g + 4f + c = -29$ . . . . . .(1)

Point (2, 3) gives $4 + 9 + 4g + 6f + c = 0$

i.e. $4g + 6f + c = -13$ . . . . (2)

Point (−3, −2) gives $9 + 4 - 6g - 4f + c = 0$

i.e. $6g + 4f - c = 13$ . . . . . .(3)

Solving these three simultaneous equations; we obtain

$$g = -2, f = +2, c = -17.$$

Hence the circle is: $x^2 + y^2 - 4x + 4y - 17 = 0$

Does (6, −5) satisfy this equation?

$6^2 + (-5)^2 - 4 \times 6 + 4 \times (-5) - 17 = 36 + 25 - 24 - 20 - 17$
$$= 61 - 61 = 0$$

∴ (6, −5) lies on the circle and all four points are concyclic.

To find the equation of a circle, given the coordinates of the ends of a diameter.

Example 6

Find the equation of the circle whose diameter is **AB** *where* **A** *is the point* (−1, 3), **B** *the point* (3, 2).

Let P (x, y) be any point on the circle (fig 19.4). Since AB is a diameter, angle APB is a right angle (angle in a semicircle).

Then (grad AP) × (grad BP) = −1 (gradients of perpendicular lines)

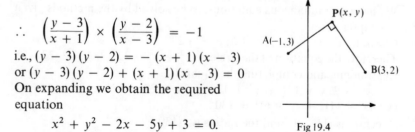

∴ $\left(\dfrac{y - 3}{x + 1}\right) \times \left(\dfrac{y - 2}{x - 3}\right) = -1$

i.e., $(y - 3)(y - 2) = -(x + 1)(x - 3)$

or $(y - 3)(y - 2) + (x + 1)(x - 3) = 0$

On expanding we obtain the required equation

$$x^2 + y^2 - 2x - 5y + 3 = 0.$$

Fig 19.4

Example 7

The equation of a circle is $x^2 + y^2 - 6x + 8y - 24 = 0$. Find the equation of the diameter of the circle which passes through the point $(-1, 1)$.

We find the equation of the straight line through the centre and the point $(-1, 1)$.

Rearrange the equation and complete the squares to find the centre:

$$x^2 - 6x + 9 + y^2 + 8y + 16 - 24 = 9 + 16$$

i.e. $$(x - 3)^2 + (y + 4)^2 = 49$$

∴   centre is $(3, -4)$.

The equation of the diameter through $(-1, 1)$ is:

$$\frac{y + 4}{x - 3} = \frac{1 + 4}{-1 - 3} = \frac{-5}{4}$$

i.e. $$4y + 16 = -5x + 15$$

or $$4y + 5x + 1 = 0.$$

Points inside and outside a circle

If a point (p, q) lies on a circle $f(x, y) = x^2 + y^2 + 2gx + 2fy + c = 0$, then $p^2 + q^2 + 2gp + 2fq + c = 0$ i.e. $f(p, q) = 0$.

If $f(p, q)$ does not equal 0, then the point (p, q) will not lie on the circle.

Consider the equation $x^2 + y^2 = 25$, i.e. a circle with centre at the origin and a radius of 5 units. The point $(3, 4)$ lies on the circle; clearly, the point $(2, 3)$ will lie inside the circle and the point $(4, 5)$ will lie outside the circle.

Let $f(x, y) = x^2 + y^2 - 25$, then $f(3, 4) = 0$ as the point is on the circle.

Now $f(2, 3) = 4 + 9 - 25 = -12$

and $f(4, 5) = 16 + 25 - 25 = +16$.

Hence if $f(x, y) = 0$ is the equation of a circle, and $f(a, b) < 0$, then the point (a, b) is inside the circle; but if $f(a, b) > 0$, then the point (a, b) is outside the circle.

Example 8

Are the points A$(1, -1)$ and B$(5, 2)$ inside or outside the circle
$$x^2 + y^2 - 3x + 4y - 12 = 0?$$

   $$f(x, y) = x^2 + y^2 - 3x + 4y - 12$$
   $$f(1, -1) = 1 + 1 - 3 - 4 - 12 = -17$$

∴ $f(1, -1) < 0$ i.e. A is inside the circle

   $$f(5, 2) = 25 + 4 - 15 + 8 - 12 = 37 - 27 = 10$$

∴ $f(5, 2) > 0$ i.e. B is outside the circle.

Exercise 19.3

1 Find the equations of the circles passing through the following points:
(a) $(-2, 2)$, $(2, 2)$, $(-2, -4)$;
(b) $(2, 0)$, $(3, -3)$, $(0, -3)$;
(c) $(-1, 1)$, $(0, -1)$, $(1, 2)$.

2 Show that the following points are concyclic:
(a) $(3, 4)$, $(0, 5)$, $(-5, 0)$, $(4, -3)$;
(b) $(2, 4)$, $(-1, 5)$, $(-2, 2)$, $(1, 1)$;
(c) $(4, 1)$, $(-3, 8)$, $(-5, 4)$, $(0, -1)$.

3 Find the equation of the circle whose diameter is the join of the points:
(a) $(1, -1)$, $(4, 1)$;   (b) $(4, -3)$, $(0, -2)$;
(c) $(-2, -3)$, $(1, 4)$;   (d) $(3, -5)$, $(-2, 0)$.

4 Show that the straight line $3x - y - 3 = 0$ is a diameter of the circle $x^2 + y^2 - 4x - 6y + 17 = 0$.

5 Show that the straight line $y = 6x + 5$ is a diameter of the circle $x^2 + y^2 - 10y - 11 = 0$.

6 Are the points $A(2, 5)$, $B(1, -3)$ and $C(1, 0)$ inside or outside the circle $x^2 + y^2 - 3x + 2y - 12 = 0$?

7 Find the equation of the diameter of the circle
$$x^2 + y^2 - 6x - 10 = 0$$
which also passes through the point $(1, -2)$.

8 Find the equation and the centre of the circle whose diameter is the join of the points $(4, 2)$ and $(-2, -1)$. Is the point $(5, 0)$ inside or outside the circle?

The Tangent to a Circle

The gradient of the tangent to a circle is found by differentiating wrt x the equation of the circle as a function of x. The equation of the tangent at any point on the circle can then be found.

Example 9

Find the equation of the tangent to the circle $x^2 + y^2 = 61$ at the point $(5, -6)$.

Differentiating the equation (wrt x) as a function of x:

$$2x + 2y\frac{dy}{dx} = 0$$

$$\therefore \frac{dy}{dx} = \frac{-x}{y}$$

$\therefore$ Gradient of tangent at $(5, -6) = \dfrac{-5}{-6}$

$\therefore$ Equation of the tangent is $\dfrac{y + 6}{x - 5} = \dfrac{5}{6}$

i.e. $5x - 6y = 61$.

Example 10

Find the equation of the tangent at the point $(7, 3)$ *on the circle*
$$x^2 + y^2 - 8x - 6y + 16 = 0.$$
Differentiate the equation as a function of x:
$$2x + 2y\frac{dy}{dx} - 8 - 6\frac{dy}{dx} = 0$$

i.e. $\dfrac{dy}{dx} = \dfrac{8 - 2x}{2y - 6}$

$\therefore$ Gradient of tangent at $(7, 3) = \dfrac{8 - 2 \times 7}{2 \times 3 - 6} = \dfrac{-6}{0}$

i.e. the tangent is a line perpendicular to the x-axis

$\therefore$ The required tangent is $x = 7$.

Conditions for tangency

A straight line, $2x + 5y + 11 = 0$, has to be tested to see whether it is a tangent to the circle $x^2 + y^2 + 2x - 8y - 12 = 0$. If the line is a tangent, then the perpendicular distance of the centre of the circle from the line is equal to the radius of the circle (See fig 19.5). Rearrange the equation of the circle and complete the square:
$$(x + 1)^2 + (y - 4)^2 = 29$$
Centre is $(-1, 4)$; radius is $\sqrt{29}$.

Perp. distance of $(-1, 4)$ from $2x + 5y + 11 = 0$ is:
$$\frac{2 \times (-1) + 5 \times 4 + 11}{\sqrt{4 + 25}} = \frac{29}{\sqrt{29}} = \sqrt{29} = \text{radius of circle}$$

$\therefore$ The line is a tangent.

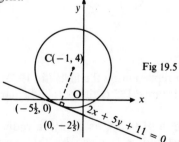

Fig 19.5

Circles touching lines

In fig 19.6 a circle touches the y-axis at P, the point of contact. The y-axis is a tangent to the circle. If C is the centre of the circle, then CP is perpendicular to the y-axis.

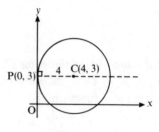

Fig 19.6

Example 11

Find the equation of the circle with centre $(4, 3)$ which touches the y-axis.

The problem is illustrated in fig 19.6. The point of contact is $(0, 3)$ as CP is perpendicular to the y-axis; the radius is 4 units.

Hence the equation of the circle is:
$$(x - 4)^2 + (y - 3)^2 = 16$$
i.e. $$x^2 + y^2 - 8x - 6y + 9 = 0.$$

Example 12

Find the equation of the smallest circle passing through $P(2, -3)$ and having its centre C on the straight line $3x + 4y - 4 = 0$.

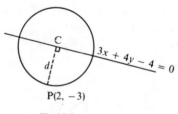

P(2, −3)

Fig 19.7

(1) Draw a sketch representing the facts (fig 19.7).
(2) The shortest distance PC is the perpendicular distance of P from the given line.
(3) For any position of C, PC is always a radius.
    ∴ the smallest radius is given by this perpendicular distance.

The perpendicular distance d is found from

$$d = \frac{ax_1 + by_1 + c}{\sqrt{a^2 + b^2}} = \frac{3(2) + 4(-3) - 4}{\sqrt{9 + 16}}$$

$$= \frac{6 - 12 - 4}{\sqrt{25}} = \frac{-10}{5}$$

$\therefore$ radius of the circle is 2.

To find C, determine the equation of PC, and find its intersection with the given line $3x + 4y - 4 = 0$.

PC is perpendicular to the given line. $\therefore$ gradient of PC is $\frac{4}{3}$.

$$\frac{y + 3}{x - 2} = \frac{4}{3}$$

i.e. $\qquad\qquad\qquad 3y + 9 = 4x - 8$

i.e. $\qquad\qquad\qquad 4x - 3y - 17 = 0$

$\left. \begin{array}{l} 4x - 3y - 17 = 0 \\ 3x + 4y - 4 = 0 \end{array} \right\}$ $\quad \begin{array}{l} 12x - 9y - 51 = 0 \\ 12x + 16y - 16 = 0 \end{array}$

$\qquad$ Subtracting $\qquad -25y - 35 = 0$

$$\therefore y = -\tfrac{7}{5}.$$

$$x = \frac{3y + 17}{4} = \frac{-\frac{21}{5} + 17}{4} = \frac{-21 + 85}{20}$$

$$= \tfrac{64}{20} = \tfrac{16}{5}$$

$\therefore$ the equation of the circle is:

$$(x - \tfrac{16}{5})^2 + (y + \tfrac{7}{5})^2 = 4$$

i.e. $\quad 25x^2 - 160x + 256 + 25y^2 + 70y + 49 = 100$

i.e. $\qquad 25x^2 + 25y^2 - 160x + 70y + 205 = 0$

i.e. $\qquad\quad 5x^2 + 5y^2 - 32x + 14y + 41 = 0$

Exercise 19.4

1 Find the equations of the tangents to the following circles at the given points:

(a) $x^2 + y^2 = 25,$ $\qquad\qquad$ $(-3, 4)$

(b) $x^2 + y^2 = 20,$ $\qquad\qquad$ $(-2, 4)$

(c) $x^2 + y^2 = 39,$ $\qquad\qquad$ $(6, -\sqrt{3})$

(d) $2x^2 + 2y^2 = 25,$ $\qquad\qquad$ $(2, -\tfrac{3}{2})$

2 Show that the tangent to the circle $x^2 + y^2 = 10$ at the point $(-3, 1)$ passes through the centre of the circle

$$x^2 + y^2 - 4x - 32y + 235 = 0.$$

3 Find the equations of the tangents to the given circles at the given points:

(a) $x^2 + y^2 + 4x - 10y - 12 = 0,$ $(3, 1)$
(b) $x^2 + y^2 - 8x - 6y - 11 = 0,$ $(-2, 3)$
(c) $x^2 + y^2 + 6x - 2y - 3 = 0,$ $(-5, -2)$
(d) $x^2 + y^2 + 6x - 4y - 7 = 0,$ $(-1, 6)$
(e) $x^2 + y^2 - 6x - 16 = 0,$ $(-2, 0)$

4 A circle has its centre at the point (a, b) and has a radius of c units. Show that the tangent at the point (s, t) is:
$$(x - a)(s - a) + (y - b)(t - b) = c^2.$$

5 Determine whether the given straight lines are tangents to the specified circles or not:

(a) $x^2 + y^2 - 2x + 6y + 5 = 0,$ $x + 2y = 0;$
(b) $x^2 + y^2 + 3x - y - 6 = 0,$ $5x + y - 6 = 0;$
(c) $2x^2 + 2y^2 - x - 5y = 7,$ $9x + y + 17 = 0.$

6 A circle touches the x-axis at the point $(2, 0)$ and has a radius of 4 units. Find the equation of the circle.

7 Find the equation of the circle which touches the y-axis at the point $(0, 3)$ and passes through the point $(8, -1)$.

8 Find the equation of the circle which has a radius of 5 units, centre on the line $x - 3 = 0$, and passes through the point $(-1, 2)$. Give the coordinates of the centre of the circle.

9 Find the equation of the circle which touches the line
$$3x + 4y - 6 = 0,$$
and passes through the points $(2, 1)$ and $(6, 1)$.

Locus Problems

As in locus problems on straight lines, the facts should first be drawn in a diagram, and then the condition expressed mathematically. The following examples illustrate locus problems on circles.

Example 13

A point P *moves so that its distance from the point* $(6, 4)$ *is twice its distance from the point* $(2, 2)$. *Find the equation of the locus of* P.

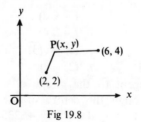

Fig 19.8

(1) Draw a diagram, as in fig 19.8, showing the facts.

(2) Let the coordinates of P be (x, y) and then write the condition as a mathematical expression.

$d_2 = \sqrt{(x - 6)^2 + (y - 4)^2}, \quad d_1 = \sqrt{(x - 2)^2 + (y - 2)^2}$

but $d_2 = 2d_1 \qquad \therefore d_2{}^2 = 4d_1{}^2$

i.e. $(x - 6)^2 + (y - 4)^2 = 4(x - 2)^2 + 4(y - 2)^2$

Hence $x^2 - 12x + 36 + y^2 - 8y + 16$

$\qquad\qquad = 4x^2 - 16x + 16 + 4y^2 - 16y + 16$

i.e. $3x^2 + 3y^2 - 4x - 8y - 20 = 0$

which is the required equation.

Note: From this equation the locus is seen to be a circle, called the Apollonius circle. The Apollonius circle is the locus of a point moving so that its distances from two fixed points are in constant ratio. If $d_1 = d_2$, the coefficients of x^2 and y^2 will each be zero, and the locus is a straight line, the perpendicular bisector. (This could be considered as a circle of infinite radius.)

Example 14

Find the locus of the centres of all circles which touch the x-axis. and pass through the point (2, 3).

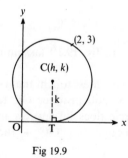

Fig 19.9

(1) Draw a diagram, as in fig 19.9. Let (h, k) be the centre of any one such circle.

(2) If T is the point of contact on the x-axis, then CT is a radius, and the length $CT = k$.

(3) Write down the equation of the circle, given its radius and centre, and substitute the point (2, 3) in the equation.

(4) An equation in h and k will be formed giving the relationship between h and k, for all possible positions of the centre.

A circle with centre (h, k) and radius k has the equation:
$$(x - h)^2 + (y - k)^2 = k^2$$
But $(2, 3)$ is a point on this circle
∴ $$(2 - h)^2 + (3 - k)^2 = k^2$$
i.e. $$4 - 4h + h^2 + 9 - 6k + k^2 = k^2$$
i.e. $$h^2 - 4h - 6k + 13 = 0.$$
This is the relationship between h and k which is true for all values of h, k when (h, k) is the centre of such a circle. Hence the equation of the required locus is: $x^2 - 4x - 6y + 13 = 0.$

Example 15
Prove that the circle $x^2 + y^2 - 4x - 6y + 9 = 0$ *touches the y-axis and also lies entirely inside the circle* $x^2 + y^2 = 36.$

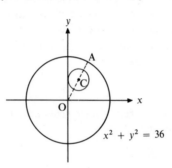

Fig 19.10

(1) Draw a diagram as in fig 19.10 (assuming that the statements are true).
(2) Prove that first circle does touch the y-axis.
(3) Join the centres O, C and produce OC to meet the outer circle at A. If the first circle does lie entirely inside the second circle, then OA will be greater than OC + the radius of the smaller circle.

The first circle is $x^2 + y^2 - 4x - 6y + 9 = 0.$ Put $x = 0.$
Then $y^2 - 6y + 9 = 0$ or $(y - 3)^2 = 0$ which gives equal roots $y = 3$ or 3.
Hence the circle touches the y-axis at the point where $y = 3.$ This proves the first part of the question.
Completing the squares, the equation of the first circle becomes:
$(x - 2)^2 + (y - 3)^2 = 4$ and so its centre C is $(2, 3)$ and its radius is 2 units.
The second circle has centre O and radius 6 units.

The distance $OC = \sqrt{2^2 + 3^2} = \sqrt{13}$.

The distance OC + the radius of the first circle $= \sqrt{13} + 2$, which is less than 6.

Hence the first circle lies entirely within the second circle.

Exercise 19.5

1 Show that the equation of the circle on the line joining $A(0, 1)$ and $B(7, 8)$ as diameter is $x^2 + y^2 - 7x - 9y + 8 = 0$.

This circle cuts the y-axis at A and C. Calculate the coordinates of the point of intersection of the tangents at A and C.   (C)

2 (i) Calculate the distance of the point $(2, 2)$ from the line
$$3x - 4y = 8.$$
Prove that this line is a tangent to the circle with centre $(2, 2)$ touching the axes.

(ii) The radius of the circle $x^2 + y^2 - 2x + 3y + k = 0$ is $2\frac{1}{2}$. Find the value of k. Find also the equation of that diameter of the circle which passes through the point $(5, 2\frac{1}{2})$.   (C)

3 (i) A circle passes through the point $(-2, 3)$ and touches the axis of x. Find the equation of the locus of the centre of the circle. Show that the locus passes through the point $(1, 3)$.

(ii) Find the equation of the tangent at the point $(1, 1)$ to the circle $x^2 + y^2 + 6x - 8y = 0$.   (C)

4 (i) Find the equation of the locus of a point which moves so that its distance from the point $(3, 4)$ is always equal to its distance from the line $y + 3 = 0$.

(ii) The coordinates of the points A and B are $(2, 1)$ and $(4, 5)$ respectively. Find the equation of the locus of a point P which moves so that the angle APB is a right angle.   (L)

5 Find the coordinates of the centre and the length of the radius of the circle
$$x^2 + y^2 - 12x - 18y + 17 = 0.$$
Obtain the equation of the circle, with centre at the point $(3, 5)$ which touches the x-axis.

Show that these circles touch one another at the point $(0, 1)$ and obtain the equation of their common tangent.   (L)

6 Prove that the circle $x^2 + y^2 - 6x - 2y + 9 = 0$ (a) touches the x-axis; (b) lies entirely inside the circle $x^2 + y^2 = 18$.   (C)

7 Calculate the coordinates of the foot of the perpendicular from the point $(-4, 2)$ to the line $3x + 2y = 5$.

Find the equation of the smallest circle passing through $(-4, 2)$ and having its centre on the line $3x + 2y = 5$.   (C)

Parametric Equations

The fact that a curve can be represented by an equation of the form
$y = f(x)$ or in an implicit form will be well known by now. For example,
$y = 3x - 2$ is the equation of a straight line, $y = 2x^2 - 3x + 4$ is the
equation of a quadratic function (a parabola), and $x^2 + y^2 - 2x + 4y$
$= 1$ is the equation of a circle. This form, connecting the variables
x and y, is called the **Cartesian** equation of the curve (after Descartes,
a 17th century French mathematician and philosopher).

This is not the only way of representing a curve. Another important
method is to use a third variable, called a **parameter**. The equation of
the curve is then given by **two** equations, called **parametric** equations,
which give the coordinates x and y of any point of the curve in terms
of this parameter.

For example, a certain curve is given by the equations $x = 2t$, $y = t^2$,
where t is a parameter.

It may be thought that this is a more complicated method (two equations instead of one) but in practice it is often easier to study the curve
from the parametric equations.

The parameter chosen is any convenient variable and may be a number,
a distance, a gradient or an angle, subject to the two conditions that:
(**1**) Every point on the curve has a unique value of the parameter.
(**2**) A particular value of the parameter must give the coordinates
of only one point of the curve.

The Cartesian equation can be found by eliminating the parameter
from the parametric equations.

Example 16
*The parametric equations of a curve
are $x = 2t$, $y = t^2$. Sketch the curve
and find its Cartesian equation.*
As $y = t^2$, y is never negative.
When $t = 0$, $x = 0$ and $y = 0$.
　　"　$t = 1$, $x = 2$ and $y = 1$.
　　"　$t = 2$, $x = 4$ and $y = 4$
and so on.

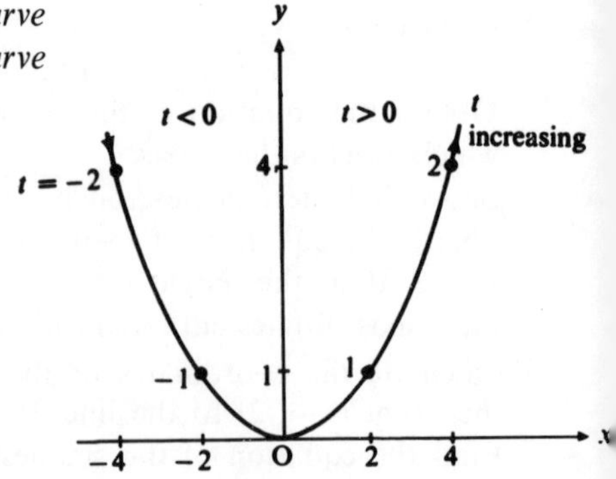

Fig 19.11

This gives an idea of the development of the curve (fig 19.11).
When t is negative, the points obtained are the reflections in the y-axis
of those for which t is positive. As t varies from $-\infty$ to $+\infty$, the
point obtained moves along the curve as shown.

If $x = 2t$, $y = t^2$, then $t = \dfrac{x}{2}$

and hence $$y = \frac{x^2}{4} \text{ or } 4y = x^2$$

which is the Cartesian equation of the curve (a parabola).

Parametric equations of a straight line

A line passes through the point A$(1, 3)$ and makes an angle of $60°$
with the x-axis (fig 19.12). Choose P, any point on the line and let r
be the distance AP. We take r as a parameter.

Then the coordinates of P will be
$$x = 1 + r\cos 60°, \quad y = 3 + r\sin 60°$$
and these are the parametric equations
of the line.

To obtain the Cartesian equation,
eliminate r.

Then $y - 3 = r\sin 60°$
and $x - 1 = r\cos 60°$

Divide: $\dfrac{y - 3}{x - 1} = \tan 60° = \sqrt{3}$

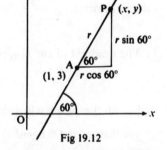

Fig 19.12

which is the required equation (equation of a line through a point
$(1, 3)$ with gradient $\sqrt{3}$).

Parametric equations of a circle.

A circle has centre A$(3, 4)$ and radius 2 (fig 19.13). Let θ be the angle
the radius to any point P (x, y) makes with the line through A parallel
to the x-axis. We take θ as a parameter. Then the coordinates of P
will be

$$x = 3 + 2\cos\theta$$
$$y = 4 + 2\sin\theta$$

and these are the parametric
equations of the circle.

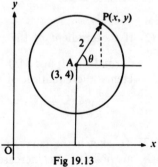

Fig 19.13

To obtain the Cartesian equation, remember that
$$(\cos \theta)^2 + (\sin \theta)^2 = 1$$
Hence $\left(\dfrac{x-3}{2}\right)^2 + \left(\dfrac{y-4}{2}\right)^2 = 1$ or $(x-3)^2 + (y-4)^2 = 4$ which
you will recognise as the equation of this circle.

Tangent and Normal

From the relationship $\dfrac{dy}{dx} = \dfrac{dy}{dt} \times \dfrac{dt}{dx} = \dfrac{dy}{dt} \div \dfrac{dx}{dt}$ (see Chapter 10),
we can find the gradient of a curve in terms of the parameter and hence
the equations of the tangent and normal.

Consider the curve given as $x = ct$, $y = \dfrac{c}{t}$ where c is a constant and t
is the parameter.

(1) The gradient of the tangent is given by $\dfrac{dy}{dx} = \dfrac{\dfrac{dy}{dt}}{\dfrac{dx}{dt}} = \dfrac{-\dfrac{c}{t^2}}{c} = -\dfrac{1}{t^2}$.

So, at the point whose parameter is 2, coordinates $(2c, \dfrac{c}{2})$, the
gradient of the tangent $= -\frac{1}{4}$.

Note: The gradient is always negative on this curve.

The equation of the tangent at the point parameter t will be
$$y - \frac{c}{t} = -\frac{1}{t^2}(x - ct) \text{ or } t^2 y - ct = -x + ct$$
i.e., $t^2 y + x = 2ct.$
At the point parameter $t = 2$, the equation of the tangent will be
$4y + x = 4c$.

(2) If the gradient of the tangent is $-\dfrac{1}{t^2}$, then the gradient of the
normal is t^2 (page 248).
Hence the equation of the normal at the point parameter t is
$$y - \frac{c}{t} = t^2(x - ct) \text{ or } ty = t^3 x + c - ct^4.$$

(3) To sketch the curve, note that when t is positive both x and y
are positive. As t increases, x increases but y decreases, approaching

zero. As t (positive) $\to 0$, $x \to 0$ but $y \to +\infty$. This enables us to sketch one branch of the curve (fig 19.14). When t is negative, both x and y are negative but with the same numerical values as for positive t. The two branches are symmetrical about the origin.

(4) If $x = ct$, $y = \dfrac{c}{t}$, by direct multiplication $xy = c^2$ which is the Cartesian equation of the curve (a hyperbola).

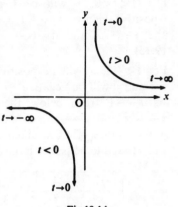

Fig 19.14

Example 17

Given the parametric equations $x = \dfrac{t}{1+t}$, $y = \dfrac{t^2}{1+t}$, sketch the curve and find its Cartesian equation.

We consider first the general pattern of values of x and y as t varies through certain ranges. Values of t which make x or y zero, positive, negative or infinite are specially noted. $\dfrac{dy}{dx}$ is also found and this will give useful clues to the shape of the curve, particularly where there are turning points. In addition, specific values of t are taken, to help locate the curve.

$$\frac{dy}{dx} = \frac{dy}{dt} \Big/ \frac{dx}{dt}$$
$$= \frac{(1+t)2t - t^2}{(1+t)^2} \div \frac{(1+t)1 - t}{(1+t)^2}$$
$$= 2t + t^2$$
$$= t(2+t)$$

So $\dfrac{dy}{dx} = 0$ when $t = 0$ or $t = -2$.

When $t = 0$, $x = y = 0$ and this is a turning point.
When $t = -2$, $x = 2$, $y = -4$ and this is also a turning point.

When $t > 0$, x and y are positive. As t increases, $x \to 1$ but y increases indefinitely. The gradient is always positive.
As $t \to -1$ both x and $y \to \infty$. We must consider values of t above and below -1.

If $-1 < t < 0$, x will be negative, y positive. Hence as $t \to -1$, $x \to -\infty$ and $y \to +\infty$ and $dy/dx \to -1$ (fig 19.15).

For $t < -1$, x will be positive and >1 and y will be negative.

As $t \to -1$ from below therefore, $x \to +\infty$ and $y \to -\infty$; $dy/dx \to -1$. A few values of t are taken and x and y found to help sketch the curve:

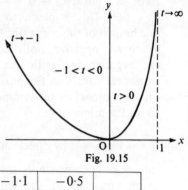

Fig. 19.15

t	-4	-1.5	-1.1	-0.5	
x	$\frac{4}{3}$	3	11	-1	etc.
y	$-\frac{16}{3}$	-4.5	-12	$+0.5$	

We can complete the sketch (fig 19.16). It has two branches and the line $x = 1$ is an asymptote.

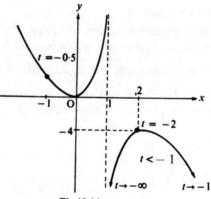

Fig 19.16

To find the Cartesian equation:

$$x = \frac{t}{1 + t}, \text{ hence } (1 + t)x = t$$

$$y = \frac{t^2}{1 + t}, \text{ hence } (1 + t)y = t^2.$$

Divide: then $\dfrac{y}{x} = t$ and substitute this in the first parametric equation.

Then
$$x = \frac{y/x}{1 + y/x} = \frac{y}{y + x}.$$

So the Cartesian equation is $xy + x^2 = y$.

Example 18

A line of variable gradient m passes through the fixed point (4, 2) *for all values of m and cuts the coordinate axes at* P *and* Q. *Find the coordinates of* P *and* Q *in terms of m. If* M *is the midpoint of* PQ, *find the coordinates of* M *and hence the locus of* M (*fig* 19.17).

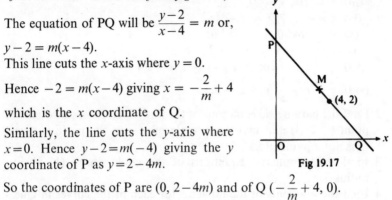

The equation of PQ will be $\dfrac{y-2}{x-4} = m$ or,

$y - 2 = m(x - 4)$.
This line cuts the x-axis where $y = 0$.

Hence $-2 = m(x-4)$ giving $x = -\dfrac{2}{m} + 4$

which is the x coordinate of Q.

Similarly, the line cuts the y-axis where $x = 0$. Hence $y - 2 = m(-4)$ giving the y coordinate of P as $y = 2 - 4m$.

Fig 19.17

So the coordinates of P are $(0, 2 - 4m)$ and of Q $(-\dfrac{2}{m} + 4, 0)$.

The coordinates of M, the midpoint, are then $(-\dfrac{1}{m} + 2, 1 - 2m)$ and these are expressed in terms of the parameter m. To obtain the Cartesian equation of the locus of M, we eliminate m between the parametric equations for M.

Now $x = -\dfrac{1}{m} + 2, \ y = 1 - 2m$.

Isolate $\dfrac{1}{m}$ and $2m$.

Then $\dfrac{1}{m} = 2 - x$ and $2m = 1 - y$.

Multiplying: $\dfrac{1}{m} \times 2m = (2 - x)(1 - y)$

giving $2 = 2 - x - 2y + xy$
or $x + 2y = xy$

which is the equation of the locus of M. The student should experiment by drawing various positions of the line PQ and marking the positions of M, to obtain a sketch of its locus. The curve is shown in fig 19.18 and is a hyperbola.
The method is typical of such locus problems.

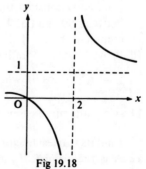

Fig 19.18

Exercise 19.6

1 Find the Cartesian equations of the following curves:
  (i) $x = 3 \cos \theta, y = 3 \sin \theta$;
  (ii) $x = 1 + 4 \cos t, y = 3 + 4 \sin t$;
  (iii) $x = 2t^2, y = 3t$;
  (iv) $x = t + 1, y = t - 1$;
  (v) $x = \cos^3 \theta, y = \sin^3 \theta$;
  (vi) $x = t^2 + 1, y = t + 2$;
  (vii) $x = 2 + r \cos 30°, y = -1 + r \sin 30°$;
  (viii) $x = t + \dfrac{1}{t}, y = t - 2$.

2 Find the parametric equations of the line which passes through the point $(-2, 3)$ and makes an angle of α with the positive x-axis (α is not a parameter).

3 Find the parametric equations of the circle centre $(1, 5)$ and radius 3.

4 Find the gradients of the tangents for each of the curves in Question **1** in terms of the parameter.

5 Find the equations of the tangent and normal at the point parameter t on the curve $x = at^2, y = 2at$ where a is a constant.

6 Find the equation of the tangent on the curve given by $x = \dfrac{t}{1 + t}$, $y = \dfrac{t^2}{1 + t}$, at the point where $t = 1$.

7 Sketch the curve given by $x = 2t - 1, y = t + 3$ and find its Cartesian equation.

8 Eliminate the parameter from the equations $x = t^2 + 1, y = 2t$ to find the Cartesian equation of the curve.

9 Sketch the curve given by $x = \cos^3 \theta, y = \sin^3 \theta$. (Take values of θ from 0° to 90°, say 0°, 30°, 45°, 60° and 90°, and calculate x and y. Use the gradient (Question **4**, part (**v**)) for additional information. Now consider the values of x and y for corresponding angles between 90° and 180°, 180° to 270° and 270° to 360° to obtain a complete sketch. This curve is called the ASTROID.)

10 Prove that the gradient of a cycloid given by $x = a(\theta - \sin \theta)$, $y = a(1 + \cos \theta)$, where a is a constant, is $\cot \dfrac{\theta}{2}$.

11 A certain curve is given by the equations $x = \dfrac{t}{\sqrt{1 + t^2}}, y = \dfrac{1}{\sqrt{1 + t^2}}$.
Find its Cartesian equation. (Consider $x^2 + y^2$). What is the curve?

12 A is the point $(t, 0)$ and B the point $(0, 2t)$ where t is a variable parameter. If M is the midpoint of AB, find the coordinates of M in terms of t, and hence find the locus of M.

13 A is the point $(t, 0)$ and B the point $(0, \frac{1}{t})$. If M is the midpoint of AB, find the locus of M. Sketch this locus.

14 Sketch the curve given by $x = t - 1$, $y = t^2 + t$ and find dy/dx in terms of t. Find the values of t **(i)** for which the tangent is parallel to the line $x + y = 5$; **(ii)** where the curve has a turning point. Obtain the Cartesian equation of the curve.

15 A curve is given as $x = 3 - 2t$, $y = t(1 - t)$. Find dy/dx in terms of t. What is the equation of the tangent which is perpendicular to the line $x - y = 2$? Find the equation of the tangent which passes through the point $(1, 0)$.

16 A is the point $(2t, 0)$ and B the point $(1, t)$ where t is a parameter. Find the locus of M, the midpoint of AB and sketch the locus.

17 The circle in fig. 19.19 touches the y-axis at O and its diameter is OB where B is $(4, 0)$. $P(x, y)$ is any point on the circle, and angle $POB = \alpha$. **(a)** Show that $OP = 4 \cos \alpha$ and then **(b)** express x and y in terms of α to obtain the parametric equations of the circle. **(c)** Show that your equations lead to the Cartesian equation $x^2 + y^2 = 4x$.

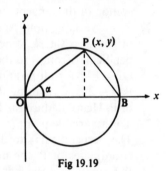

Fig 19.19

18 Find the gradient of the tangent to the curve $x = t + \frac{1}{t}$, $y = t - \frac{1}{t}$. Find the equation of the tangent at the point where $t = 2$ and the Cartesian equation of the curve.

19 Find the equations of the tangent and normal to the curve $x = 4t - 1$, $y = 2t^2$ at the point where $t = 1$.

20 P is the point $(2t, t)$ and O is the origin. The line through P perpendicular to OP cuts the axes at A and B. Find the coordinates of A and B in terms of t. Hence find the locus of the midpoint of AB.

21 The parametric equations of a curve are $x = t^2 + t$, $y = 2t - 1$. What are the values of the parameter at the points where the line $y = 2x - 3$ meets the curve? Give the coordinates of these points.

22 Find the equations of the tangent and normal to the curve given by $x = (t - 1)^2$, $y = t^2 - 1$ at the point where $t = -2$.

23 A is the fixed point $(1, 0)$ and T the variable point $(0, t)$. A line is drawn through T perpendicular to AT and cuts the x-axis at Q. Find the coordinates of Q in terms of t. Hence find the locus of M, the midpoint of TQ and sketch this locus.

24 P is the variable point $(2t, t)$. The line through P with gradient -1 cuts the x-axis at A and the y-axis at B. A second line through P with gradient 1 cuts the x-axis at C and the y-axis at D. BC and AD intersect at T. Find the coordinates of T in terms of t and the locus of T.

25 A is the point $(2, 0)$ and a line of varying gradient m passes through A, cutting the y-axis at P. From P a line is drawn at right angles to AP to cut the x-axis at Q. Find the coordinates of M, the midpoint of PQ, in terms of m and hence find the locus of M.

26 Find the equation of the tangent to the curve $x = 3t^2$, $y = 6t$ at the point parameter t. Find the coordinates of the point of intersection of the tangents to the curve at the points $t = 2$ and $t = -1$.

27 P is a variable point such that $OP = r$ units long, where O is the origin. A is the fixed point $(-2, 0)$ and M is the foot of the perpendicular from P to the line $x = -2$ through A. P moves so that $OP = PM$ for all positions of P. Show that the x coordinate of P is $(r - 2)$ and by using Pythagoras' Theorem find the y coordinate of P. Hence deduce the locus of P.

28 T is a variable point $(t, 0)$ on the x-axis and R is the point $(3, 1)$. The perpendicular to TR at T meets the y-axis at Q. Find the coordinates of P, the midpoint of TQ, in terms of t. Hence find the equation of the locus of P as t varies.   (C)

29 The parametric equations of a curve are $x = 3t^2$, $y = 2t^3$ where t is the parameter. Find the equation of the tangent at the point parameter t. Sketch the curve and obtain its Cartesian equation.

30 A circle is drawn with centre $(0,1)$ and radius 1. A line OAB is drawn, making an angle θ with the x-axis to cut the circle at A and the tangent to the circle at $(0, 2)$ at B. Lines are now drawn through A and B parallel to the x-and y-axes respectively to intersect at P. Prove that **(i)** $OA = 2 \sin \theta$ and **(ii)** the coordinates of P are $(2 \cot \theta, 2 \sin^2 \theta)$. Hence find the Cartesian equation of the locus of P.

Applied Mathematics

20 Displacement, Velocity, Acceleration, Relative Velocity

This Chapter will be concerned with the measurement of motion, the branch of Applied Mathematics called **Kinematics**. In Chapter 11 we first met the idea of velocity as a rate of change of distance in a specific direction and acceleration as the rate of change of velocity. Also in Chapter 11 we discussed simple problems involving s or v as a function of t. The present chapter will deal in more detail with the relations between distance, velocity and acceleration.

Graphical Representations

The two most important graphs are the s-t (distance or displacement-time) and the v-t (velocity-time) graphs which were mentioned in Chapter 11.

s-t graphs

Fig 20.1 shows the s-t graph of the motion of a body. We already know that the gradient of the graph ($\dfrac{ds}{dt}$) is a measure of the velocity in m s^{-1}. The following facts can be seen from the graph:

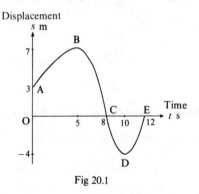

Fig 20.1

(1) The starting or initial position of the body is 3 m from a reference position O. The initial velocity is given by the gradient at O and is positive. The body is travelling away from O in the positive direction.

(2) From A to B the body travels further away from O but its velocity is decreasing, as the gradient is decreasing.

(3) At B, 7 m from O, the velocity is 0. The body is momentarily at rest. The time is 5 s from the start.

(4) From B to C the gradient is negative. Hence the velocity is negative, i.e., the body is travelling back towards O and arrives there after 8 s from the start. Note that the distance travelled up to this time is $4 + 7 = 11$ m but the **displacement** from O at this time is 0. As this graph always shows the displacement of the body at any time, it is better described as a displacement-time graph.

(5) The body reaches position D (4 m from O in the opposite direction to B) after 10 s, and is momentarily at rest.

(6) From D to E the body is again travelling in the positive direction and reaches O again after 12 s from the start.

If s is given as a function of t, the velocity can be found by obtaining the value of $\dfrac{ds}{dt}$. If not, an approximate method must be used, by fitting a tangent to the curve and measuring its gradient.

v-t graphs

A more useful graph is the v-t graph, relating velocity to time. Fig 20.2 shows a specimen graph.

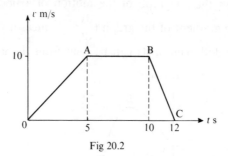

Fig 20.2

We know that the gradient of this graph $\left(\dfrac{dv}{dt}\right)$ measures the acceleration in m s^{-2}. Note that there is no indication of the displacement of the body.

(1) The body starts at $t = 0$, from rest (i.e., with zero velocity). From O to A the velocity increases until it reaches 10 m s^{-1} at time 5 s. Since OA is a straight line, the acceleration is **uniform** or constant and equal to $\frac{10}{5} = 2$ m s^{-2}. At A the acceleration ceases.

(2) From A to B the body travels with uniform velocity (10 m s^{-1}).

(3) From B to C the velocity decreases steadily and the body comes to rest again at $t = 12$ s. From B to C the acceleration is negative or alternatively, the body has a uniform **retardation** of $\frac{10}{2} = 5$ m s^{-2}.

Area under the *v-t* graph

The area under a *v-t* graph supplies additional information. In the centre part of the graph, the body is travelling with constant velocity of 10 m s^{-1} for 5 s. Hence the distance covered is 50 m. The area of this part is 10 m s^{-1} × 5 s = 50 m. So the area under a *v-t* graph is numerically equal to the distance covered by the body (strictly speaking, the displacement of the body). If *v* is measured in m s^{-1} and *t* in s, the area will give the displacement in m; if *v* is in km h^{-1} and *t* in h, the area represents the distance measured in km.

(In calculus, the area under the graph would be

$$\int v \, dt = \int \frac{ds}{dt} \, dt = \int ds = s$$

with the appropriate limits taken).
This fact makes the *v-t* graph very useful. Care must be taken to work in consistent units, not for example m s^{-1} and h.

Example 1

A train starts from station P *and accelerates uniformly for* 2 min *reaching a speed of* 48 km/h. *It continues at this speed for* 5 min *and then is retarded uniformly for a further* 3 min *to come to rest at station* Q. *Find* (**i**) *the distance* PQ *in* km; (**ii**) *the average speed of the train*; (**iii**) *the acceleration in* m s^{-2}; (**iv**) *the time taken to cover half the distance between* P *and* Q.

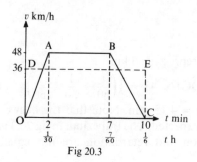

Fig 20.3

The *v-t* graph is shown in fig 20.3. OA is a straight line as the acceleration is uniform. AB is a straight section parallel to the *t*-axis as the velocity is constant at 48 km/h. BC is also straight, with negative gradient, as the retardation is uniform. *v* is marked in km/h and to be consistent *t* is also marked in h.

(i) Then the area under the graph represents the distance travelled. OABC is a trapezium of area $= \frac{1}{2}$ (sum of parallel sides) × height

$$= \frac{1}{2}(AB + OC) \times 48$$
$$= \frac{1}{2} \times (\frac{5}{60} + \frac{1}{6}) \times 48 = 6.$$

So the distance PQ = 6 km.

(ii) The average speed $= \dfrac{\text{total distance travelled}}{\text{time taken}} = \dfrac{6}{\frac{1}{6}} = 36$ km/h.

This is shown by the dotted line DE, and ODEC also represents 6 km.

(iii) To find the acceleration over section OA, work in m and s.

$$48 \text{ km/h} = \frac{48 \times 1000}{60 \times 60} \text{ m s}^{-1} = \frac{40}{3} \text{ m s}^{-1}.$$

Hence the acceleration $= \dfrac{40}{3} \div 120 = 0\cdot11 \text{ m s}^{-2}.$

(iv) Let half the distance be covered in t min $\left(\dfrac{t}{60}\text{h}\right)$ (fig 20.4.).

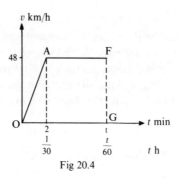

Fig 20.4

The area OAFG represents 3 km.

$$\therefore 3 = \tfrac{1}{2}(AF + OG) \times 48 = \tfrac{1}{2}\left(\frac{t}{60} - \frac{1}{30} + \frac{t}{60}\right) \times 48 \text{ which leads to}$$

$7\cdot5 = 2t - 2$ or $t = 4\cdot75$ min. Note that the answer is not 5 min as might have been expected. The trapezium in fig 20.3 is not symmetrical as the acceleration and retardation are not equal.

Example 2

Car A is travelling at a steady speed of 60 km/h and passes a stationary car B. The driver of car B immediately accelerates, reaching 80 km/h in 20 s, and continues at this speed until he overtakes car A. Find the distance travelled by car B when this happens.

PQ is the uniform speed graph of car A (fig 20.5).

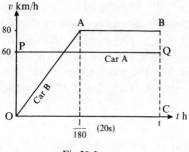

Fig 20.5

OABC is the graph of car B. Since they travel the same distance, the area OABC = the area OPQC. Let t (in hours) be the time taken to overtake.

Then $\frac{1}{2}\left(t + t - \frac{1}{180}\right) \times 80 = 60 \times t$ which gives $t = \frac{1}{90}$ h = 40 s.

Hence the distance travelled by B $= 60 \times \frac{1}{90} = 0.67$ km.

Example 3
A bus, travelling at 15 m/s and accelerating at 0·2 m/s² passes a stationary car. The bus accelerates for 5 s and then continues at a steady speed. Thirty seconds after being passed the car starts with acceleration 0·5 m/s² until it reaches a speed of 30 m/s, with which speed it continues to travel. (i) What is the distance between the bus and the car $\frac{1}{2}$ min later? (ii) When will the car overtake the bus?

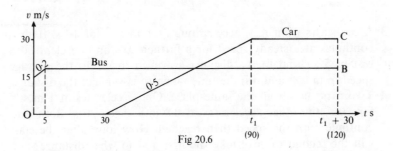

Fig 20.6

The *v-t* graphs are shown in fig 20.6.
Note that the points B and C are simultaneous in *time* but not in displacement. Let the car reach its maximum speed at time t_1 (measured from when the bus passes the car).
Then in $(t_1 - 30)$ s the speed of the car increases from 0 to 30 m/s.

Hence $30 = 0.5 (t_1 - 30)$ or $t_1 = 90$ s.

The distance travelled by the car is given by the area under its graph
$= \frac{1}{2}(90 + 30) \times 30 = \frac{1}{2} \times 120 \times 30 = 1800$ m.

The steady speed reached by the bus after 5 s is $15 + 0.2 \times 5 = 16$ m/s.

The distance travelled by the bus is given by the area of the trapezium
+ the rectangle $= \frac{1}{2}(15 + 16) \times 5 + 16(115) = 15.5 \times 5 + 16 \times 115$
$= 1917.5$ m.

Hence the bus is 117.5 m ahead of the car (**i**).

To answer (**ii**) the car is gaining 15 m/s $(45 - 30)$ every s on the bus.

To overtake the bus in a distance of 117.5 m will take a further
$$117.5/15 = 7.8 \text{ s.}$$
So the car overtakes the bus at time $90 + 30 + 7.8 = 127.8$ s from
the start.

Exercise 20.1

1 A car starts from rest, accelerates at 0.8 m s^{-2} for 10 s and then
continues at a steady speed for a further 20 s. Draw the v-t graph
and find the total distance travelled.

2 Interpret the v-t graph shown in
fig 20.7 and find (**i**) the acceleration;
(**ii**) the retardation, both in m s^{-2};
(**iii**) the distance travelled in km.

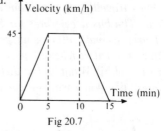

Fig 20.7

3 A car starts from rest, accelerating at 1 m s^{-2} for 10 s. It then
continues at a steady speed for a further 20 s and decelerates to
rest in 5 s. Find (**i**) the distance travelled in m, (**ii**) the average
speed in m s^{-1} and (**iii**) the time taken to cover half the distance.

4 Two cars start from the same place. One accelerates at 1 m s^{-2}
for 10 s, the other accelerates at 0.8 m s^{-2} for 20 s. Both cars
continue with the speed then reached. How long after the start
will the second car overtake the first and in what distance?

5 A car accelerates from rest to reach a certain speed in 10 min. It
then continues at this speed for another 10 min and decelerates
to rest in a further 5 min. The total distance covered is 17.5 km.
Find the steady speed reached.

6 The table gives the speeds at 1 s intervals for a car which started
from rest:

Time s	1	2	3	4	5	6	7	8	9	10
Speed m/s	5	9·5	13·5	17	20	22·5	24·5	26	27	27·5

Draw the speed-time graph and use it to find as accurately as you can (i) the acceleration after 2 s; (ii) the total distance travelled in the 10 s. (The area under the curve can be found by counting squares. Count each square which is equal to or greater than $\frac{1}{2}$ square, reject the others, at the curved boundary).

7 Two trains A and B, starting together from rest, arrive together at rest 10 min later. Train A accelerates uniformly at 0.125 m s^{-2} for 2 min, continues at the steady speed reached for another 4 min and then retards uniformly to rest. Train B accelerates uniformly for 5 min and then retards uniformly to rest.
Draw both journeys on the same v-t graph and find (a) the distance (in m) travelled, (b) the acceleration of train B, (c) the distance between the two trains after 3 min.

8 A car travelling at a constant velocity of 20 m s^{-1} passes a stationary sports car. Ten seconds afterwards the sports car accelerates uniformly at 3 m s^{-2} to reach a speed of 30 m s^{-1} with which it continues. Draw the v-t graphs together and find when and where the sports car overtakes the first car.

9 In rising from rest to rest in 8 s, a lift accelerates uniformly to its maximum speed and then retards uniformly. The retardation is one-third the acceleration and the distance travelled is 20 m. Find the acceleration, the retardation and the maximum speed reached.

10 An electric train takes 3 min to travel between two stations 2970 m apart. The train accelerates uniformly to a speed of 18 m s^{-1} and then travels for a time at this speed before retarding uniformly to rest at the second station.
If the acceleration and retardation are in the ratio $2:3$ calculate the times for which the train was accelerating and travelling at steady speed.

Straight Line Motion with Constant Acceleration

A basic type of motion is that of a body travelling in a straight line with constant or uniform acceleration (or retardation) (fig 20.8). For this motion we can derive useful equations. Let the body have initial velocity u at zero time and move with uniform acceleration a. At time t let the velocity be v and the displacement at that time be s (all in consistent units).

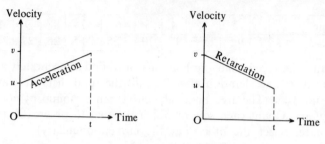

Fig 20.8

(1) Since the velocity increases by a units in unit time

$$v = u + at$$

(2) The area under the v-t graph represents the displacement.

Hence $$s = \tfrac{1}{2}(u + v) \times t$$

(3) Substituting for v in this second equation, $s = \tfrac{1}{2}(u + u + at) \times t$

which gives $$s = ut + \tfrac{1}{2}at^2.$$

(4) From the first two equations, $s = \tfrac{1}{2}(u + v) \times t = \tfrac{1}{2}(u + v)(\dfrac{v - u}{a})$

or $2as = v^2 - u^2$

i.e. $$v^2 = u^2 + 2as$$

These four equations for uniformly accelerated motion in a straight line can be used to solve problems instead of using a v-t graph or in addition to doing so. Solutions will often involve solving simultaneous equations.

Example 4

A body is moving in a straight line with constant acceleration. One second after passing a point A it is 10 m from A. In the next second it travels a further 14 m. Find (i) the velocity on passing A, (ii) the acceleration, (iii) the velocity after travelling 100 m from A, (iv) the time taken to cover this distance.

A v-t sketch is shown in fig 20.9. Let the initial velocity (at A) be u m s^{-1} and the acceleration be a m s^{-2}.

Then the velocities 1 s and 2 s after passing A are $u + a$ and $u + 2a$.

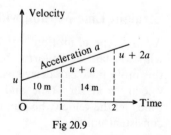

Fig 20.9

Using equation **2** we have $10 = \frac{1}{2}(u + u + a) \times 1$

or $2u + a = 20$

and $14 = \frac{1}{2}(u + a + u + 2a) \times 1$

or $2u + 3a = 28.$

Solving we obtain $u = 8$ and $a = 4.$

Hence **(i)** the velocity at A is 8 m s^{-1} and **(ii)** the acceleration is 4 m s^{-2}.

To answer **(iii)** use equation **4**, $v^2 = u^2 + 2as.$

Then $v^2 = 8^2 + 2 \times 4 \times 100 = 864$

and $v = 29\cdot4 \text{ m s}^{-1}.$

For part **(iv)** use equation **1**, $v = u + at$ or $t = \dfrac{v - u}{a}.$

Hence $t = \dfrac{29\cdot4 - 8}{4} = 5\cdot35 \text{ s.}$

Example 5

A, B *and* C *are three points on a straight road. A car passes* A *with a speed of* 5 m/s *and travels from* A *to* B *with a constant acceleration of* 2 m/s². *From* B *to* C *the car has a constant retardation of* 3·5 m/s² *and comes to rest at* C. *If the total distance from* A *to* C *is* 475 m, *find* **(a)** *the speed of the car at* B, **(b)** *the distance of* B *from* A.

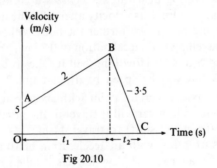

Fig 20.10

The v-t graph is sketched in fig 20.10.

Using equation **1**, the speed at B is given by

$v = 5 + 2t_1$ (i)

From B to C, using the same equation,

$0 = v - 3\cdot5t_2$ (ii)

This gives a relation between t_1 and t_2:

$v = 3\cdot5t_2 = 5 + 2t_1$ (iii)

The distance from A to B (s_1) is given by equation **2**.

Thus $s_1 = \frac{1}{2}(5 + 5 + 2t_1)t_1 = 5t_1 + t_1^2$

and the distance from B to C (s_2) is similarly given by
$$s_2 = \tfrac{1}{2}(3 \cdot 5 t_2) t_2 = 1 \cdot 75 t_2{}^2.$$
Now the total distance is $475 = s_1 + s_2 = 5 t_1 + t_1{}^2 + 1 \cdot 75 t_2{}^2$.
Substitute for t_2 in terms of t_1, using equation (iii).

Then
$$475 = 5 t_1 + t_1{}^2 + 1 \cdot 75 \left(\frac{5 + 2 t_1}{3 \cdot 5} \right)^2$$

$$= 5 t_1 + t_1{}^2 + \frac{25 + 20 t_1 + 4 t_1{}^2}{7} \quad \text{leading to}$$

$$3325 = 35 t_1 + 7 t_1{}^2 + 25 + 20 t_1 + 4 t_1{}^2$$

or
$$11 t_1{}^2 + 55 t_1 - 3300 = 0$$

or
$$(11 t_1 - 165)(t_1 + 20) = 0$$

giving the solution $\qquad t_1 = 15$ s.

Hence the speed of the car at $\mathbf{B} = v = 5 + 2 t_1 = 35$ m/s and the distance of B from A is $s_1 = \tfrac{1}{2}(5 + 35) \times 15 = 300$ m.

Exercise 20.2

(All accelerations are to be taken as uniform and in a straight line).

1 A particle starts with velocity 3 m s^{-1} and accelerates at 0·5 m s^{-2}. What is its velocity after (i) 3 s, (ii) 10 s, (iii) t s? How far has it travelled in these times?

2 A body, decelerating at 0·8 m s^{-2}, passes a certain point with a speed of 30 m s^{-1}. Find its velocity after 10 s, the distance covered in that time and how much further the body will go until it stops.

3 A particle travelling with acceleration of 0·75 m s^{-2} passes a point O with speed 5 m s^{-1}. How long will it take to cover a distance of 250 m? What will its speed be at that time?

4 If a particle passes a certain point with speed 5 m/s and is accelerating at 3 m/s^2 how far will it travel in the next 2 s? How long will it take (from the start) to travel 44 m?

5 A body falls from rest with an acceleration of 10 m s^{-2}. What is its velocity 5 s afterwards? How far has it fallen by then?

6 A car, retarding uniformly, passes over three cables P, Q, R set at right angles to the path of the car and 11 m apart. It takes 1 s between P and Q and 1·2 s between Q and R. Find (a) its retardation, (b) its velocity when it crosses P, (c) its distance beyond R when it comes to rest. (C)

7 A car, slowing down with uniform retardation, passes a stationary man. 20 s after passing the man the car is 40 m from him and 30 s after passing him it is 50 m away. Find the speed of the car as it passed the man. Find the distance from the man where the car comes to rest. Draw a sketch of the v-t graph.

8 A body passes a certain point A of the straight line on which it is moving with uniform acceleration. One second afterwards it is 11 m beyond A and in the next second it travels a further 13 m. Find the velocity it had when passing A, and how far it travels in the third second after passing A.

9 A motor car travelling at 30 km/h on a straight road is overtaken by another car travelling at 40 km/h. The first car immediately accelerates uniformly and after travelling 0·5 km it overtakes the second car which has kept its speed constant. Calculate the time taken, the magnitude of the acceleration and the greatest distance between the cars during this time. Illustrate your answer by a speed-time graph. (L)

10 A particle starting from rest moves with constant acceleration x m s^{-2} for 10 s; travels with constant velocity for a further 10 s and then retards at $2x$ m s^{-2} to come to rest 300 m from its starting point. Find the value of x. (C)

Acceleration due to Gravity

A very important case of motion with uniform acceleration is that of a body moving in space under the influence of a gravitational force. This will be discussed further in the next chapter. For the present the following facts may be stated.

(1) The earth (or moon or any planet) attracts bodies near it, the force of attraction being called the gravitational force.

(2) This force produces an acceleration towards the centre of the planet in any body free to move.

(3) This acceleration due to gravity depends on distance from the centre of the planet. Near the surface of the earth it is about 9·8 m s^{-2} (or approximately 10). It is slightly less near the equator than near the poles, as the earth is not a perfect sphere.

On the moon the acceleration due to gravity is about 1·62 m s^{-2}.

(4) The symbol g is always used to denote the acceleration due to gravity.

(5) At the same place, g is the same for all bodies. Ignoring air resistance, a piece of paper and a stone would each fall with the same acceleration g.

In all our work we ignore the effect of air resistance.

Example 6

A stone is dropped from the top of a building 25 m high. How long does it take to reach the ground and what is its velocity on arrival?
$(g = 9·8 \text{ m s}^{-2})$.

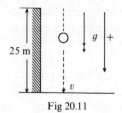

Fig 20.11

The initial velocity $u = 0$ and the final velocity is v. The positive direction is taken vertically downwards. (fig 20.11)

Using the equation $s = ut + \frac{1}{2}at^2$,

$$25 = 0 + \frac{1}{2} \times 9{\cdot}8 \times t^2$$

or $$t^2 = \frac{50}{9{\cdot}8} = 5{\cdot}1$$

and hence $$t = 2{\cdot}26 \text{ s.}$$

Also using $v = u + at$, $v = 0 + 9{\cdot}8 \times 2{\cdot}26$

$$= 22{\cdot}1 \text{ m s}^{-1}.$$

Example 7

A stone A is thrown vertically upwards with velocity 20 m s^{-1}. At the same time and 30 m vertically above, a second stone B is let fall. After what time and at what height do they collide? Take $g = 10$ m s^{-2}.

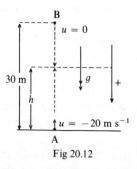

Fig 20.12

Take the downward vertical as the positive direction. For stone A, $u = -20$, $g = -10$ m s^{-2} and the displacement is $-h$ m. (fig 20.12) For stone B, $u = 0$, $g = 10$ and the displacement is $(30 - h)$ m. The time taken for both stones to arrive at the collision point, t, is the same.

For stone A, using $s = ut + \frac{1}{2}at^2$,

$$-h = -20t + 5t^2$$

or $\qquad\qquad\qquad h = 20t - 5t^2 \qquad$ (i)

For stone B, similarly, $30 - h = 5t^2 \qquad$ (ii)

Add these two equations to eliminate h:

$$30 = 20t \text{ giving } t = 1.5 \text{ s.}$$

Then $\qquad\qquad h = 20(1.5) - 5(1.5)^2 = 18.75 \text{ m.}$

The stones collide after 1·5 s at a height of 18·75 m.

Example 8

A rocket is fired vertically upwards with a speed of 60 m s^{-1}. *Find*
(i) *the maximum height reached,* **(ii)** *the time taken to return to the point
of projection,* **(iii)** *the time taken to reach a height of* 145 m. *Take*
$g = 10$ m s^{-2}.

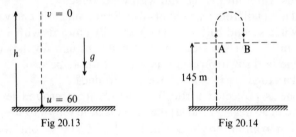

Fig 20.13   Fig 20.14

Take the positive direction as vertically upwards (fig 20.13).

$u = +60, g = -10, v = 0, s = h.$

(i) At the maximum height, $v = 0$ and using $v^2 = u^2 + 2as$ we obtain

$$0 = 60^2 - 2 \times 10 \times h \text{ giving } h = 180 \text{ m.}$$

(ii) Time taken to reach the maximum height, using $v = u + at$, is

$$0 = 60 - 10t \text{ or } t = 6 \text{ s.}$$

As the downward path is subject to the same acceleration, the time
of descent will also be 6 s. (Check this using $s = ut + \frac{1}{2}at^2$). Hence
the total time to return to the point of projection is 12 s.

This can be done more neatly by noting that the total displacement
is zero.

Hence using $s = ut + \frac{1}{2}at^2$, $\quad 0 = 60t - 5t^2$

which gives $t = 0$ or $t = 12$ s.

The second answer is the time taken over a displacement of 0 m
though the particle actually travels a distance of 360 m.

(iii) The displacement of the particle is 145 m (fig 20.14).

Then $145 = 60t - 5t^2$ which gives $t = 3$ or $t = 9$ s.

$t = 3$ s is the time to reach 145 m on the upward journey (A) and
$t = 9$ s is the time to reach 145 m on the downward or return journey
(B).

Exercise 20.3

Take $g = 10$ m s^{-2}.

1 A ball is thrown upwards with a velocity of 12 m s^{-1}. How high does it reach and how long does it take to get there?

2 A particle, thrown vertically upwards, reaches a height of 5 m. What was its initial velocity?

3 A stone is let fall from the top of a building 20 m high. How long does it take to reach the ground and with what velocity does it strike the ground?

4 From the top of a cliff 50 m above sea level a stone is thrown vertically upwards with a velocity of 15 m s^{-1}. After how many seconds will it hit the sea and with what velocity?

5 A ball is let fall from a height of 40 m. Simultaneously and vertically below it, a second ball is thrown vertically upwards with a velocity of 20 m s^{-1}. At what distance from the ground do they collide?

6 A stone is thrown vertically upwards with a speed of 25 m s^{-1}. At what times is it 5 m above the ground?

7 A stone is projected vertically upwards from ground level with a speed of 30 m s^{-1}. Find (a) the time taken to return to the ground, (b) the maximum height reached and (c) for how long the stone was more than 40 m above the ground.

8 A ball is dropped from a height of 5 m onto a concrete floor and rebounds with a speed 0·8 times the downward speed on arrival. Find the height reached from the rebound.

9 A stone is thrown vertically upwards with a speed of 20 m s^{-1}. One second later a second stone is thrown vertically upwards with a speed of 30 m s^{-1}. At what height above the ground do they collide? (Hint: the first stone has a certain displacement in t s; the second stone has the same displacement in $(t - 1)$ s).

10 Two rockets are fired vertically from launching pads side by side. The first rocket moves vertically upwards with an acceleration of $6g$ and the second with an acceleration of $8g$. If the second rocket is fired 1 s after the first, find how long after its launching the second rocket overtakes the first.   (C)

Vectors

A velocity is an example of a **vector** quantity, i.e. a quantity whose specification requires a magnitude and a direction. Other examples of vectors are displacements, accelerations, forces. The important point about vectors for our work is that they are combined or 'added' by a special rule, the **parallelogram law** of addition of vectors. Quantities which are not vectors, called **scalar** quantities, are added according to the usual rules of arithmetic. For example, a mark of 50 added to a mark of 30 produces a mark of 80. But a velocity of 50 km/h combined with a velocity of 30 km/h does not necessarily produce a velocity of 80 km/h. The directions have to be considered as well.

A vector can be represented by any line whose length is a scale multiple of its magnitude and whose direction is parallel to that of the vector.

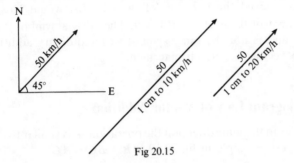

Fig 20.15

In fig 20.15 each line will represent the same velocity, 50 km/h NE, the direction being indicated by the arrow.

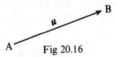

Fig 20.16

A vector is symbolized either by a single letter **u** (printed in black as here, written as u̲) or as **AB** (written either as A̲B̲ or $\overline{AB}$) (fig 20.16). In the second symbol the direction is also clear, from A to B.

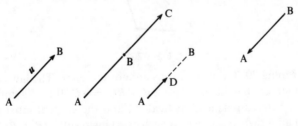

Fig 20.17

A vector can have scalar multiples. Fig 20.17 shows a vector **u** ($\overline{AB}$) and multiples $AC = 2u = 2AB$, $AD = \frac{1}{2}u = \frac{1}{2}AB$, $BA = -u = -AB$.
Note: The direction is unchanged for positive multiples but reversed for the negative multiple.

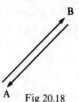

Fig 20.18

From fig 20.18 it is seen that vector AB followed by vector $BA = 0$.
A particle moved through $\overline{AB}$ followed by $\overline{BA}$ would have a final
displacement of 0, i.e. $AB + BA = 0$. The + is a symbol for vector
addition and is not the same symbol as for ordinary addition. 0 is
the zero vector, i.e. no displacement.

Parallelogram Law of Vector Addition

For vectors in the same direction the operation + is similar to ordinary
addition. For example in fig 20.19, $AB + BC = AC$.

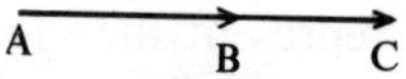

A $\qquad$ B $\qquad$ C

Fig 20.19

A displacement of 3 km in a certain direction followed by a displacement of 5 km in the same direction = a displacement of 8 km in
that direction.

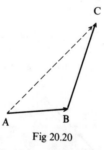

Fig 20.20

However in fig 20.20 the directions are not the same. The law of vector
addition states that in this case $AB + BC = AC$, the + symbol here
no longer behaving as arithmetical +, and the student must note its
wider meaning when applied to vectors. This result, $AB + BC = AC$,
can also be seen as being derived from the parallelogram in fig 20.21.
AD is an equivalent vector to BC as AD = BC in magnitude and has
the same direction. Therefore ABCD is a parallelogram of which AC
is a diagonal. The vector AC is called the **resultant** of the vectors AB
and BC (or AD). This is the **parallelogram law of vector addition**. The

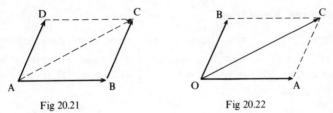

Fig 20.21   Fig 20.22

resultant of two vectors is the diagonal of the parallelogram based on the two given vectors, where the three originate from a common vertex. In fig 20.22, $OC = OA + OB$.

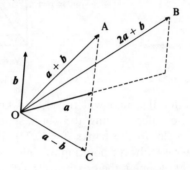

Fig 20.23

In fig 20.23 two vectors a and b are given. From these are constructed, using the parallelogram law and scalar multiplication, $OA = a + b$, $OB = 2a + b$, $OC = a - b$ i.e. $a + (-b)$.

Exercise 20.4

1 Draw a vector a. Then draw the vectors $2a$, $\frac{1}{2}a$, $-a$.

2 Using graph paper, draw two axes of coordinates Ox, Oy. Mark the points $A(3, 4)$, $B(5,5)$, $C(-2, 4)$ and $D(-3, -5)$. Draw the vectors OA, AB, CD, OD. Starting from the point $(-1, -1)$ draw vectors equivalent to $OA, AO, OA + AB, \frac{1}{2}(OA + AB), AO + OB$.

3 Mark two vectors a and b on your paper. Now sketch vectors equivalent to $a + b$, $2a + b$, $2a - b$, $2a + 3b$.

4 A, B, C, D are 4 points in a plane. If $AB = CD$ what can you say about the four points?

5 If a and b are vectors and $a + b = 0$ what can you say about b?

6 a is a vector length 4 cm in direction NE, b is a vector length 3 cm in direction N, c is a vector length 5 cm in direction E. Construct accurate diagrams to show the vectors: $a + b$, $a + b + c$ (add c to $a + b$), $a - b$, $b - c$, $a + b - c$, $2a + b$, $a + 2b$.

Combining Velocities

A simple example will illustrate the vector combination law.

A river is flowing at 3 km/h. *A man sets out, at right angles to the banks, to row across. He can row at* 4 km/h *in still water. What is his actual velocity across the river?*

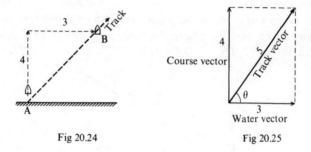

Fig 20.24   Fig 20.25

Relative to the water the boat will travel at 4 km/h perpendicular to the bank; but the water, moving at 3 km/h, simultaneously carries the boat along parallel to the banks (fig 20.24). The boat therefore moves crabwise across, always pointing at right angles to the bank. During any element of time t, the boat travels a distance proportional to $4t$ at right angles to the banks and simultaneously a distance $3t$ parallel to the banks. Hence its path will be *AB*. This is the resultant of the two vectors and has magnitude 5 km/h at an angle θ to the bank where $\tan \theta = \frac{4}{3}$ ($\theta = 53°$ approx.). This is the **track** of the boat, the path it actually takes (fig 20.25). The **course** is the direction in which the boat is headed, relative to the water.

Hence   **course vector + water vector = track vector.**

This deals with the velocities. To find the time of crossing and where the boat lands, see fig 20.26: The boat's track is *AB* (speed 5 km/h). If there was no current the boat would travel straight across (*AC*) at 4 km/h. Hence the time taken $= \frac{0.5}{4} = 0.125$ h $= 7\frac{1}{2}$ min, i.e. the same time as AB would be covered at 5 km/h.

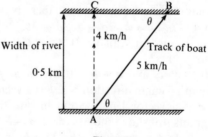

Fig 20.26

The distance downstream from C of B is given by $\tan \theta = \dfrac{0 \cdot 5}{BC}$ where $\tan \theta = \frac{4}{3}$ (from fig 20.25). Therefore $BC = 0 \cdot 375$ km.

This calculation was easy as the two given velocities were at right angles and they formed a rectangle. In general, calculation of the resultant of two vectors will involve using the sine and cosine rules for a triangle.

Alternatively a solution can be obtained by accurate scale drawing. Problems will either require (i) finding the resultant of two given vectors or (ii) finding one of the vectors if the other and the resultant are known.

Example 9

A helicopter has an airspeed of 100 km/h and flies on a course bearing 150°. A wind is blowing steadily at 30 km/h from a bearing of 080°. Find the track and ground speed.

Note the terms used: course and track have been mentioned above. The **airspeed** is the speed along the course, the speed relative to the air. The **groundspeed** is the speed along the track, the speed relative to the ground. A wind is always described as coming **from** a certain direction. If the wind is blowing from 080° it is actually blowing in the direction 260°.

Now **wind vector + course vector = track vector.**

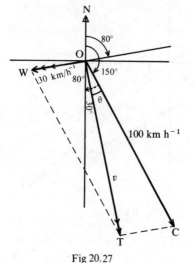

Fig 20.27

1 *By calculation*
Draw a sketch of the various vectors (fig 20.27).

From the triangle OCT using the cosine rule:
$$v^2 = 100^2 + 30^2 - 2 \times 100 \times 30 \times \cos 70°$$
$$= 10\,000 + 900 - 2052 = 8848$$
and hence $v = 94.06$ i.e. 94 km/h

Using the sine rule, $\dfrac{30}{\sin \theta} = \dfrac{94}{\sin 70°}$ or $\sin \theta = \dfrac{30 \sin 70°}{94}$ which

gives $\theta = 17° \, 27'$.

Hence the track is $167\frac{1}{2}°$ and the ground speed is 94 km/h.

2 *By drawing*

Select a suitable scale, say 1 cm for 10 km/h. Draw OW and OC to scale, 110° apart. Complete the parallelogram OWTC and take the necessary measurements from your figure. Compare your results with the calculated values as a guide to the accuracy achieved by drawing.

The next example shows how to find a course when the track is known.

Example 10

An aeroplane leaves an airfield **A** *at 1 p.m. to fly to another airfield* **B**, *400 km away on a bearing of* 130°. *There is a steady wind of* 30 km/h *coming from the north-east. The aeroplane has an airspeed of* 200 km/h. *Find the course the pilot must take and his time of arrival.*

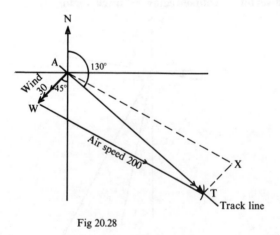

Fig 20.28

1 *By drawing*

Fig 20.28 shows the sketch which should be made first. Take a scale of say 1 cm for 20 km/h. Draw the N/E framework and then **AW** at 45° to the S line, to represent the wind vector. Draw *AT*, the track line, bearing 130°. The resultant vector lies along this line. *AW* is one side of the vector parallelogram. Now with centre W and radius 10 cm (representing the airspeed) cut the track line at T. Next complete the parallelogram AWTX and *AX* is the course vector. Measure its bearing. This gives the course to be taken. The ground speed is given by the length of *AT*. Divide 400 by this speed to find the time taken (see below). Compare your results with those from the calculations below.

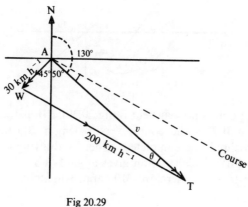

Fig 20.29

2 *By calculation*
All the facts are shown in fig 20.29. v is the ground speed. We solve the triangle AWT, given two sides and 1 angle.

First, using the sine rule, $\dfrac{30}{\sin \theta} = \dfrac{200}{\sin 95°}$ which gives $\theta = 8° \, 36'$.
Hence the course to take is $130° - \theta = 121° \, 24'$.
To find v, use the cosine rule:
$$\text{angle AWT} = 180° - (95° + \theta) = 76° \, 24'.$$
Then $v^2 = 200^2 + 30^2 - 2 \times 200 \times 30 \times \cos 76° \, 24'$
$= 40\,900 - 12\,000 \times 0.2351 = 38\,079$
and hence $v = 195.1 \text{ km h}^{-1}$.
The time taken $= \dfrac{400}{195.1} = 2.05$ h or 2 h 3 min and so he will arrive at 3.03 p.m.

Example 11
What course should the pilot in Example 10 take for the return journey, if the wind continues as before?

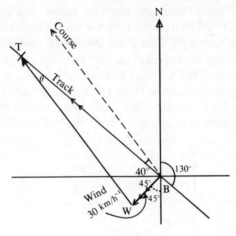

Fig 20.30

The starting point is now airfield B (fig 20.30). **BW** is the wind vector and therefore **WT** is the course vector, length 200 (the airspeed). The solution should be completed, either by drawing or calculation, using triangle BWT. Calculation gives angle WTB = θ = 8° 36′. The course to be taken is therefore 319° approximately.

Exercise 20.5

1 Find the resultants, in magnitude and direction, of the following velocities by drawing or calculation:
   (a) 5 km/h due north and 3 km/h due east; (b) 8 m/s north-east and 5 m/s north-west; (c) 10 m/s N 60° E and 4 m/s east; (d) 40 km/h on a bearing 050° and 40 km/h on a bearing 110°.

2 A steamship is sailing due north at 20 km/h. There is a wind of 15 km/h from the west. Find the direction in which the smoke from the funnel is drifting.

3 A ship is travelling at 10 m/s. A ball is rolled across the deck (at right angles to the motion of the ship) at 4 m/s. Find the actual velocity of the ball relative to the ground.

4 An aeroplane pilot sets a course due north. The airspeed is 150 km/h. There is a wind of 40 km/h from the west. Find the track of the plane (i.e. ground speed and its direction).

5 A man sets out to swim across a river 0·4 km wide at right angles to its parallel banks. He can swim at a steady speed of 2 km/h in still water but there is a currect flowing at 1 km/h. Find (a) his track across the river and (b) how far downstream he is carried when he lands.

6 An aeroplane has an airspeed of 200 km/h and is headed on a course of 330°. There is wind of 50 km/h from a bearing of 210°. Find the track of the plane by drawing or calculation.

7 A river flows at 3 km/h and its parallel banks are 400 m apart. A man wishes to cross in a motorboat which can make 5 km/h in still water, to reach the point directly opposite. Find the direction in which he must steer and the time taken to cross the river.

8 An aeroplane pilot wishes to fly from an airfield P to another airfield Q, 300 km due east. The airspeed is 200 km/h. There is a steady wind of 40 km/h from the north-east. Find (**a**) the course he must take, (**b**) the actual ground speed, (**c**) the time taken to reach Q, (**d**) the course required for the return journey, the wind being the same and (**e**) the time taken over the return journey.

9 A ship is steering due south in still water at a speed of 24 km/h. It then gets into a current running north-east at 4 km/h without the captain being aware of the current's existence. Find, by calculation or drawing, the magnitude and direction of the ship's new velocity. After a short time the captain recognises the existence of the current and estimates its direction and velocity correctly. If he alters helm so that the ship now moves due south, while the engine speed remains unchanged, find the course he has to steer.

(C)

10 A ship is steaming at 60 km/h on a bearing of 012°. It is known that a wind of strength 15 km/h is blowing. A boy flying a kite from the ship observes that the kite is due south of him. Find, by drawing or calculation, the possible directions of the wind. Give your answers to the nearest degree. (C)

11 An aeroplane, flying at a steady airspeed of 200 km/h, flies in a straight line from A to B, where B is due east of A and then returns along the same route. There is a constant wind of 60 km/h blowing from the north-east throughout the double flight. Find, by drawing or calculation, the ground speed of the aeroplane on both trips.

12 A motorboat, capable of travelling at 40 km/h is to be sailed from a point A to a point B where B lies 25 km on a bearing of 070° from A. There is a steady current of 10 km/h running due S. Find the course to be taken and the time required for the journey.

Relative Velocity

Two aeroplanes are flying horizontally at the same height. Aeroplane A is flying due north at 200 km/h, aeroplane B is flying at 300 km/h on a course 060°. These are the velocities relative to the ground, the fixed reference. To the pilot of aeroplane A however, B will appear to have a different velocity, as he himself is moving. We call this apparent velocity, the velocity of B **relative** to A, which we shall write as **B'A** and we now show a method of finding it.

First we find a meaning for subtraction of vectors. Consider fig 20.31. Then $b = a + c$ by the usual rule.

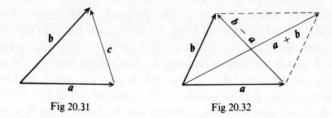

Fig 20.31   Fig 20.32

Then $b - a = (a + c) - a = c$ which gives us the rule for subtracting one vector from another. Note that this result is the second diagonal of the parallelogram formed by a and b (fig 20.32) and note carefully the direction of the vector $b - a$: from the end of vector a to the end of vector b.

To connect this rule with relative velocity, return to the example, p. 314, of the man rowing across a river. The man rows at 4 km/h relative to the water, vector **M'W** and the water flows at 3 km/h relative to the ground, vector **W'G**. We found that
$$M'G = M'W + W'G \text{ (fig 20.33)}.$$

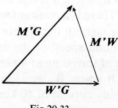

Fig 20.33

So $M'W = M'G - W'G$ i.e.

man relative to water = man rel. to ground − water rel. to ground.
Applying this result to the problem of the two aeroplanes,
B rel. to A = B rel. to G − A rel. to G
i.e., $B'A = B'G - A'G$ (fig 20.34).

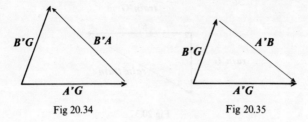

Fig 20.34   Fig 20.35

If the velocity of A relative to B is required, then similarly (fig 20.35).
$$A'B = A'G - B'G.$$

Hence if we draw the two vectors representing the velocities of the two aeroplanes (relative to the ground) (fig 20.36), the third vector $B'A$ can be found by calculation or drawing.

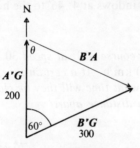

Fig 20.36

By calculation, $(B'A)^2 = 200^2 + 300^2 - 2 \times 200 \times 300 \times \cos 60°$
$$= 70\,000$$
and $\qquad B'A = 265$ km/h.

Also $\qquad \dfrac{\sin \theta}{300} = \dfrac{\sin 60°}{265}$ giving $\theta = 78° \, 36'.$

Hence from the point of view of A, B appears to be flying at 265 km/h on a course 101° 24′ and this is the velocity of B relative to A. Further, the velocity of A relative to B, $A'B$, is 265 km/h on a course 101° 24′ + 180° = 281° 24′, i.e. the vector $B'A$ reversed.

Example 12

Rain is falling vertically at 5 km/h. A man is travelling in a train whose velocity is 60 km/h. In what direction do the raindrops cross the windows of the train?

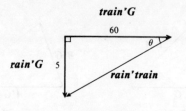

Fig 20.37

We require the direction of the vector **rain' train**. We know
train' ground = 60 km/h horizontally and
rain' ground = 5 km/h vertically (fig 20.37).
Then from the figure, tan $\theta = \frac{5}{60}$ and $\theta = 4°\ 46'$.
The rain crosses the windows at 4° 46' to the horizontal.

Example 13

*Ship A is sailing on a course 010° at speed 30 km/h. Ship B is sailing
on a course 340° at 20 km/h. At a certain time, B is 40 km from A on
a bearing 060°. After what time will they be closest together and what
will be their minimum distance apart?*

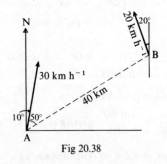

Fig 20.38

Fig 20.38 shows the positions of the two ships at the start, in a space-
diagram.
Now find the velocity of B relative to A, **B'A**. Fig 20.39 shows the
velocity diagram. The 40 km distance between A and B is irrelevant
here.
Then $(\textbf{B'A})^2 = 20^2 + 30^2 - 2 \times 30 \times 20 \times \cos 30° = 261$

giving   **B'A** = 16·12

and $\dfrac{\sin\theta}{20} = \dfrac{\sin 30°}{16·12}$

giving $\theta = 38°\ 21'.$

Fig 20.39

Hence **B'A** is 16·12 km/h in a direction $180° + 10° + 38°\ 21'$
$$= 228°\ 21'.$$
The position relative to A is as shown in fig 20.40 where K is the point on the relative track of B nearest to A (AK being perpendicular to this track). AB = 40 km and hence AK = 40 sin 38° 21' = 24·8 km. This is the nearest distance B ever approaches A.

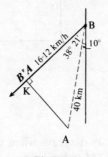

Fig 20.40

The time taken to reach the point K will be the time taken by B to cover the distance BK at the (relative) speed of 16·12 km/h.

$$BK = 40\cos 38°\ 21'$$

and therefore the time taken is $\dfrac{40\cos 38°\ 21'}{16·12}$

$$= 1·946\ h$$

$$= 1\ h\ 57\ min.$$

Example 14

When a man cycles due north at 15 km/h *the wind appears to blow from the direction* 030°. *When he cycles due east at* 10 km/h *the same wind appears now to blow from direction* 060°. *Find the true direction of the wind.*

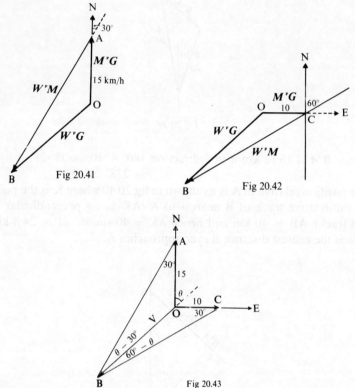

Fig 20.41

Fig 20.42

Fig 20.43

Fig 20.41 shows the velocity diagram when the man is cycling north. **OA** = **M'G** and **AB** is **W'M** (whose length is unknown) making an angle of 30° with **OA**. The position of B is not known but **OB** = **W'G**. Fig 20.42 shows the velocity diagram when the man is cycling east. **OB** is the same vector as in fig 20.41. Hence if we put the two diagrams together to the same scale, B will be located at the intersection of **AB** and **CB** (fig 20.43). The angle $\theta°$ can then be measured.

Calculation here is rather difficult and an acceptable solution can be found by scale drawing. The calculation is outlined here.

Using the sine rule in triangles OAB and OBC (fig 20.43) we obtain:

$$\frac{OB}{\sin 30°} = \frac{15}{\sin (\theta - 30)°}$$

and
$$\frac{OB}{\sin 30°} = \frac{10}{\sin (60 - \theta)°}$$

Therefore $15 \sin (60 - \theta)° = 10 \sin (\theta - 30)°$.

Both sides are now expanded and the equation solved for tan θ. We obtain

$$\tan \theta = \frac{3\sqrt{3} + 2}{2\sqrt{3} + 3} = \frac{7·196}{6·464} \text{ giving } \theta = 48° 4'.$$

Hence the wind is blowing from the direction 048°.

The next problem shows an alternative method of dealing with relative velocities which is very useful, especially in problems where one object has to intercept another.

Example 15

A pirate ship is sailing due north at 10 km/h. *25 km away on a bearing* 130° *a police launch, which can travel at* 20 km/h, *is ready to intercept the pirate ship. What course should it take and how long will it take to catch up with the pirate ship (assuming the pirate ship does not alter its velocity).*

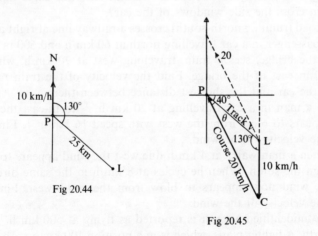

Fig 20.44

Fig 20.45

Fig 20.44 is the space diagram (P = pirate ship, L = launch).
If the pirate ship were at rest, the launch would travel along the line **LP** which is the shortest distance to the pirate. The relative velocity of the launch to the pirate will not be altered by giving each a second velocity 10 km/h due south. P will then be stationary and L must take a course parallel to **CP** (starting from position L) (fig 20.45).

By calculation, $\theta = 22° \ 31'$ and the track speed v (LP) = 12·05 km/h. The student should perform the calculation or complete the problem by drawing to check these results.

The time taken to reach the pirate ship is then $\dfrac{25}{12·05} = 2·074$ h = 2 h 4 min. In many such problems this method of reducing the object to be overtaken to rest provides a straightforward solution.

Exercise 20.6

1 Aeroplane A is flying due N at 150 km/h. Aeroplane B is flying due E at 200 km/h. Find $B'A$.

2 Two cars are travelling on roads which cross at right angles, one over the other. The speeds of the cars are 60 km/h and 40 km/h. Find the velocity of the second car relative to the first.

3 A passenger is on the deck of a ship sailing due east at 25 km/h. The wind is blowing from the north-east at 10 km/h. What is the velocity of the wind relative to the passenger?

4 A car is travelling at 50 km/h and rain is actually falling at an angle of 10° to the vertical at 4 km/h. At what angle does the rain cross the side windows of the car?

5 A road (running north-south) crosses a railway line at right angles. A passenger in a car travelling north at 60 km/h and 600 m south of the bridge, sees a train, travelling west at 90 km/h, which is 800 m east of the bridge. Find the velocity of the train relative to the car and the shortest distance between them.

6 To a man in a car travelling at 20 km h^{-1} north-east the wind appears to blow from the west with speed 16 km h^{-1}. Find the true velocity of the wind.

7 When a man walks at 4 km/h due west the wind appears to blow from the south. When he cycles at 8 km/h in the same direction the wind now appears to blow from the south-west. Find the true velocity of the wind.

8 An unidentified aircraft is reported as flying at 500 km h^{-1} due north. A fighter plane, which is in a position 100 km on a bearing of 225° from the unknown plane, is ordered to contact it. If the fighter can fly at a speed of 600 km h^{-1} what course should the pilot take? After what time will the fighter be in contact? (It is assumed that the plane does not alter velocity).

9 A man on a ship steaming due south at 12 km/h sees a balloon apparently travelling due west at 15 km/h. Find the true velocity of the balloon.

10 A ferry boat travelling at a constant speed, crosses a river 1·5 km wide in 10 min, going straight across at right angles to the banks. If the river flows at 2 km/h, find the course taken by the ferryboat and its speed in still water.

11 An aeroplane is flying on a course 060° at a speed of 300 km h⁻¹. 100 km away due east, a helicopter is flying at 120 km h⁻¹ on a course 330°. How near do they get to each other and after what time?

12 Two ships are sailing, one due west at 12 km/h and the other due south at 18 km/h. The second ship is 15 km north-west of the first. Find how close the two ships approach each other and after what time.

13 A ship A is heading due north at 20 km/h and at 12·00 h is 50 km south-east of a ship B. If B steers at 25 km/h so as to *just* intercept A, find (**a**) the direction in which B must travel and (**b**) the time when the interception takes place.   (L)

14 A helicopter flies from a point A to a point B due north of A on a course of 350°. From B the helicopter flies due east to C on a course of 080°. Assuming that the airspeed of the helicopter is constant and that the wind is steady, find the direction from which the wind is blowing.   (C)

21 Force, Acceleration: Newton's Laws

In the previous chapter, acceleration was studied as a feature of motion in kinematics. This Chapter will study the connection between force and acceleration, which is at the heart of **dynamics**, the study of the causes and effects of motion. The three fundamental laws were finally stated by Newton (17th Century) though Galileo had previously gone some way towards evolving them.

Forces

We all have some idea of a **force** and are familiar with many varieties of force. We push some object with a force or pull on a rope which transmits our pull. These forces act directly on a body or in contact with it. We have already mentioned gravitational force which attracts bodies towards the earth. Magnets exert a magnetic force of attraction. These forces act at a distance without any apparent contact. It is also fairly obvious that a force is a vector quantity. The direction in which I push a box is just as important as the strength of my push. Hence we can represent a force by a straight line (fig 21.1). The length of the line represents the magnitude of the force F and the direction of

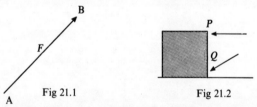

Fig 21.1   Fig 21.2

AB is the direction of the force. A third important feature is the point or line of application. It will make some difference if I push the box in fig 21.2 at P or Q in the directions shown. A force is an example of a *tied* or bound vector where the position of the vector is important.

Unit of Force
The unit of force is 1 **Newton** (1 **N**) which we shall discuss later.

Types of Force

Certain types of force have particular names and we introduce them here.

Weight

If I hold a brick, I feel it exerting a downward force on my hand. If I let go, the brick will fall to the ground. The force I felt on my hand is now free to pull the brick towards the ground. This force, the gravitational attraction of the earth on the brick is called the **weight** of the brick. It always acts vertically downwards (towards the centre of the earth). The weight of a brick is about 20 N (fig 21.3). This will give some idea of the size of the unit. Note that weight is the one inescapable force on earth. All bodies have weight.

Fig 21.3

The weight of a body acts through a point called the **centre of gravity (CG)**. We shall discuss the position of the CG of various bodies in Chapter 24. For symmetrical bodies the CG is at the centre of symmetry. For example, the CG of a cube is at its centre, the CG of a long *uniform* rod (same cross section throughout) is at the geometrical centre of the rod. The CG of a thin rectangular plate (called a **lamina**) is at the intersection of the diagonals and similarly for a parallelogram.

Reaction

If I place the brick, weight 20 N, on a horizontal table, it does not move. We say that the brick is in *equilibrium.* Now it would be unrealistic to assume that the gravitational attraction has ceased, so there must be an opposing force exactly equal to 20 N, acting vertically upwards and also through the CG (fig 21.4(i)).

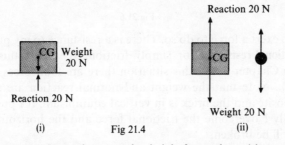

(i)   Fig 21.4   (ii)

This is the **normal reaction** on the brick from the table, normal in the geometrical sense of being at right angles to the common surface. Sometimes this force is called the normal contact force. In diagrams we show only the forces acting *on* a body, so fig 21.4(ii) shows the forces on the brick. Note also that the dimensions of the body are not relevant in this context. The brick is treated as a **particle**.

Tension

If I hold the brick suspended by a string or spring and the string does not break, the brick is again in equilibrium (fig 21.5).

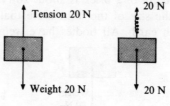

Fig 21.5

In this case the force which equalises the weight passes along the string or spring and is called the **tension** in the string. The weight will also stretch the spring until the equilibrium position is reached. (We regard the string as being inextensible, i.e. any stretch is too small to be noted).

Friction

I place the brick on a horizontal table and try to push it along (fig 21.6).

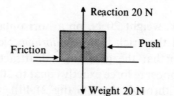

Fig 21.6

I have to exert a force to do so. There is a resistance to my push, called the **frictional resistance** or simply friction. We study this in more detail in Chapter 22. In this situation there are four forces acting on the brick. Note that the weight and normal reaction are still equal and opposite and the brick is in vertical equilibrium. If I push harder, eventually I overcome the frictional force and the horizontal equilibrium will be broken.

Thrust

If the brick is placed on a large strong spring, the spring will resist the weight with a force called a **thrust** (fig 21.7). This is a similar force

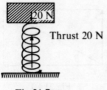

Fig 21.7

to a tension, but acts in the opposite direction. Each of the above are examples of forces, but they need not *all* operate together in any given problem.

Resolving a Vector: Components

As we saw, two vectors are combined using the parallelogram law (fig 21.8) $AB + AD = AC$. As a force is also a vector this law can be applied to two forces and we shall use it to produce a single resultant force.

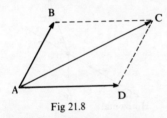

Fig 21.8

The reverse process is very important. A single vector can be **resolved** into two vectors called **components**. The given single vector will be the diagonal of a parallelogram and the components the adjacent sides. If AC is the given vector (fig 21.8) then a parallelogram ADCB can be constructed and AD and AB are its components.

This parallelogram can be drawn in an infinite number of ways but the most useful case is when the components are perpendicular. The parallelogram is then a rectangle (fig 21.9). We are then free to choose the angle θ. So if $\mathbf{F}$ is the magnitude of the vector AC its perpendicular components are $\mathbf{F} \cos \theta$ (along AD) and $\mathbf{F} \sin \theta$ (along AB).

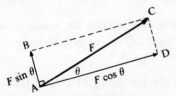

Fig 21.9

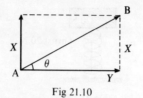

Fig 21.10

Again, if X and Y are the magnitudes of the perpendicular components of a vector AB (fig 21.10) then the magnitude of AB is given by

$$(AB)^2 = X^2 + Y^2 \text{ and } \tan \theta = \frac{X}{Y}$$

which gives the direction of AB (the direction of Y being known).

Example 1

Resolve the forces shown in fig 21.11, where W acts vertically downwards, (a) *horizontally and vertically,* (b) *along and normal to the plane.*

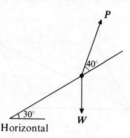

Fig 21.11

(a) Resolving vertically (fig 21.12a, b) the component of W is W itself

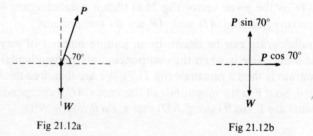

Fig 21.12a   Fig 21.12b

and the component of P is $P \sin 70°$ upwards. Resolving horizontally, the component of W is zero and the component of P is $P \cos 70°$ (to the right).

(b) Resolving along the plane (fig 21.13a, b) the component of P is $P \cos 40°$. The component of W is $W \sin 30°$ down the plane.

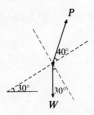

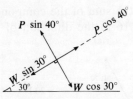

Fig 21.13 *a*   Fig 21.13b

Resolving normal to the plane, the component of *W* is *W* cos 30°
downwards and the component of *P* is *P* sin 40° upwards.

Exercise 21.1

1 Resolve the forces shown in each diagram (fig 21.14) in the directions
of the dotted lines.

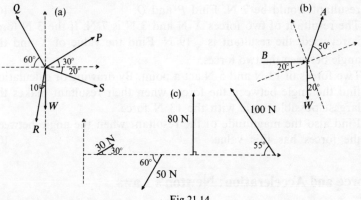

Fig 21.14

2 Find (by drawing or calculation) the resultant of the following forces,
in magnitude and direction (fig 21.15).

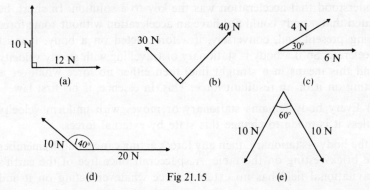

Fig 21.15

3 Find the sum of the components of the forces acting as shown in fig 21.16 in the direction of the dotted lines.

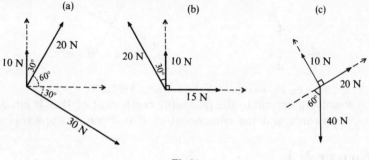

Fig 21.16

4 Two forces P N and Q N include an angle of 120° and their resultant is $\sqrt{19}$ N. If the included angle between the forces were 60°, their resultant would be 7 N. Find P and Q.   (C)

5 The resultant of two forces X N and 3 N is 7 N. If the 3 N force is reversed, the resultant is $\sqrt{19}$ N. Find the value of X and the angle between the two forces.

6 Two forces of 13 N and 5 N act a point. By drawing or calculation find the angle between the forces when their resultant makes the largest possible angle with the 13 N force.

Find also the magnitude of the resultant when the angle between the forces has this value.   (C)

Force and Acceleration: Newton's Laws

In the early days of science, a vexed question was "what is the cause of motion?" The main difficulty in answering this question was the lack of clear distinction between velocity and acceleration. Newton understood that acceleration was the key to a solution. In short, he stated that a body could not have an acceleration without some force being present and, conversely, if a force acted on a body, it must accelerate. So if a body is stationary or travelling with steady velocity (and this means in a straight line) then either no force whatever is acting on it or no resultant force. This, in essence, is his first law.

(1) Every body remains stationary or moves with uniform velocity unless it is made to change this state by external forces.

If the body is stationary, then any forces acting cancel out. Remember the brick resting on the table. A spacecraft once free of the earth's gravitational field has no external force whatever acting on it and

will continue to travel in a straight line with steady speed until affected
by the pull of another planet. It is not possible to attain this state on
earth, as friction or air resistance is always acting and gradually slows
down any moving body.

It follows that if a body is not at rest or moving with uniform velocity,
it has an acceleration (or retardation) and some force is acting on it.
So acceleration is linked with any external force or forces acting.
Newton's second law now leads to the precise relation between force
and acceleration.

(2) If a force acts on a body and produces a certain acceleration, then
the force is proportional to the product of the *mass* of the body (assumed
constant in our work) and the acceleration. Also the acceleration
takes place in the direction of the force.

An important concept here is the mass of a body. If I push a certain
truck with a certain force, an acceleration will be produced. If I push
two similar trucks joined together with the same force, the acceleration
will be halved. I am pushing an object of twice the mass. It is difficult
to define mass: we might say it is the amount of 'matter' in the body
but the idea is reasonably easy to grasp. The standard unit of mass
is 1 kg (kilogram) = 1000 g (gram).

Now the second law states that $P \propto ma$ (P = force, m = mass and
$$a = \text{acceleration})$$
or $P = kma$ where k is some constant.

For a standard mass of 1 kg having standard unit acceleration 1 m s^{-2}
the force acting will be $P = k \times 1 \times 1 = k$ units.

If we now take the unit of force to be that which will produce an
acceleration of 1 m s^{-2} in a mass of 1 kg, then $P = 1$ and $k = 1$. This
unit of force is 1 **Newton (1 N)**. The equation of motion then takes
the direct form

$$P = m\ a$$

or **Force = mass × acceleration**
(in N)   (in kg)   (in m s^{-2})

This unit of force (1 N) is an absolute unit and is invariable. For large
forces the kN (1 kilonewton = 1000 N) can be used.

(Other units of force have been used in the past, called gravitational
units, which depend on the value of g (see below) such as kgf (kilo-
gram-force) or gf (gram-force). 1 kgf is the force equal to the weight
of a mass of 1 kg. 1 gf is the force equal to the weight of a mass of
1 g. The British system had a similar unit called lbf (pound force).
These units have all been superseded by the Newton).

Mass and Weight

It is known by experiment that all bodies are attracted to the earth by a gravitational force. Near the surface of the earth this force produces an acceleration symbolised as g m s^{-2}. g is about 9·8, but varies slightly over the surface of the earth.

Hence a mass of m kg if free to move would fall with an acceleration g m s^{-2}. The force acting on the body is thus mg N and this is the weight of the body. This weight is not constant but varies with locality. There is a slight variation over the earth's surface, so the weight of a mass of 1 kg is about 9·8 N. In space, where $g = 0$ or nearly so, its weight would be zero (weightlessness). On the moon, where $g = 1·6$, its weight would be about 1·6 N though its mass is still 1 kg.

To compare the masses of two bodies we 'weigh' them, which compares the gravitational forces acting on the bodies. If the masses are m and x kg then (fig 21.17) $F_1 = mg$ N and $F_2 = xg$ N. So $\dfrac{F_1}{F_2} = \dfrac{m}{x}$.

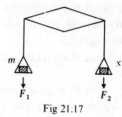

Fig 21.17

Hence if the two masses are suspended at equal distances on a balance which rests horizontally, $F_1 = F_2$ and thus $m = x$. In this way we can copy the standard kilogram, which is a block of platinum kept in France, and subdivisions of the unit of mass can be made. In everyday life the distinction between mass and weight is blurred and a packet of tea is usually labelled 'net weight: 0·25 kg' which really should read 'net mass'. The labelling really means that the weight of the tea is that of a mass of 0·25 kg. In our work the distinction will be carefully noted between mass and weight and the correct units used for each. The relation is that the weight (in N) of a mass m (in kg) is weight = mg. (Hence 1 kgf = g N)

The remainder of the chapter gives worked solutions for various problems using the equation $F = ma$ and also introduces Newton's Third Law.

In diagrams forces are represented by $\rightarrow$ and accelerations by $\rightarrowtail$ and velocities by $\rightarrow$.

Example 2

What force will produce a velocity of $1 \cdot 2$ m s^{-1} *in 6 s on a body of mass* $0 \cdot 5$ kg?

To reach a velocity of $1 \cdot 2$ m s^{-1} in 6 s a uniform acceleration of $0 \cdot 2$ m s^{-2} is required. Hence $F = ma = 0 \cdot 5 \times 0 \cdot 2 = 0 \cdot 1$ N.

Example 3

A horizontal force of $0 \cdot 6$ N *acts on a body of mass* $0 \cdot 3$ kg. *There is a resistance of* $0 \cdot 15$ N *opposing the first force. What acceleration will be produced?* (Fig 21.18)

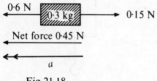

Fig 21.18

The resultant or net force acting is $0 \cdot 45$ N. The acceleration will take place in the direction of this resultant.

$$F = ma \text{ gives } 0 \cdot 45 = 0 \cdot 3a$$

i.e. $a = 1 \cdot 5$ m s^{-2}.

Example 4

Two forces F_1 *and* F_2 *act on a body of mass* $2 \cdot 5$ kg. F_1 *has magnitude* 5 N *in direction* N 30° E *and* F_2 *has magnitude* 8 N *in direction* E. *Find the acceleration of the body.*

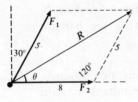

Fig 21.19

First find the resultant of the two forces by drawing or calculation (fig 21.19).

The results are: $R = 11 \cdot 36$ N and $\theta = 22°\ 24'$.

Then by the equation of motion, $11 \cdot 36 = 2 \cdot 5 \times a$

and $a = 4 \cdot 54$ m s^{-2} in the direction of R.

Example 5

A parcel of mass 4·5 kg *is suspended from a spring balance in a lift. What does the balance read if the lift is* (a) *moving with uniform speed,* (b) *accelerating upwards at* 0·5 m s^{-2}, (c) *accelerating downwards at* 0·5 m s^{-2}? ($g = 9·8$ m s^{-2})

Fig 21.20

A spring balance consists of a strong spring with a pointer and scale attached. Using a result in physics that the extension of a spring is proportional to the tension in the spring, the scale can be calibrated to show the force extending the spring.

We consider the forces acting on the parcel, i.e. its weight 4·5g N acting vertically downwards and the tension in the spring vertically upwards.

(i) Since the parcel is moving with uniform speed, there is no net force acting on it. Therefore $T = 4·5g$ N $= 44$ N which is what the balance should read.

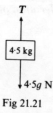

Fig 21.21

(ii) The parcel is accelerated upwards. There must be a resultant upward force, i.e. $T > 4·5g$.

Then $T - 4·5g = 4·5 \times 0·5$
or $T = 4·5 \times 0·5 + 4·5 \times 9·8$
   $= 4·5 \times 10·3$
   $= 46$ N

which will be the new reading on the balance.

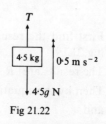

Fig 21.22

(iii) Similarly, $4\cdot5g - T = 4\cdot5 \times 0\cdot5$
or $T = 4\cdot5 \times 9\cdot8 - 4\cdot5 \times 0\cdot5$
    $= 42$ N.

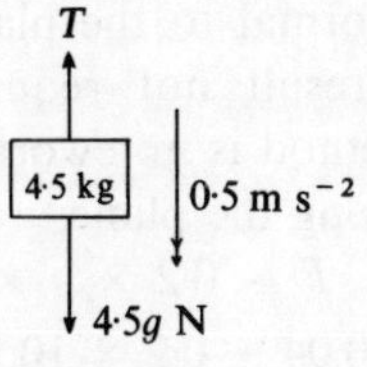

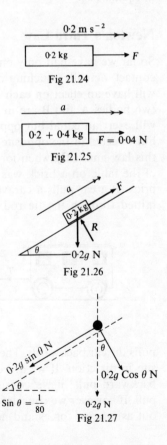

Fig 21.23

Example 6
On a level table, a toy car of mass $0\cdot2$ kg can reach a speed of $0\cdot6$ m s^{-1} in 3 s from rest. What is the effective pull of the engine? If the car is loaded with another $0\cdot4$ kg how long would it take to reach the same speed?
The car is now placed by itself on an inclined surface, which has a slope of θ where $\sin\theta = \frac{1}{80}$. How long will the car now take to reach the same speed from rest? (Take $g = 10$ m s^{-2}).

To reach $0\cdot6$ m s^{-1} in 3 s requires a uniform acceleration of $0\cdot2$ m s^{-2}.
(i) Hence (fig 21.24) $F = 0\cdot2 \times 0\cdot2$
    $= 0\cdot04$ N
which is the effective pull of the engine after overcoming any resistances.

Fig 21.24

(ii) The mass is now $0\cdot6$ kg (fig 21.25) but F is as before.
Therefore $F = 0\cdot04 = 0\cdot6 \times a$ giving $a = \frac{1}{15}$ m s^{-2}.
To reach a speed of $0\cdot6$ m s^{-1} with this acceleration requires t s where $0\cdot6 = \frac{1}{15}t$ or $t = 9$ s (using $v = u + at$).

Fig 21.25

In each of these cases the weight being at right angles to the motion plays no part in the solution.

Fig 21.26

(iii) In this case (fig 21.26) the forces on the body are $F = 0\cdot04$ N, R the normal reaction and the weight ($0\cdot2g$ N). We resolve the weight along and and normal to the plane (fig 21.27) and the forces now acting in these two directions, with the acceleration, are shown in fig 21.28.

$\sin\theta = \frac{1}{80}$

Fig 21.27

(Normal to the plane $R = 0\cdot2g \cos \theta$,
a result not required here, but the
method is noteworthy).

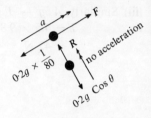

Along the plane,

$$F - 0\cdot2 \times g \times \tfrac{1}{80} = 0\cdot2 \times a.$$

$$\therefore\ 0\cdot04 - 0\cdot2 \times 10 \times \tfrac{1}{80} = 0\cdot2 \times a$$

which reduces to $a = 0\cdot075 \text{ m s}^{-1}$.

Fig 21.28

To reach a speed of $0\cdot6$ m s^{-1} from rest therefore will require a time
t s given by $0\cdot6 = 0\cdot075t$ i.e. $t = 8$ s.

*The method of resolving the weight in this solution should be carefully
noted. As a general rule, all forces should be resolved into components
parallel and normal to any acceleration. The equation of motion is then
applied in the direction of the acceleration, the other components being
in equilibrium.*

Newton's Third Law

So far we have had only one body in problems. If two bodies are in
contact, actually touching or connected by a string or tie rod, they
will have an effect on each other. Newton's Third Law states that if
two bodies A and B are in contact, A will exert a force on B and B
will exert an equal but opposite force on A, i.e. equal in magnitude
but directed in the opposite sense along the same line. We have used
this law implicitly when for example we concluded that the reaction
of the table on a brick was equal and opposite to the weight of the
brick. If a car pulls a caravan (fig 21.29) the pull of the car is trans-
mitted through the tie rod to the caravan but the caravan equally

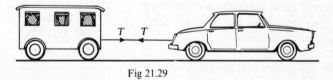

Fig 21.29

pulls the car backwards. The two pulls are the same size (T) but opposite in direction. If we are considering the car, we must include the
backward pull; if we consider the caravan we include the forward
pull. If however we consider the two as *one* body, the two pulls cancel
out as *internal* forces and need not be considered.

Again, if two masses are suspended by a
string over a frictionless (smooth) pulley,
the string transmits a tension which
pulls A (fig 21.30) upwards when considering A but pulls B upwards when
considering B.

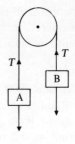

Fig 21.30

Example 7

*A string (assumed to have no weight and not to stretch) is placed over
a smooth pulley. To the ends of the string are attached masses of 2 kg
(A) and 1 kg (B) and both parts of the string are vertical. With what
acceleration does the system move   What is the reaction at the axle
of the pulley? ($g = 10\,\mathrm{m\,s}^{-2}$)*

The system is shown in fig 21.31 with
the weights of the masses. Let the acceleration of the 2 kg mass be a m s^{-2}
downwards and hence the 1 kg mass will
have the same acceleration upwards. Now
consider each mass and the pulley separately (fig 21.32). The string transmits
a tension T and the reaction at the axle
of the pulley is R.

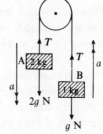

Fig 21.31

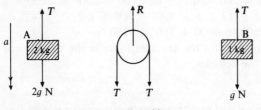

Fig 21.32

For the mass A, since the acceleration is downwards: $2g - T = 2a$ (i)
For the pulley, since it has no acceleration vertically: $R = 2T$ (ii)
For the mass B, acceleration upwards: $T - g = a$ (iii)
Now solve these equations for a, T and R.
From (i) and (iii) by addition, $g = 3a$ or $a = \frac{10}{3}$ m s^{-2}.
From (iii) $T = a + g = \frac{10}{3} + 10 = 13 \cdot 3$ N.
Finally from (ii) $R = 26 \cdot 6$ N. (Note that this is less than the total
weight of the masses, 30 N).

This problem illustrates the important point that each mass must be considered in isolation with all the forces acting on that mass taken into account.

Example 8

A car of mass 800 kg *is towing a caravan of mass* 500 kg *at a steady speed. Road resistances etc. are estimated at 5 N per kg of mass for each vehicle. What is the tractive force of the car? If the car and caravan accelerate at* 0·2 m s^{-2} *what is the new tractive force?*

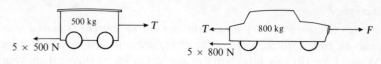

Fig 21.33

Fig 21.33 shows the forces acting on each body.

By Newton's third law if the pull of the car on the caravan is *T*, the pull of the caravan on the car is also *T* (reversed in direction). The resistance to the motion of the caravan is 5 × 500 N and to that of the car 5 × 800 N.

For the caravan, *since the speed is steady*: $T = 5 \times 500 = 2500$ N.

For the car, similarly: $F = T + 5 \times 800 = 6500$ N.

The weights and road reactions, being perpendicular to the line of acceleration, do not enter into the solution.

When the vehicles accelerate, we have for the caravan:

$T - 5 \times 500 = 500 \times 0.2$ or $T = 100 + 2500 = 2600$ N.

For the car: $F - (T + 5 \times 800) = 800 \times 0.2$

giving $F = 2600 + 4000 + 160 = 6760$ N.

The 260 N increase in the tractive force is the extra force necessary to accelerate the vehicles.

Example 9

Fig 21.34 *shows a wedge with a smooth surface and two masses connected by a light string over a smooth pulley. The strings run parallel to the wedge surface. Find the acceleration of either mass and the tension in the string.* $(g = 9.8$ m s$^{-2})$

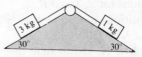

Fig 21.34

Clearly the 3 kg mass will slide downwards. Consider each mass separately, marking the forces acting on it (fig 21.35).

 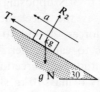

Fig 21.35

Now resolve in the direction of the acceleration of each mass and at right angles, i.e. along and normal to the surface (fig 21.36).

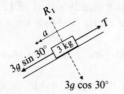 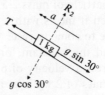

Fig 21.36

For the 3 kg mass:

$3g \sin 30° - T = 3a$ (i)

(Also $R_1 = 3g \cos 30°$)

For the 1 kg mass:

$T - g \sin 30° = a$ (ii)

(Also $R_2 = g \cos 30°$)

Add equations (i) and (ii): $2g \sin 30° = 4a$ which gives $a = 2\cdot5$ m s^{-2}.

From (ii) $T = a + g \sin 30°$

$= 2\cdot5 + 4\cdot9$

$= 7\cdot4$ N.

Exercise 21.2

Take $g = 10$ m s^{-2} unless otherwise instructed.

1 If a force of 20 N acts on a mass of 2 kg what is the acceleration produced?

2 If a force of 5 N acts on mass of 750 g what acceleration results?

3 If a force of 4 kN acts on a mass of $2\cdot2$ t (1 tonne $= 10^3$ kg) what is the acceleration in m s^{-2}?

4 If $g = 1\cdot6$ m s^{-2} on the moon, what is the weight of a packet of tea labelled: net mass 250 g?

5 A mass of $1\cdot5$ kg has an acceleration of $0\cdot8$ m s^{-2}. What force is acting on it?

6 What is the ratio of the weight of an astronaut on the moon to his weight on earth? ($g_{earth} = 9\cdot8$, $g_{moon} = 1\cdot6$ m s^{-2}).

7 If a force of 2 N acts on a mass of 1·5 kg at rest initially, what is its velocity after 6 s?

8 A mass of 5 kg is dragged across a rough surface (frictional force equal to 3 N opposing the motion) by a horizontal force of 20 N. What is the acceleration produced?

9 Two forces, 20 and 10 N, act on a body of mass 0·5 kg at right angles to each other. What is the acceleration of the mass, in magnitude and direction?

10 A body, of mass 10 kg, slides down a smooth slope whose inclination to the horizontal is 30°. What is its acceleration? ($g = 9\cdot8$)

11 A force of 20 N is applied at an angle of 60° to the horizontal to a mass of 4 kg on a smooth horizontal table. What is the acceleration of the mass?

12 A particle of mass 2·5 kg is moving at a steady speed of 12 m s^{-1} when it meets with a fixed resistance of 10 N. How long does it take to come to rest?

13 Find the constant force which would give a body of mass 5 kg at rest a velocity of 6·4 m s^{-1} in 4 s.

14 A mass of 2 kg is at rest on a rough horizontal table. A force of 20 N is applied to the mass, the force making an angle of 30° with the table. Frictional resistance is equal to 5 N. What is the acceleration of the mass?

15 A toy engine of mass 350 g exerts a driving force of 0·1 N. With what acceleration could it climb a smooth slope of $\frac{1}{98}$? (i.e. a slope making angle θ with the horizontal where $\sin \theta = \frac{1}{98}$) ($g = 9\cdot8$).

16 Masses of 5 kg and 3 kg are connected by a light string over a smooth pulley. Find (i) the acceleration of the masses (ii) the tension in the string and (iii) the reaction at the axle of the pulley.

17 A man of mass 80 kg stands in a lift. What is the reaction from the floor of the lift if the lift (i) moves upwards with steady speed, (ii) moves upwards with acceleration 0·5 m s^{-2}, (iii) moves downwards with acceleration 0·4 m s^{-2}? ($g = 9\cdot8$)

18 A mass of 3 kg rests on a smooth horizontal table connected by a light string passing over a smooth pulley at the edge of the table to another mass of 2 kg hanging vertically. When the system is released from rest, with what acceleration do the masses move and what is the tension in the string?

19 A truck of mass 4·5 kg is being pulled along a smooth horizontal floor by two forces, one of 10 N at an angle of 60° to the floor and the other of 20 N at an angle of 20° to the floor. Find the acceleration of the truck.

20 A parcel weighing 45 N is carried by a man in a lift moving upwards with acceleration $0\cdot4$ m s^{-2}. What is the apparent weight of the parcel?

21 A lorry, mass 1000 kg, pulling a trailer of mass 450 kg, accelerates on the level at $0\cdot6$ m s^{-2}. Resistances to motion on either vehicle are 4 N per kg of mass. What is the tension in the tow bar? What is the tractive force of the engine?

22 A trolley of mass 25 kg rolls down a hillside of slope θ to the horizontal, where $\sin\theta = \frac{1}{70}$, at a steady speed. Find the resistance to motion. $(g = 9\cdot8)$

23 A car of mass 750 kg is accelerating up a slope of θ to the horizontal where $\sin\theta = \frac{1}{70}$ at $1\cdot5$ m s^{-2}. Ignoring any road resistances, find the tractive force of the engine. $(g = 9\cdot8)$

24 A barge is towed by two tugs at a steady speed. One tow rope makes an angle of $20°$ to the line of motion and its tension is 500 N; the other tow rope makes an angle of $17°$ and its tension is 600 N. Find the resistance to the motion of the barge.

25 Fig 21.37 shows three masses connected by two pieces of light inextensible string as shown. The horizontal surface is smooth and the pulleys are free from friction. Find the acceleration with which the masses move and the tensions in the strings. $(g = 9\cdot8)$

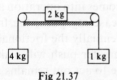

Fig 21.37

26 A body of mass 10 kg lies on a smooth inclined plane. A light string attached to this body passes over a smooth pulley at the top of the plane and supports a mass of 2 kg hanging freely. If the inclination of the plane is θ to the horizontal where $\sin\theta = \frac{1}{14}$, find the acceleration of the masses. $(g = 9\cdot8)$

27 Fig 21.38 shows two masses connected by a light inextensible string on a smooth incline where $\sin\theta = \frac{1}{40}$. Find the acceleration of the masses when the system is released from rest. $(g = 9\cdot8)$

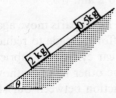

Fig 21.38

22 Friction

Though we have specified 'smooth' surfaces in problems, it is impossible to obtain such a surface in practice. A 'smooth' surface is a convenient approximation in applied mathematics. When one body moves over another, there is always some resistance to motion, the frictional resistance between the two surfaces. Clearly the nature of the two surfaces in contact is important in determining the size of this resistance. Experiments can be made to study this frictional resistance and we summarise the experimental results.

(1) The frictional force only comes into operation when an attempt is made to move one body relative to another in contact with it and this force always acts to oppose the relative motion.

If I push the block on the table with a small force P (fig 22.1),

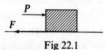

Fig 22.1

the frictional force F comes into operation and opposes the possible movement. As I increase my push, the frictional force increases to equal my push. Eventually the frictional force reaches its maximum and a slightly greater push will make the block move. The friction does not cease to act but remains at this maximum value. (This is not strictly true but we shall ignore any variation).

(2) The frictional force F has a value given by $F \leqslant \mu R$ where R is the normal reaction between the surfaces and μ (Greek letter, read mu) is the **coefficient of friction**. The maximum value of F (μR) is reached only on the point of sliding and is called the **limiting friction**. Until sliding takes place F is less than μR and its exact value is indeterminate. The value of μ depends only on the nature of the surfaces in contact and is independent of the areas in contact or the forces present. For two wooden surfaces for example μ has a value between 0·2 and 0·5.

In many machines, where metal parts move against each other, friction is a nuisance and every effort is made to reduce it by lubrication. For example, the piston in a car cylinder is lubricated by the oil pumped up from the sump. On the other hand, without friction we could not move. We rely on the friction between our shoes and the ground to push us forward. On smooth ice or a polished floor we tend to slip as this frictional force is small. Similarly the frictional force between the tyres of a car and the road is essential to motion. In solving problems, first mark and find the normal reaction between the two surfaces. Then provided sliding is taking place or about to take place, put $F = \mu R$. Otherwise $F < \mu R$.

Example 1

A block of mass 10 kg rests on a horizontal floor, coefficient of friction 0·4. (1) What force is required just to make the block move when (a) pulling horizontally, (b) pulling at an angle of 60° to the horizontal? (2) If the block is pulled with a horizontal force of 50 N, with what acceleration does it move? ($g = 9·8 \text{ m s}^{-2}$)

1 (a) The normal reaction $R = 10g$ N (fig 22.2) as the block is in equilibrium vertically. On the point of sliding,
$$F = \mu R = 0·4 \times 10g \approx 39 \text{ N}.$$
and this is the minimum force required to make the block begin to move.

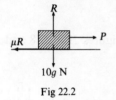
Fig 22.2

(b) In this case the normal reaction R is not equal to $10g$ as we must take into account the vertical component of P (fig 22.3).
Fig 22.4 shows the situation when P is resolved.
Resolving vertically,
$$10g = R + P \sin 60° \qquad \text{(i)}$$
Since the block is in limiting equilibrium,
$$P \cos 60 = F = \mu R \qquad \text{(ii)}$$
From (ii) $P \times 0·5 = 0·4R$ or $R = 1·25P$.
Hence from (i)
$$10g = 1·25P + 0·866P = 2·116P$$
giving $\qquad\qquad\qquad P \approx 46 \text{ N}.$
Note that this is greater than the force required in part **(a)**.

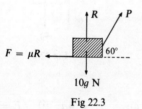
Fig 22.3

2 The pulling force (50 N) is greater than the value of P required to just move the block (fig 22.5), as found in part **(a)** above. Hence there is a net forward force of
$$50 - 39 = 11 \text{ N}$$
and thus $\qquad\qquad 11 = 10a$
giving $\qquad\qquad a = 1·1 \text{ m s}^{-2}.$

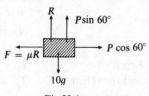
Fig 22.4

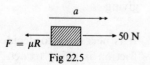

Fig 22.5

Example 2

Three masses A, B, C *are connected by light inextensible strings as shown in fig 22.6, where* B *is held on a rough horizontal plane (coefficient of friction* 0·6*). The pulleys are smooth. When* B *is released what will be the acceleration of the masses?* $(g = 10 \text{ m s}^{-2})$

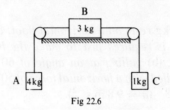

Fig 22.6

Figures 22.7, 22.8 and 22.9 show the forces acting on each mass.

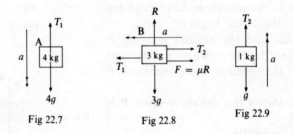

Fig 22.7   Fig 22.8   Fig 22.9

$4g - T_1 = 4a$ (i) $F = \mu R$ (as B moves) $T_2 - g = 1 \times a$ (iii)
and acts against
the motion.
$R = 3g, F = 0.6 \times 3g$
$T_1 - (T_2 + F) = 3a$ (ii)

From (i) $T_1 = 4g - 4a$
and from (iii) $T_2 = g + a$
Substituting in (ii),
     $4g - 4a - (g + a + 0.6 \times 3g) = 3a$
which simplifies to $1.2g = 8a$
giving $a = 1.5 \text{ m s}^{-2}$.

It is necessary to decide in certain problems in which direction the friction is going to act. The next example illustrates this point.

Example 3

A mass of 10 kg *is placed on a plane inclined at an angle of* 30° *to the horizontal. What force parallel to the plane is required* **(a)** *to hold the mass at rest;* **(b)** *to make the mass move steadily up the plane?*
$\mu = 0.5, g = 9.8.$

(a) Fig 22.10 shows the forces acting. As the block is just being held at rest, it is on the verge of slipping *down*. Hence $F(= \mu R)$ acts upwards. Now resolve the weight parallel and normal to the surface (fig 22.11).

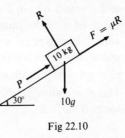

Fig 22.10

Then $R = 10g \cos 30°$ (i)
and $P + \mu R = 10 g \sin 30°$ (ii)
From (i) $R = 5g\sqrt{3}$
and therefore in (ii)
      $P + 0.4 (5g\sqrt{3}) = 10g\ 0.5$
or $P = 5g - 2\sqrt{3}g = 15$ N.

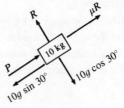

Fig 22.11

(b) In this case (fig 22.12) the mass is moving steadily (i.e. with uniform speed) *up* the plane, $F(= \mu R)$ acts *down* the plane.

Then resolving as before,
      $R = 10g \cos 30°$ (i)
and $P = 10g \sin 30° + \mu R$ (ii)
Then $P = 10g \times 0.5 + 0.4 (5g\sqrt{3})$
       $= 5g + 2\sqrt{3}g = 83$ N

Note the considerable difference in the two forces.

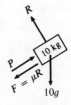

Fig 22.12

Angle of Friction

Consider a mass resting on a rough horizontal table (fig 22.13). A gradually increasing force is applied (horizontally) to the mass. Then the frictional force F also increases. $F < \mu R$ until sliding commences and then $F = \mu R$.

If F and R are combined into one force, their resultant T (fig 22.14) will make an angle λ (Greek letter: read lambda) with the normal reaction R. This resultant is sometimes called the **total reaction**.

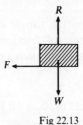

Fig 22.13

From the diagram $\tan \lambda = \dfrac{F}{R}$.

But $F \leqslant \mu R$ or $\dfrac{F}{R} \leqslant \mu$.

Hence $\tan \lambda \leqslant \mu$. So the maximum value
of λ is given by $\tan \lambda = \mu$. This maximum
value of λ is the **angle of friction** and is
sometimes used as an alternative way of
specifying the coefficient of friction. It

Fig 22.14

also gives the direction of the total reaction in limiting equilibrium.
The next example shows an experimental method of finding μ (or λ)
for two surfaces.

Example 4

*A block of mass m kg rests on a rough inclined plane whose angle of tilt
can be altered slowly. The angle of inclination is increased until the
block is just on the point of sliding (rather difficult to determine exactly).
We can deduce a result for μ.*

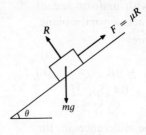

Fig 22.15

Fig 22.15 shows the position just at the point of limiting friction.
Resolving the weight along and normal to the plane we obtain:
$$R = mg \cos \theta$$
and $F = \mu R = mg \sin \theta$
Therefore $\mu mg \cos \theta = mg \sin \theta$
or $\mu = \tan \theta.$
As $\mu = \tan$ (angle of friction), the angle of friction $(\lambda) = \theta$. Hence
the angle of friction between the two surfaces equals the angle of
inclination of such a plane when the mass is on the point of sliding on it.

Example 5

*A parcel of mass 10 kg rests on a lorry. When the lorry is accelerating
at 1·5 m s⁻² the parcel is just on the point of sliding backwards. What*

is the coefficient of friction between the parcel and the lorry?
$(g = 9.8 \text{ m s}^{-2})$

Fig 22.16 shows the forces acting on the
parcel. Vertically the parcel is in equilibrium and therefore $R = 10g$ N.
Horizontally: $F = 10 \times 1.5 = 15 (F = ma)$
Therefore $\mu R = 15$
or $\mu \times 10g = 15$
which gives $\mu = \dfrac{15}{10 \times 9.8} = 0.15.$

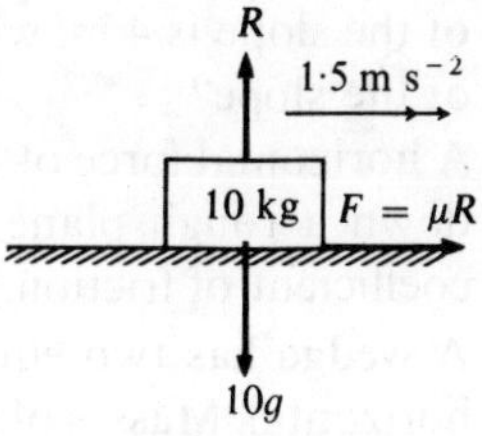

R

1.5 m s^{-2}

10 kg $F = \mu R$

10g

Fig 22.16

Exercise 22

$(g = 9.8 \text{ m s}^{-2})$

1 A mass of 4 kg rests on a rough table $(\mu = 0.4)$. Find the least
   force sufficient to make it move.
2 If a force of 10 N is just sufficient to move a mass of 2 kg resting
   on a rough horizontal table, find the coefficient of friction.
3 A block is placed on an inclined plane at angle 30° to the horizontal
   and just remains at rest. Find the coefficient of friction and the
   angle of friction.
4 A block of mass 10 kg is placed on an inclined plane of angle 30°
   to the horizontal, where $\mu = 0.4$. Find (a) the acceleration of the
   block down the plane, (b) the least force parallel to the plane
   required to keep the block at rest, (c) the force required to make
   the block begin to move up the plane.
5 A mass of 8 kg is placed on a horizontal table $(\mu = 0.3)$ connected
   by a light inextensible string placed over a smooth pulley at the
   edge of the table to another mass of 4 kg hanging freely. Find the
   acceleration of the masses when released from rest.
6 A block of mass 1 kg is placed on an inclined plane of angle 60°
   and is just held there at rest by a horizontal force P. If the coefficient
   of friction is 0.4 find P.
7 A body of mass 5 kg can just rest on a plane when the plane is
   inclined at 60° to the horizontal. Find the force parallel to the
   plane required to push the body up the plane if the inclination
   is reduced from 60° to 30°.   (C)
8 A horizontal force of 2000 N is pulling a block of mass 150 kg
   over a horizontal surface. The angle of friction is 20°. Find the
   acceleration of the block.

9 A box of mass 20 kg starts from rest and slides down a slope inclined to the horizontal at an angle of 30°. If the coefficient of friction is 0·4, find the acceleration of the box. If the length of the slope is 4 m, with what velocity does it arrive at the bottom of the slope?

10 A horizontal force of 10 N just prevents a mass of 2 kg from sliding down a rough plane inclined at 45° to the horizontal. Find the coefficient of friction.

11 A wedge has two equally rough faces each inclined at 30° to the horizontal. Masses of 5 kg and 2 kg, one on each face, are connected by a light string passing over a smooth pulley at the top of the wedge. The coefficient of friction μ, between each mass and the surface of the wedge is 0·2. Find the acceleration of the masses when they are released.

12 The least force required to move a block of mass 10 kg on a rough horizontal plank is 25 N. If the plane is tilted to an angle of 45° to the horizontal, the coefficient of friction being unchanged, find the force, parallel to the plank, required to give the block an acceleration of 1 m s^{-2} up the slope.

13 A particle of mass 0·1 kg is projected along a rough horizontal table with a speed of 0·42 m s^{-1}. If the coefficient of friction between the particle and the table is $\frac{1}{7}$, find the total distance travelled by the particle. (L)

14 Two bodies A and B, joined by a light inextensible string, are placed on a plane which is inclined to the horizontal at an angle whose tangent is 0·75 so that the string is taut and lies along a line of greatest slope and B is higher up the plane than A. The body A is smooth and its mass is 9 kg. The mass of B is 3 kg and the coefficient of sliding friction between B and the plane is 0·5. The system is allowed to slide down the plane. Calculate (**a**) the frictional resistance to the motion of B, (**b**) the acceleration of the system, (**c**) the tension in the string. (L)

15 A wedge has equally rough faces inclined at 30° and 60° respectively to the horizontal. A mass of 5 kg on the 30° face is connected by a light string passing over a smooth pulley at the top of the wedge of a mass of 8 kg on the 60° slope. The masses are on the point of sliding. Calculate the coefficient of friction.

23 Statics: Moments, Forces in Equilibrium

We now look at problems involving forces acting on a body in equilibrium. This branch of applied mathematics is called **Statics**. First we introduce an important idea, the turning effect of a force.

The Moment of a Force

In dynamics we concentrate on the force-acceleration relationship. The shape and dimensions of the body have not been considered and we have treated it as a particle. This is correct provided the body is travelling in a straight line without rotation. In statics however the shape and dimensions of the body and exactly where the forces act are important.

ABCD is a uniform beam of wood of weight W balanced on a smooth pivot at its midpoint B with weights W_1 and W_2 ($W_1 \neq W_2$) attached at A and C (fig 23.1). The beam rests horizontally in equilibrium and its CG (B) is directly above the pivot. R is the normal reaction from the pivot on the beam. The weights W_1 and W_2 are vertical forces acting on the beam.

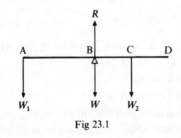

Fig 23.1

By common experience it is known that if the distance AB is greater than BC, then W_1 must be less than W_2. If W_2 was made smaller, the beam would tilt anticlockwise. On the other hand if W_2 was moved to the right (nearer D) the beam would tilt clockwise. Each force exerts a turning effect or **moment** about an axis through the pivot, round which the beam could turn. This turning effect depends on the size of the force and its distance from the axis. The beam balances only if the anticlockwise and clockwise turning effects are equal about the axis. So we define the moment of a force about an axis as:

Moment of force = magnitude of force × *perpendicular* distance of
about axis   line of force from axis.

Anticlockwise moments ⌒ are usually taken as +, clockwise moments
⌒ as −.

The axis is the line about which the body would turn under the action
of the force. In two-dimensional problems the axis will be perpendicular to the plane of the paper, and the axis cuts this plane in a point,
such as B above. Hence, in such problems, we usually speak of taking
moments about a point, the axis being understood to pass through
this point. Fig 23.2 shows forces acting on a body and the various
moments of these forces are given below.

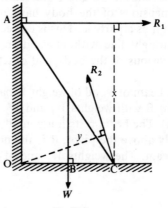

Fig 23.2

Moment of R_1 about C = $-R_1 \times x = -R_1 \times$ AO
Moment of W about C = $+W \times$ BC
Moment of R_2 about C = 0
Moment of R_2 about O = $+R_2 y$

If the force is measured in N and distance in m, then the unit of moment
will be 1 N m, but other non-standard units such as N cm or N mm
could be used. If a consistent system of units is used in a problem the
actual unit of moment is not important.

The Principle of Moments

By simple experiments such as illustrated in fig 23.1 it is found that
the sum of the clockwise moments about any point equals the sum
of the anticlockwise moments about the same point, provided the
body is in equilibrium.

For example, in fig 23.1 if we take moments about B, it is found that $W_1 \times AB = W_2 \times BC$ (R and W have zero moments about this point). If moments are taken about A then $R \times AB = W \times AB + W_2 \times AC$ (as W_1 has zero moment).

Note: The moments of all the forces must be included.

Such experiments lead to the **principle of moments**:

If a body is in equilibrium under the action of forces, then the algebraic sum of the moments of the forces about any point is zero.

Alternatively we can say, *the sum of the clockwise moments = the sum of the anticlockwise moments, about the same point.*

This principle is very important and will be used a great deal. In using it is essential to count the moment of every force acting and to note its sign.

Example 1

A uniform beam of mass 10 kg is pivotted as in fig 23.3 with the weights attached. $AB = BD = 1.5$ m *and* $BC = 1.2$ m. *Can the beam rest in equilibrium horizontally?*

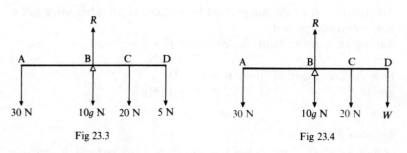

Fig 23.3   Fig 23.4

The weight ($10g$ N) and the normal reaction R act through the CG which is above the pivot at B. Take moments about B.

Sum of anticlockwise moments $= 30 \times 1.5 = 45$ N m.

Sum of clockwise moments $= 20 \times 1.2 + 5 \times 1.5 = 31.5$ N m.

Hence the beam will not rest in equilibrium, but will tilt anticlockwise. Find the new weight to be attached to D to keep the beam balanced on the pivot. Let this weight be W N (fig 23.4).

Taking moments about B,

$$30 \times 1.5 = 20 \times 1.2 + W \times 1.5$$

or $\qquad\qquad 45 = 24 + 1.5 W$

giving $\qquad\qquad W = 14$ N

Note that when this weight is attached and the beam is in equilibrium,

$R = 30 + 10g + 20 + 14 = 164$ N approx ($g \approx 10$) as there is no acceleration vertically. R must be vertical as otherwise it would have a horizontal component along the beam.

Example 2

A non-uniform beam length 2·5 m and weight 80 N is attached to a hinge at a wall (fig 23.5). Its CG is 1 m from the wall. What vertical upward force applied at the other end will keep the beam horizontal?

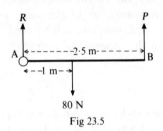

Fig 23.5

The reaction R at the hinge must be vertical as no other force has a horizontal component.

Taking moments about A, $80 \times 1 = P \times 2·5$

or $P = 32$ N.

If R is also required, then $R + P = 80$

giving $R = 48$ N.

Example 3

A uniform plank, 2 m long and weight 90 N, rests inclined at an angle of 30° to the horizontal with one end on a rough table. It is supported in this position by a force applied to its other end at right angles to the plank. Find this force.

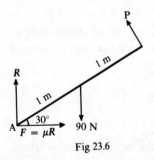

Fig 23.6

Fig 23.6 shows the forces acting on the plank. At the end A on the table, R is the normal reaction and F ($\leqslant \mu R$) acts to prevent the plank slipping to the left.

Take moments about A. R and F will then be absent from the equation.

$P \times 2 = 90 \times 1 \times \cos 30°$, all distances from A being at right angles to the forces.

Then $P = 39$ N.

Example 4

In fig 23.7 the beam **ABCDEF** *is shown resting on two pivots at* **B** *and* **D**. *The weight of the beam is* 300 N *and weights are attached at* **A** *and* **E** *at distances (in* m) *shown. Find the reactions at the pivots.*

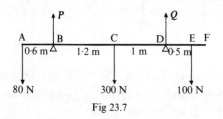

Fig 23.7

There are two unknowns, P and Q, so two equations will be required. Take moments about two different points. Any two can be chosen but if we take points B or D only one of P or Q will appear in the equations.

We can also use the vertical equilibrium equation:

$P + Q = 80 + 300 + 100 = 480.$ (no vertical acceleration)

Taking moments about B gives:

$$80 \times 0.6 - 300 \times 1.2 + Q \times 2.2 - 100 \times 2.7 = 0$$

or $\qquad 2.2Q = 270 + 360 - 48 = 582$

So $\qquad Q = 260$ N (2 sig figs)

To find P take moments about D or use the equilibrium equation above, $P + Q = 480$.

$$\therefore P = 480 - 260 = 220 \text{ N}.$$

Example 5

A beam **AB**, 5 m *long, is balanced horizontally on two supports* **P** *and* **Q** *so that* **AP** = 0.6 m *and* **QB** = 1 m. *It is found that, if a child of mass* 25 kg *stands on the beam at either end, the beam is on the point of toppling. Find* **(a)** *the weight of the beam,* **(b)** *the distance of its CG from* **A** *and* **(c)** *the distance from* **A** *at which the child must stand for the reactions at the supports to be equal* $(g = 9.8 \text{ m s}^{-2})$.

Let the CG of the beam (weight W N) be at G where AG = x m.
(a) When the child stands at end A the beam is on the point of turning about P and the reaction at Q will be zero (fig 23.8).

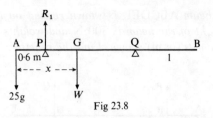

Fig 23.8

Taking moments about P, then $25g \times 0.6 = W(x - 0.6)$ (i)

(b) When the child stands at end B, the beam will tilt about Q and the reaction at P will be zero (fig 23.9).

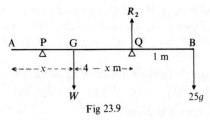

Fig 23.9

Taking moments about Q, $25g \times 1 = W(4 - x)$.   (ii)

Now solve these two equations.

Dividing (i) by (ii), $0.6 = \dfrac{x - 0.6}{4 - x}$

or $2.4 - 0.6x = x - 0.6$

giving $x = 1.875 \approx 1.9$ m.

Substituting in (ii), $25g = W(4 - 1.9)$

giving $W = 116$ N.

(c) Let the child stand y m from A (fig 23.10).

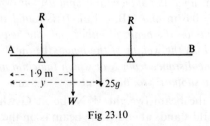

Fig 23.10

The reactions are equal, both R.

Considering vertical equilibrium, $2R = 116 + 25g = 361$ and so
$$R = 180 \text{ N (2 sig figs)}.$$

Taking moments about A:
$$R \times 0·6 - 116 \times 1·9 - 25g \times y + R \times 4 = 0$$
i.e.,
$$245y = R \times 4 + R \times 0·6 - 116 \times 1·9$$
$$= 4·6 \times R - 116 \times 1·9$$
$$= 608$$
and
$$y = 2·5 \text{ m}.$$

Exercise 23.1

Where required take $g = 9·8 \text{ m s}^{-2}$.

1 Find the moments of each of the forces shown in fig 23.11 about
the points A, B and C.

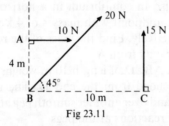

Fig 23.11

2 A light aluminimum straight rod ABC, 2 m long, is balanced
about a point B in its length, where AB = 0·5 m. A weight of
20 N is suspended from A. What weight suspended from C will
keep the rod in equilibrium?

3 A uniform beam AB, length 80 cm, weight 20 N, rests on a support
30 cm from one end A. At A a weight 30 N is suspended. What
weight suspended from B will keep the beam in equilibrium?

4 The rod ABCD in fig 23.12 is hinged at D and is in equilibrium
under the action of the forces shown. Find the value of P.

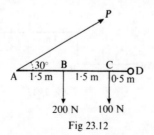

Fig 23.12

5 A rod ACD, length 4 m, is suspended horizontally by two vertical strings attached at A and D, the ends of the rod. The weight of the rod is 100 N and its CG is 1·7 m from A. A weight of 150 N is attached at C where CD = 1·2 m. Find the tensions in the strings.

6 A horizontal beam ABCD of length 3 m is supported at B and C where AB = 1·2 m, CD = 0·8 m. A force of 200 N applied vertically downwards at A just causes the beam to turn about B: a similar force of 300 N applied at D just causes the beam to turn about C. Find the weight of the beam and the position of the CG.

7 A non-uniform rod AB of length 2 m is supported horizontally on two supports at A and B. The reactions at A and B are 60 N and 40 N respectively. Find the weight of the rod and the position of its CG.

8 A straight rod AB, which is 40 cm long, is in equilibrium in a horizontal position when supported at the point C, 10 cm from A with masses of 9 kg and 2 kg attached to the rod at A and B respectively. It is also in equilibrium in a horizontal position when supported at the midpoint with masses of 4 kg and 8 kg attached at A and B respectively. Find the mass of the rod and the distance of its centre of gravity from A.   (L)

9 A uniform plank ABCD of length 10 m weighs 40 N and a weight of 20 N is suspended from the end D. The rod passes under a smooth peg at B and over another smooth peg at C, where AB = BC = 2 m. Find the reactions at the pegs.

10 ABC is an equilateral triangle of side 2 m. Forces of 10, 5 and 8 N act along the sides BC, CA, and BA respectively in the order of the letters. Find the sum of the moments of these forces about (i) the centre of the triangle, (ii) the point D, where D lies on BC produced and CD = 1 m.

11 In fig 23.13 two horizontal uniform bars AB, CD are freely hinged at A and E and connected by a light vertical cord BC which will break if the tension exceeds 8 N. The bar CD weighs 24 N. If AB = 3·5 m, CE = 3 m, ED = 2 m, find the maximum weight

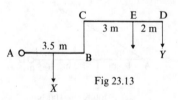

Fig 23.13

X of the bar AB and the corresponding value of the load Y which must be suspended from D to maintain equilibrium.   (C)

12 ABCD is a rectangle where AB = 4 m and BC = 3 m. Forces of 5 N, 10 N, 8 N, 12 N and 6 N act along the lines AB, BC, CD, AD and AC respectively in the order of the letters. Find the sum of the moments of these forces about **(i)** A, **(ii)** B and **(iii)** E where E is on DA produced and AE = 1 m.

Two Forces

If two forces act on a body they must either be parallel or their lines of action intersect.

I Parallel forces

If the two forces are parallel and the body is in equilibrium then the forces must be in the same line and (fig 23.14) $F_1 = F_2$.

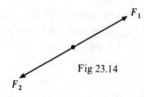

Fig 23.14

Otherwise the body will not be in equilibrium and the parallel forces will have a resultant (with one exception, see 3 below).

1 *Like parallel forces*

Parallel forces in the same direction are called **like** parallel forces (fig 23.15). The resultant of F_1 and F_2 will be R, parallel to and between F_1 and F_2 and $R = F_1 + F_2$. Since it replaces the two forces the moment of R about any point must be the same as the sum of the moments of F_1 and F_2 about the same point. If R cuts a perpendicular line at B, then the moment of R about B = 0, the moment of F_1 about B = $- F_1 \times$ AB and the moment of F_2 about B is $+ F_2 \times$ BC.

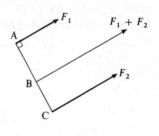

Fig 23.15

Then $- F_1 \times$ AB $+ F_2 \times$ BC $= 0$

or $\dfrac{\text{AB}}{\text{BC}} = \dfrac{F_2}{F_1}$.

This gives the point at which the line of action of the resultant divides the distance between two parallel forces.

2 *Unlike parallel forces*

Parallel forces in opposite directions are called **unlike** parallel forces (fig 23.16).

Suppose $F_2 > F_1$. Then the resultant $R = F_2 - F_1$ and acts parallel to F_1 and F_2 in the direction of the larger force (F_2) at a point B on the farther side of the larger force. Once again taking moments about B,

$$F_1 \times AB = F_2 \times BC \text{ giving } \frac{AB}{BC} = \frac{F_2}{F_1}$$

which gives the position of B.

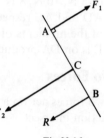

Fig 23.16

3 *Couple*

There is one exception to the above. If the unlike forces are equal, they have no resultant but they are NOT in equilibrium (fig 23.17). This pair of forces is called a **couple** and its effect is one of pure rotation of the body affected. No single force is equivalent to a couple. We shall not use this type of force in this work.

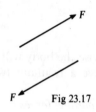

Fig 23.17

II Non-parallel forces

If the two forces acting on a body are not parallel the lines of action must intersect (fig 23.18). The parallelogram law gives the resultant R, which also acts at A. The two forces can be replaced by this resultant

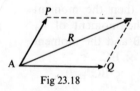

Fig 23.18

and the moment of the resultant about any point will also equal the (algebraic) sum of the moments of the two original forces, about the same point.

Example 6

Find the resultant of two forces 30 *and* 50 N *if the angle between them
is* 110° (fig 23.19).

By calculation, using the cosine and
sine rules, R = 48·7 N and θ = 35° 23'.
The student should also check these
answers by a scale drawing.

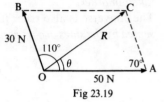

Fig 23.19

If the given forces are at right angles
(fig 23.20) the resultant is quickly found.

$$R = \sqrt{P^2 + Q^2} \text{ and } \tan \theta = \frac{P}{Q}.$$

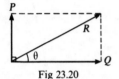

Fig 23.20

Three Forces: the Triangle of Forces

If two (non-parallel) forces act on a body the resultant *R* is the diagonal
OC of the parallelogram based on the vectors *OA* and *OB* which
represent *P* and *Q* respectively (fig 23.21).

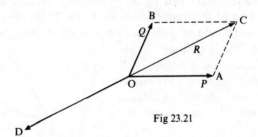

Fig 23.21

Hence *OD* will represent a force equal and opposite to this resultant
R where OD = OC and D lies on CO produced. These two forces
will be in equilibrium.

This means that the three forces represented by *OA*, *OB* and *OD* are
in equilibrium. Now these three forces can equally well be represented
in magnitude and direction by *OA*, *AC* and *CO* which form a triangle,
where the sides are taken in order (following on one after the other).

This leads to an important result, **the triangle of forces**.

If three forces act at a point and are in equilibrium, then they can be represented in magnitude and direction by the three sides of a triangle taken in order.

The converse is also true: *if three forces acting at a point can be represented by the sides of a triangle taken in order, then they are in equilibrium.*

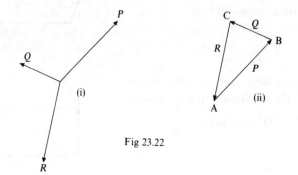

Fig 23.22

The three forces P, Q, R in fig 23.22i can be represented by the sides **AB**, **BC** and **CA** respectively of the triangle ABC (fig 23.22ii) which must be drawn to scale.

The triangle of forces is important in solving problems in statics in view of the following:

If any three forces are in equilibrium they must either be parallel or be concurrent (meet at a point).

Clearly three parallel forces can be in equilibrium. If they are not parallel suppose two of them meet at point A (fig 23.23). Since they are in equilibrium the sum of their moments about A must be zero.

The moments of P and Q about A are both zero and hence the moment of R about A must also be zero. This means that R must also pass through A. Hence the three forces must act at a point.

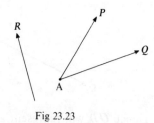

Fig 23.23

Therefore if three forces act on a body which is in equilibrium, the three forces can be represented by the sides of a triangle (taken in order) in magnitude and direction. This gives what is known as the **graphical** method for solving problems involving three forces as it depends on geometry (using the triangle of forces).

Lami's Theorem

If three forces are in equilibrium, Lami's theorem (which is a version of the sine rule) relates them to the angles between their directions.

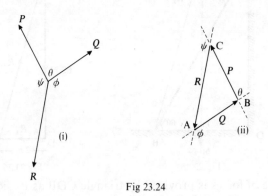

Fig 23.24

P, Q, R are three forces (fig 23.24i) and ABC is a triangle of forces (fig 23.24ii). If the angle between P and Q is θ then

$$\text{angle ABC} = 180° - \theta$$

and similarly for the other angles.

By the sine rule, $\dfrac{\text{AC}}{\sin(180° - \theta)} = \dfrac{\text{AB}}{\sin(180° - \psi)} = \dfrac{\text{BC}}{\sin(180° - \phi)}$

i.e. $\dfrac{R}{\sin\theta} = \dfrac{Q}{\sin\psi} = \dfrac{P}{\sin\phi}$ as P, Q, R are proportional to BC, AB, CA

respectively. Lami's theorem is useful if the angles between the three forces are known.

Example 7

A uniform ladder rests at an angle of 60° with the horizontal against a smooth vertical wall and on rough ground. The ladder weighs 60 N and its length is 8 m. Find the reactions at the wall and at the ground.

In fig 23.25 AB is the ladder and the weight (60 N) acts vertically through G, halfway along the ladder. At A, the reaction R_1 is normal to the wall as it is smooth. (No frictional force along the wall). The lines of these two forces intersect at C and hence the line of the third force (R_2) must also pass through C. R_2 cannot be vertical as the ground is rough.

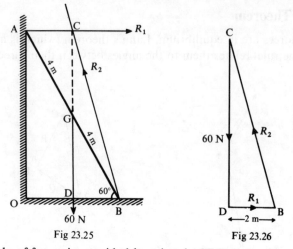

Fig 23.25   Fig 23.26

A triangle of forces is provided by triangle CDB as its sides are each parallel to one force (fig 23.26). The length of DB = 4 cos 60° = 2 m and the length of CD = AO = 8 sin 60° = $4\sqrt{3}$ m. *CD* represents the force of 60 N (the weight of the ladder).

This concludes the statical part of the solution. We have found a suitable triangle of forces CDB. The solution can now be completed either by a scale drawing of the triangle or by trigonometry.

1 *By drawing*
Draw a right-angled triangle similar to triangle CDB with DB proportional to 2 m and CD proportional to $4\sqrt{3}$ (6·93) m. Measure the length of CB.

Then $\dfrac{R_2}{CB} = \dfrac{60}{4\sqrt{3}} = \dfrac{R_1}{2}$ which gives the reactions R_1 and R_2. Also measure the angle CBD.

2 *By calculation* $CB^2 = CD^2 + DB^2 = (4\sqrt{3})^2 + (2)^2 = 52$
                          Then CB = 7·21.

Hence $\dfrac{R_2}{7\cdot21} = \dfrac{60}{6\cdot93} = \dfrac{R_1}{2}$

which gives $R_1 = 17\cdot3$ N, $R_2 = 62\cdot4$ N

and angle CBD is given by tan $\widehat{CBD} = \dfrac{4\sqrt{3}}{2} = 2\sqrt{3}$

and   angle CBD = 73° 54′.
The solution is now complete.

Example 8

A uniform rod AB of weight 30 N is jointed at a hinge on a vertical wall and is held in a horizontal position by a string attached to the end B and to a point C of the wall vertically above A, where angle ACB = 30° Find the tension in the string and the reaction at the hinge.

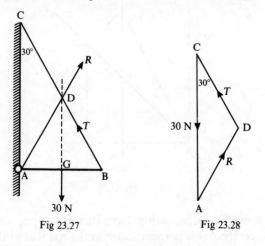

Fig 23.27   Fig 23.28

Fig 23.27 shows the three forces acting on the rod. The weight acts through G (the midpoint of the rod), the tension T along the string and so the reaction R must pass through D.

Triangle CAD is a triangle of forces. Since G is the midpoint of AB, D is the midpoint of CB. Also, since $\widehat{BAC}$ is a right angle, the semicircle on BC as diameter, with D as centre passes through A, and CD = DB = DA (fig 23.28).

Hence $T = R$ and by the sine rule,

$$\frac{30}{\sin 120°} = \frac{T}{\sin 30°}$$

giving $\qquad T = 10\sqrt{3} = 17{\cdot}32 \text{ N}$

$R = 17{\cdot}32$ N and makes an angle of 30° with the upward vertical at A.

Example 9

A uniform rod ABC, weight 50 N and length 4 m, rests with one end A on rough horizontal ground and is supported by a smooth peg at B where AB = 2·5 m. The peg is 2 m from the ground. Find the reactions at A and at the peg.

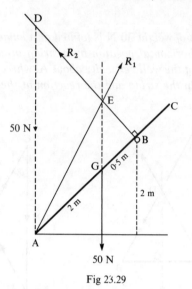

Fig 23.29

Fig 23.29 shows the forces acting. Note that as the peg is smooth, the reaction R_2 on the rod is perpendicular to the rod (i.e., to the common surface). The weight acts vertically through G and so R_1 passes through E. To solve, find a suitable triangle of forces. If BE is produced to meet the vertical through A at D, triangle DAE is a triangle of forces.

Then
$$\frac{50}{DA} = \frac{R_1}{AE} = \frac{R_2}{ED}$$

Now draw the figure to scale and measure the length of DA, AE and ED, thus finding R_1 and R_2 from the above ratios.

To complete the solution by calculation is rather long in this case: a better method using calculation is shown later.

In solving three force problems by the graphical method, note:

(1) The three forces will be concurrent. Two are usually fixed by the conditions of the problem (e.g., weight always acts vertically through CG, tension along a string, reaction perpendicular to a smooth wall, etc) and so the third force passes through their point of intersection.

(2) Find a suitable triangle of forces. There may be one already in the figure or one may be adapted by extension of lines.

(3) For a solution by drawing, draw the triangle of forces to a suitable scale. The lengths of the sides will be proportional to the forces

acting along these sides. The direction of at least one force will be known. The others will follow in order round the triangle and so their directions can be found.

(4) For a solution by calculation, find the lengths of the sides of the triangle of forces. If the angles can be found more easily, use the sine rule (this is using Lami's theorem in effect).

(5) Such problems can also be solved by an alternative method, the analytical method, described later.

Exercise 23.2

1 Two parallel forces 50 N and 30 N, act in lines which are 40 cm apart. Find their resultant and where it acts if the forces are (a) like, (b) unlike.

2 ABC is a horizontal straight line, AB = 30 cm, BC = 20 cm. Forces of 10, 20, 40 N act at A, B, C respectively perpendicular to the line ABC. The forces at A and C act in one direction, the force at B in the opposite direction. Find the resultant of the three forces.

Solve the remaining problems by the graphical method, completing the solution either by drawing or calculation.

3 A uniform ladder, weight 150 N and length 6 m, is placed on rough horizontal ground at an angle of 60° with its top resting against a smooth vertical wall. Find the reactions at the ends of the ladder.

4 A uniform rod AB, length 3 m and weight 100 N is freely hinged to a vertical wall at A. It is held horizontally by a string BC, 5 m long, attached to the wall at C, vertically above A. Find the tension in the string and the reaction at the wall in magnitude and direction.

5 A sphere of weight 100 N and radius 10 cm rests against a smooth vertical wall and is supported by a string 50 cm long attached to to the surface of the sphere and to the wall. Find the tension in the string and the reaction at the wall.

6 Find the unknown forces and angles in fig 23.30.

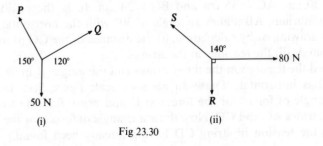

(i)   (ii)

Fig 23.30

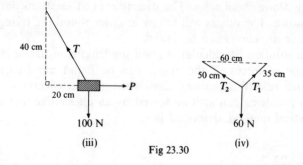

(iii)

Fig 23.30

(iv)

7 Three forces acting at the origin O can be represented by the vectors OA, OB, OC where A, B, C have coordinates $(5, 2), (-3, 8), (-2, -10)$ respectively. Show that the forces are in equilibrium. (L)

8 A block of weight 30 N rests on a smooth plane inclined at an angle of 20° to the horizontal. Find the least horizontal force required to keep it in equilibrium and the reaction of the plane.

9 A uniform rod ABC of weight 60 N and length 2 m rests over a smooth peg at B and with the end A against a smooth vertical wall. The rod makes an angle of 30° with the upward vertical wall. Find the distance AB and the reactions at the peg and the wall.

10 A uniform rod AB of length 80 cm and weight 20 N is suspended by two strings of lengths 1 m and 50 cm attached to its ends A and B and to a point C above the rod. Find the tensions in the strings and the angle the rod makes with the horizontal.

11 A heavy uniform beam AB is hinged to a wall at its lower end A and is kept in equilibrium at 60° to the horizontal by a string BC attached to the end B and to a point C on the wall vertically above A. If angle BCA = 60° and the tension in the string is 30 N find the weight of the beam and the magnitude and direction of the reaction at A. (C)

12 A non-uniform rod AB weighs 30 N and hangs in a vertical plane supported by two strings AC and BC attached to a peg C. AB = 30 cm, AC = 36 cm and BC = 24 cm. If, in the position of equilibrium, AB makes an angle of 30° with the horizontal, find, by drawing or by calculation, (**i**) the distance of the CG of the rod from A, (**ii**) the tensions in the strings. (L)

13 Find the tensions in the three strings and the weight W in fig 23.31. AB is horizontal. (Draw an accurate scale figure first. Draw a triangle of forces for the forces at C and hence find the tensions in strings AC and CD. Now draw a triangle of forces for the point D, the tension in string CD having already been found.)

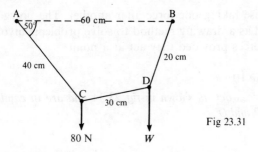

Fig 23.31

14 Find the value of *W* and the tensions in the three strings in fig 23.32. CF and DE are horizontal.

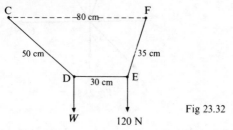

Fig 23.32

The Polygon of Forces

The triangle of forces can be extended to more than three forces provided they *act at a point and lie in the same plane*. This extension is called the **polygon of forces**. If a set of forces acting at a point are in equilibrium then they can be represented by the sides of a polygon taken in order. The sides are proportional and parallel to the forces.

Fig 23.33i shows forces *P*, *Q*, *R*, *S*, *T* which can be represented by the polygon ABCDE in fig 23.33ii. If the forces are in equilibrium the polygon must be closed. The sides of the polygon may cross each other if this is necessary by the layout of the forces, but must follow each other in order. Start with any force and proceed clockwise or anti-

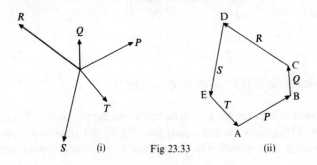

(i)   Fig 23.33   (ii)

clockwise taking one force after another. The polygon of forces can be used as a drawing method to solve problems involving more than three forces provided they act at a point.

Example 10

Five forces act as shown in fig 23.34 and are in equilibrium. Find the forces P and Q.

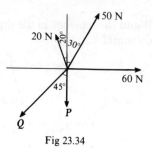

Fig 23.34

Construct a polygon to a suitable scale (say 10 N to 1 cm) (fig 23.35). ABCD is drawn and then the line DE of unlimited length. The angles are taken from fig 23.34. AE is drawn perpendicular to AB to intersect DE at E. The lengths of DE and EA give the values of Q and P. Note that P is actually directed upwards to close the polygon.

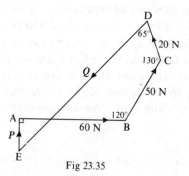

Fig 23.35

You should find that $P = 16$ N, $Q = 110$ N.

If the forces at a point are not in equilibrium the polygon is not closed. Suppose ABCDE is an open polygon (fig 23.36) obtained from four forces acting at a point. The forces are not in equilibrium and *EA*

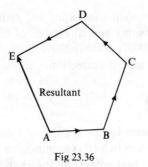

Fig 23.36

represents the fifth force necessary for equilibrium. Hence **AE** will represent the resultant of the original four forces in magnitude and direction. The resultant must of course pass through the same point as the original forces.

Exercise 23.3

Find the unknown forces in magnitude and direction in these systems of forces which are in equilibrium.

1   **2**   **3**

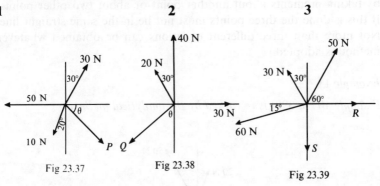

Fig 23.37   Fig 23.38

Fig 23.39

4 Find the resultant in magnitude and direction of forces 10, 20, 30, 40 N acting in directions 060°, 120°, 180°, 270° respectively.

5 Referred to rectangular axes with O as the origin, A is the point (4, 3), B is the point (12, 5), C is the point (0, −6) and the same scale is used for the x and the y-coordinates. Forces of magnitude 75 N, 65 N and 35 N act at O towards A, B and C respectively. Find, by calculation or drawing, the magnitude of the resultant force and the angle it makes with the x-axis.   (L)

6 Five strings are attached to a point and radiate horizontally from this point. The tensions and directions of four of the strings are: 50 N, 060°; 40 N, 090°; 100 N, 270°; 20 N, 330°. Find the tension in the fifth string and its direction.

Analytical Method

The solution of problems by the triangle of forces is obviously limited to the case of three forces (which is common enough) and the polygon of forces can be only used where the forces meet at a point. A more general method is now outlined, based on the following principles.

A system of forces acting on a body is in equilibrium provided **(1)** *the sum of the components of all the forces in two perpendicular directions is zero in both directions, and* **(2)** *the algebraic sum of the moments of the forces about any point is also zero.*

This will provide three equations connecting the forces of the system from which up to three unknowns can be found.

The resolution equations show that the body cannot move in any linear direction and the moment equation shows that the body cannot rotate. (It is possible to replace either or both of the resolution equations by taking moments about another point or about two other points. If this is done the three points must not lie in the same straight line. Not more than three different equations can be obtained whatever method is adopted.)

Example 11

Example 10 (p. 390) is reworked by the analytical method.

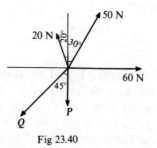

Fig 23.40

There are two unknowns P and Q. Resolve along and perpendicular to the 60 N force. (fig 23.40)

Resolving in direction →,
$$60 + 50 \cos 60° - 20 \cos 70° - Q \cos 45° = 0$$
(P has no component in this direction.)
Therefore $Q \cos 45° = 60 + 50 \times 0·5 - 20 \times 0·3420 = 78·16$.
Hence $Q \times 0·7071 = 78·16$ N giving $Q = 110$ N as before.

Now resolving in direction ↑,
$$50 \cos 30° + 20 \cos 20° - P - Q \cos 45° = 0$$
or
$$P = 50 \times 0·8666 + 20 \times 0·9397 - 78·16$$
$$= 43·3 + 18·79 - 78·16 = -16 \text{ N}.$$
(P actually acts upward, as we found before.)

Example 12

Example 9 (p. 385) solved by the analytical method.

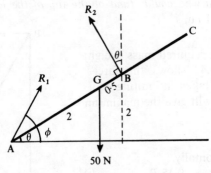

Fig 23.41

First from fig 23.41 note that $\sin \theta = \dfrac{2}{2·5} = 0·8$ and $\cos \theta = \frac{3}{5} = 0·6$.

The direction of the reaction at A is unknown, so introduce angle ϕ as a third unknown with R_1 and R_2.

Take moments about A, to find R_2 without introducing R_1.

Therefore
$$50 \times 2 \cos \theta = R_2 \times 2·5$$
or
$$50 \times 2 \times 0·6 = 2·5 R_2$$
giving
$$R_2 = 24 \text{ N}.$$

Resolving horizontally,
$$R_1 \cos \phi = R_2 \sin \theta = 24 \times 0·8 = 19·2 \quad \text{(i)}$$
Resolving vertically,
$$R_1 \sin \phi + R_2 \cos \theta = 50$$
or $R_1 \sin \phi = 50 - 24 \times 0·6 = 50 - 14·4 = 35·6$ (ii)

Note the following method for finding R_1 and ϕ from two such equations.

Squaring and adding (i) and (ii), remembering that $\sin^2\phi + \cos^2\phi = 1$.

$$R_1{}^2 = (19\cdot2)^2 + (35\cdot6)^2 = 1636$$

giving $R_1 = 40\cdot4$ N

Dividing (ii) by (i), $\tan\phi = \dfrac{35\cdot6}{19\cdot2}$ giving $\phi = 61°\ 40'$.

The solution is now completed.

Example 13

A ladder is 8 m long and weighs 400 N and its CG is 3 m from its foot. It stands on rough horizontal ground making an angle of 60° with the horizontal and leans against a smooth vertical wall. If the coefficient of friction between the ladder and the ground is 0·45, find the mass of the heaviest man who could stand at the top of the ladder without it slipping. $(g = 10\ \text{m s}^{-2})$

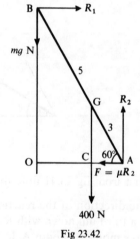

Fig 23.42

Let m kg be the required mass (weight mg N). Fig 23.42 shows the forces, with the ladder taken in limiting equilibrium, as this will give the maximum value of m.

Resolving horizontally,

$$R_1 = \mu R_2 = 0\cdot45\ R_2 \qquad \text{(i)}$$

Resolving vertically,

$$400 + mg = R_2 \qquad \text{(ii)}$$

Taking moments about A,

$$mg \times \text{OA} + 400 \times \text{CA} = R_1 \times \text{OB}$$

i.e., $m \times 10 \times 8\cos60° + 400 \times 3\cos60° = R_1 \times 8\sin60°$

or $40m + 600 = 4\sqrt{3}\,R_1$ $\qquad$ (iii)

Eliminate R_1 and R_2 to find m.

From (i) using (ii), $R_1 = 0\cdot45(400 + mg) = 180 + 4\cdot5\,m$

In (iii), $40m + 600 = 4\sqrt{3}(180 + 4\cdot5m) = 720\sqrt{3} + 18\sqrt{3}m$

$\therefore \qquad 40m - 18\sqrt{3}m = 720\sqrt{3} - 600$

or $\qquad 8\cdot82m = 647$

giving $m = 73\cdot5$ kg.

Exercise 23.4

The problems in Exercise 23.2 and 23.3 and Examples 7 and 8 should be reworked by the analytical method as well.

1 A sphere of weight 40 N and radius 15 cm rests against a smooth vertical wall and is held by a string of length 15 cm attached to a point on its surface. Find the reaction at the wall and the tension in the string.

2 A rod of length 50 cm and weight 60 N is freely hinged to a vertical wall. Its CG is 30 cm from the wall along the rod. The rod is held horizontally by a string 80 cm long attached to the other end of the rod and to the wall at a point vertically above the hinge. Find the reaction at the wall and the tension in the string.

3 A uniform rod AB, 2 m long and weight 60 N, is freely hinged at A to a vertical wall. It is supported by a string BC attached to a point of the wall 2 m vertically above A. In equilibrium the rod makes an angle of 60° with the downward vertical. Find the tension in the string and the reaction at the hinge.

4 A uniform rod AB, 3 m long and of weight 120 N, rests in equilibrium with end A on a rough horizontal table and supported by a force P applied at its end B at right angles to the rod. The rod makes an angle of 30° with the table. Find the force P and the reaction at the floor.

5 A uniform rod AB of weight 100 N and 2 m long is hinged at A. A weight of 50 N is hung from the end B. The rod is held horizontally by a string attached to B and to a point C, 1·5 m vertically above A. Find the tension in the string and the reaction at the wall.

6 PQ is a heavy, non-uniform bar, 4 m long and weighing 10 N. It is supported in a horizontal position by two strings at P and Q which are inclined at 25° and 55° respectively to the horizontal. Calculate the tension in each string.
If the weight of the bar acts at the point G, calculate the length PG.   (C)

7 A straight uniform rod ABC of length 1 m and weight 10 N rests with its end A on rough horizontal ground. It is supported in limiting equilibrium at an angle of 40° to the horizontal by a string at right angles to the rod at B where AB = 80 cm. Find the tension in the string and the coefficient of friction between the rod and the ground.

8 A uniform rod AB of weight 30 N and length 2 m is hinged at A to a vertical wall and is supported in a horizontal position by a string BC 4 m long to a point C of the wall vertically above A. If the string can only sustain a maximum tension of 200 N find the

greatest weight which can be hung from B.

9 In fig 23.43, AB represents a uniform beam of weight 150 N which rests with the end B on a rough plane inclined at an angle of 30° to the horizontal. DBE is a line of greatest slope of the plane. The beam is supported by a light rope AC attached to a point C vertically above B, so that angle ABE is 30° and angle CAB is a right angle. The points A, B, C, D and E are in a vertical plane. Find, by drawing or calculation, **(a)** the tension in the rope, **(b)** the inclination to the vertical of the reaction at B.　　(C)

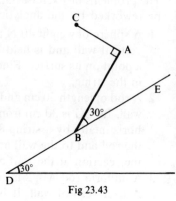

Fig 23.43

(A 'line of greatest slope' is a line directly up the plane, i.e. at an angle of 30° to the horizontal in the above.)

10 A body of weight 5 N hangs by a strong cord from a hook in the ceiling. To the body is attached a piece of string which will break under a tension of 3 N. Calculate the largest angle which the cord can make with the vertical if the body is pulled aside by the string **(a)** if the string remains horizontal, **(b)** if the string may be pulled in any direction.　　(C)

11 A ladder length 4 m and weight 200 N rests on rough horizontal ground with the other end against a smooth vertical wall. The foot of the ladder is 2 m from the wall and the ladder is on the point of slipping. The CG of the ladder is 1 m from its foot. Find the coefficient of friction between the ladder and the ground.

12 A uniform rod of length 2·8 m and weight 150 N rests on rough horizontal ground (coefficient of friction 0·5) and against a smooth vertical wall, inclined at an angle of 60° to the ground. Find the greatest weight which can be hung from the top end of the ladder without the ladder slipping.

13 A uniform sphere of mass 2 kg, resting on a rough plane which is inclined at an angle of 37° to the horizontal, is held in position by a horizontal string attached to the highest point of the sphere. Find the least possible value of the angle of friction between the sphere and the plane, for equilibrium to be maintained. Find also the tension in the string.　　(L)

14 A uniform rod AC, inclined at 60° to the horizontal, rests with its lower end against a smooth vertical wall. It is supported by

a smooth peg B which is 10 cm away from the wall. Find, by drawing or calculation, the length of the rod.
The peg will just break under a force of 24 N. What is the maximum weight of the rod if the peg is not to break? (C)

15 A uniform ladder of weight 250 N and length 5 m rests on a smooth horizontal floor and against a smooth vertical wall. The foot of the ladder is 3 m from the wall. A man of weight 800 N ascends the ladder. Find the horizontal force to be applied at the foot of the ladder when the man is **(a)** at the middle of the ladder; **(b)** at top of the ladder, to maintain the ladder in equilibrium.

16 A uniform ladder 6 m long and of weight 100 N stands on a smooth horizontal floor and rests against a smooth vertical wall, making an angle of 60° with the floor. It is held in this position by a horizontal rope tied to the ladder and fixed to the wall 1 m above the floor. Find the tension in the rope and the reactions at the wall and floor.

Systems of Forces

A system of forces (not in equilibrium) in the same plane is given and it is required to find the resultant and its line of action. The principles above can be used in a modified form.

(1) The components of the resultant in two perpendicular directions will equal the sum of the components of the forces in the same directions.

(2) The moment of the resultant about any point will equal the sum of the moments of the forces about the same point. Hence if the resultant passes through a certain point P, its moment about P is zero and so is the sum of the moments of the original forces.

Example 14

A square ABCD is of side 5 m. Forces of 10, 5, 12, 15 N act along the sides AB, BC, CD, DA respectively (in the direction of the letters) (fig 23.44). Find the resultant (in magnitude and direction) and where its line of action cuts AB and AD.

(1) Resolve in directions AB and AD.
In direction AB the sum of the components is: $10 - 12 = -2$ N
In direction AD the sum of the components is: $5 - 15 = -10$.

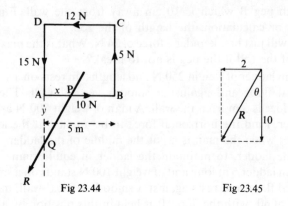

Fig 23.44   Fig 23.45

Hence the components of the resultant are as shown in fig 23.45
and $R^2 = 10^2 + 2^2 = 104$ giving $R = 10 \cdot 2$ N
Also $\tan \theta = \frac{2}{10}$ giving $\theta = 11° \ 19'$.
The resultant is $10 \cdot 2$ N acting in a direction making an angle of
$11° \ 19'$ with DA.

(2) To find where the resultant cuts AB and AD.
Let the resultant cut AB at P (x m from A) and DA produced at Q
(y m from A).
Take moments about P: the moment of R about P is zero and so
the sum of the moments of the original forces about P is also zero.
Therefore $5 \times (5 - x) + 12 \times 5 + 15x = 0$ giving $x = -8 \cdot 5$.

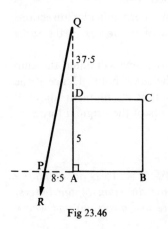

Fig 23.46

So the resultant cuts AB $-8 \cdot 5$ m from
A i.e. on BA produced.
Repeat, taking moments about Q.
Then $-10y + 5 \times 5 + 12(5 + y) = 0$
giving $y = -42 \cdot 5$.
So the resultant cuts AD (produced)
$42 \cdot 5$ m from A (fig 23.46).

(This second result could have been
obtained from the fact that

$$\frac{AP}{AQ} = \frac{2}{10} \ (\tan \theta \text{ above}).$$

So $AQ = 5 \times AP$
             $= 5 \times 8 \cdot 5$
             $= 42 \cdot 5$ m).

Example 15
Forces of 3, 4, 5 N act along the sides **AB, BC, CA** *respectively of an*

equilateral triangle ABC *of side* 1 m (*in the directions shown by the letters*). *Find the single force required to keep the triangle in equilibrium in magnitude and direction and where its line of action will cut* AB.

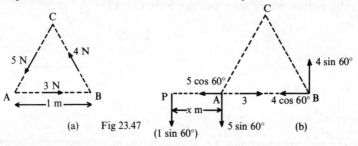

(a)   Fig 23.47   (b)

Fig 23.47a shows the three forces acting along the sides of the triangle. The 3 forces will be combined into a single resultant R. A single force equal and opposite to R will therefore keep the triangle in equilibrium. We find R and where it cuts AB. The 5 N force acts through A, the 3 N force along AB and the 4 N through B. Resolve the 5 N and the 4 N forces at A and B respectively as shown in fig 23.47b.

Resolving along BA,

$$5 \cos 60° - 3 + 4 \cos 60° = 1.5 \text{ N}.$$

The resultant of the other two components (unlike parallel forces) gives 1 sin 60° N acting as shown at P (on the further side of the larger force).

Taking moments about P, $5 \sin 60° \, x = 4 \sin 60° \, (x + 1)$ and so $x = 4$ m.

Hence the final resultant of the original three forces will pass through P where AP = 4 m.

We now have reduced the three forces to two components 1.5 N and

$1 \sin 60° \, (= \dfrac{\sqrt{3}}{2})$ N acting through P, as shown in fig 23.48.

Then $R^2 = (1.5)^2 + (\dfrac{\sqrt{3}}{2})^2 = 3$

and $R = \sqrt{3}$ N.

Also $\tan \theta = \dfrac{1.5}{\sqrt{3}/2} = \sqrt{3}$. Hence $\theta = 60°$.

So $R = \sqrt{3}$ N and acts at 60° as shown.

Hence the single force required to keep the triangle in equilibrium must be $\sqrt{3}$ N, acting through P in the direction opposite to that of R (i.e. at 30° with AB).

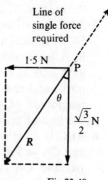

Fig 23.48

Exercise 23.5

Find (i) the resultants of each of the following systems of forces and (ii) the points where they cut the axes Ox, Oy.

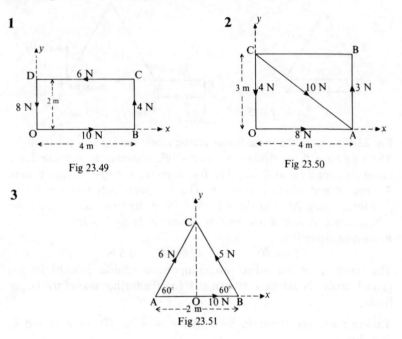

Fig 23.49

Fig 23.50

Fig 23.51

4 The side AB of a rectangle ABCD is 24 cm and the side BC is 12 cm. M is the midpoint of AB. The point P in BC is such that BP = 9 cm and Q is the point on AD such that AQ = 5 cm. Forces of 5 N and 13 N act along MP and MQ respectively. Prove that the resultant of these two forces passes through D and calculate the magnitude of this resultant. (C)

5 ABC is a triangle in which AB is horizontal and of length 12 m. BC is vertical and of length 16 m. D is the point in BC, 5 m from B. A force of 80 N acts at A in the direction AC and a force of 91 N acts at A in the direction AD. By resolving these forces into horizontal and vertical components, or otherwise, calculate (a) the magnitude of their resultant, (b) the distance from B at which the line of action of this resultant cuts BC. (C)

6 ABC is a right-angled triangle with sides AB = 3 m, BC = 4 m and CA = 5 m. Forces of 10, 20 and 30 N act along the sides AB, BC and CA respectively. Find the magnitude and direction of the resultant and where its line of action cuts AB and BC.

7 Find the resultant of the following forces acting along the sides of a square ABCD of side 2 m : 20 N along AB, 30 N along CB, 15 N along CD and 20 N along AD. Find where the resultant cuts AB and AD.

8 Forces of 5, 5 and 8 N act along the sides AB, BC and CD respectively of a square ABCD of side 2 m. A force *P* maintains the square in equilibrium. Find the magnitude and direction of *P*, and where its line of action cuts AD.

9 Forces of 5, 10, 8 and 4 N act along the sides AB, BC, CD and DE of a regular hexagon ABCDEF of side 1 m. Calculate the magnitude and direction of the single force which will keep the hexagon in equilibrium. Find where this single force cuts AB.

10 ABCD is a square of side 20 cm. Forces of 2, 5, 6, 3 and 5 N act along the lines AB, BC, CD, AD and AC respectively. Find the resultant of these forces in magnitude and direction and where its line of action cuts AB and AD.

11 Forces of 4, 5 and 5 N act along the sides AB, CB and CA respectively of an equilateral triangle ABC. Show that it is possible to keep the triangle in equilibrium by a single force. Find the point in which the line of action of this force cuts AB and calculate its magnitude and direction.   (C)

24 Centre of Gravity

The centre of gravity (CG) has been mentioned briefly as the point through which the weight of a body acts. For example,
CG of a uniform rod (straight with constant cross-section) is at the midpoint of the rod and at the centre of the cross-section;
CG of a rectangular lamina (negligible thickness) is at the intersection of the diagonals;
CG of a circular disc or lamina is at its centre;
CG of a sphere is at its centre;
CG of a cylinder is at the midpoint of its axis.
For other shapes or bodies, the position of the CG can sometimes be found by treating the body as built up of parts whose separate centres of gravity are known. Otherwise more advanced methods involving calculus have to be used. Note that if a body is symmetrical about an axis, the CG will lie on that axis.

Composite Bodies

Fig 24.1 shows a body which is divided into two parts, weights W_1 and W_2, whose centres of gravity G_1 and G_2 are known. The total weight of the body $(W_1 + W_2)$ is the resultant of the two parallel forces W_1, W_2 and hence acts through a point G on the line joining $G_1 G_2$.

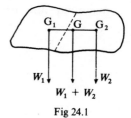

Fig 24.1

By taking moments about G,
$$W_1 \times G_1 G = W_2 \times G_2 G$$
which will give the position of G on the line $G_1 G_2$.

Alternatively, if we have a suitable axis Oy (fig 24.2) and if the distances of G_1, G, G_2, from Oy are x_1, $\bar{x}$, x_2 respectively, the moment of $(W_1 + W_2)$ about Oy equals the sum of the moments of W_1 and W_2.
$(W_1 + W_2) \bar{x} = W_1 x_1 + W_2 x_2$ thus giving $\bar{x}$. This method is especially useful in dealing with two-dimensional lamina.

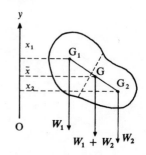

Fig 24.2

Example 1

Find the position of the CG *of the lamina* ABCDEF *made from thin sheet metal shown in fig* 24.3. *If the lamina is suspended freely from* F, *what angle will* AF *make with the vertical? What horizontal force applied at* A *will maintain the lamina in equilibrium when suspended from* F *with AF vertical? Mass of lamina* = 0·8 kg.

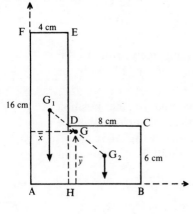

Fig 24.3

Divide the lamina into two rectangles AHEF and HBCD.
If the lamina is of uniform thickness, the weight of any part is proportional to its area. Take axes AF and AB. It is best to set out the details in a table.

Part	Weight proportional to	Co-ordinates of CG from		
			AF	AB
AHEF	64	G_1	2	8
HBCD	48	G_2	4 + 4 = 8	3
Whole Lamina	112	G	$\bar{x}$	$\bar{y}$

Taking moments about AF, $112\bar{x} = 64 \times 2 + 48 \times 8 = 512$
$$\therefore \quad \bar{x} = 4\cdot57 \text{ cm.}$$
,,   ,,   ,,   AB, $112\bar{y} = 64 \times 8 + 48 \times 3 = 656$
$$\therefore \quad \bar{y} = 5\cdot85 \text{ cm.}$$
The co-ordinates of the CG are (4·57, 5·85) from AF and AB respectively.

If the lamina is suspended from F (fig 24.4) and can swing freely, the reaction at F must act in the same line as the weight through G. Hence the lamina will hang so that FG is vertical.

From fig 24.4 it is seen that

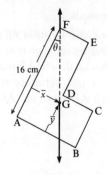

$$\text{Tan } \theta = \frac{\bar{x}}{16 - \bar{y}} = \frac{4 \cdot 57}{10 \cdot 15}$$
$$\text{giving } \theta = 24° \ 14'.$$

Hence when hanging freely from F, AF will make an angle of 24° 14' with the vertical through G.

Fig 24.4

To restore the lamina to the position where AF is vertical, a horizontal force P is applied at corner A (fig 24.5).

Taking moments about F in the equilibrium position,

$$\text{Weight} \times \bar{x} = P \times 16$$
$$\therefore \ 0 \cdot 8g \times 4 \cdot 57 = P \times 16$$
$$\text{giving } P = 2 \cdot 2 \text{ N.}$$

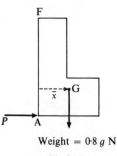

Weight = 0·8 g N

Fig 24.5

Example 2

ABC *is a uniform rod of mass* 8 kg *and length* 3·5 m. AB *and* BC *form two sides of a right-angled triangle with* AB = 1·5 m. *What is the least force applied at* C *which will maintain the rod in equilibrium when suspended freely from* A *so that* BC *is vertical?* ($g = 9 \cdot 8 \text{ m s}^{-2}$)

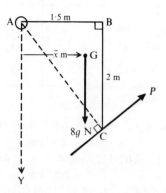

Fig 24.6

We first find the position of the CG as measured from AY (fig 24.6).

Part	Weight (N)	Distance of CG from AY
AB	$\frac{3}{7} \times 8g$	0·75
BC	$\frac{4}{7} \times 8g$	1·5
Whole rod	$8g$	$\bar{x}$

Taking moments about A,

$$8g \times \bar{x} = \tfrac{3}{7} \times 8g \times 0\cdot 75 + \tfrac{4}{7} \times 8g \times 1\cdot 5$$

$$\therefore \qquad 7\bar{x} = 2\cdot 25 + 6 = 8\cdot 25$$

or $\qquad \bar{x} = 1\cdot 18$ m.

Let P N be the required force acting through C. The moment of this force about A = the moment of the total weight ($8g$ N) about A = $8g\bar{x}$, which is constant. So the moment of P about A is constant, i.e., the product of P and the perpendicular distance of its line of action from A is constant. To obtain the least value of P we must take the *greatest* perpendicular distance, i.e., AC. Hence P must act perpendicular to AC, where AC = 2·5 m (3 − 4 − 5 triangle).

$$\therefore \qquad P \times 2\cdot 5 = 8g \times 1\cdot 18$$

or $\qquad P = 37$ N.

Example 3

A solid is constructed by glueing a hemisphere radius 8 cm onto a cylinder of equal radius and length 16 cm made of the same material. (CG of a hemisphere is $\frac{3}{8}$ of radius from the base on the central axis). Find the position of the CG of the combined solid.

G_1, the CG of the cylinder, is 8 cm from the common face. (fig 24.7).
G_2, the CG of the hemisphere, is $\frac{3}{8} \times 8$ = 3 cm from the common face.
Let the required CG (G) be $\bar{x}$ cm from the common face. The weight of each solid is proportional to its volume.

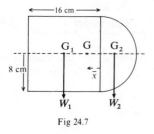

Fig 24.7

So W_1 (the weight of the cylinder) is proportional to $\pi r^2 h = \pi 8^2 \times 16$ and W_2 (the weight of the hemisphere) is proportional to $\frac{2}{3}\pi r^3 = \frac{2}{3}\pi 8^3$.
Then taking moments about G, $W_1 \times G_1G = W_2 \times GG_2$
$$\therefore \pi \times 64 \times 16 \times (8 - \bar{x}) = \tfrac{2}{3}\pi \times 8 \times 8 \times 8 \times (3 + \bar{x})$$

$$\therefore \qquad (8 - \bar{x}) = \tfrac{1}{3}(3 + \bar{x})$$

or $\qquad\qquad\qquad 24 - 3\bar{x} = 3 + \bar{x}$

giving $\qquad\qquad\qquad\qquad \bar{x} = 5\cdot2$ cm

The CG is $5\cdot2$ cm inside the cylinder from the common face, or $10\cdot8$ cm from the base of the cylinder.

The method can also be adapted to deal with a body from which a part has been removed.

Example 4

A square plate of side 120 mm *has a hole of radius* 40 mm *bored in it. The centre of the hole lies on a diagonal and is* 40 mm *from the centre of the square. Find the position of the* CG *of the remainder.*

G is the CG of the original plate (fig 24.8). G_1 is the CG of the hole (circular disc removed). G_2 is the CG of the remainder.

By symmetry G_2 will also lie on the diagonal, to the left of G where $G_2G = x$ mm.

The weights of all parts are proportional to their areas.

Then the moment of the circle removed about G will equal the moment of the remainder: the two parts "balance" about G.

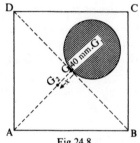

Fig 24.8

Part	Weights (proportional to areas)	Distance from G
Square	120^2	0
Circle	$\pi \times 40^2$	40
Remainder	$120^2 - \pi \times 40^2$	x

Then taking moments about G,

$$\pi \times 40^2 \times 40 = (120^2 - \pi \times 40^2)x$$

or

$$x = \frac{\pi \times 40^2 \times 40}{120^2 - \pi \times 40^2}$$

$$= \frac{\pi \times 40}{9 - \pi} \text{ (dividing by } 40^2)$$

$$= \frac{40 \times 3\cdot142}{5\cdot858} = 21 \text{ mm}$$

As $AG = 120 \times \cos 45° = 85$ mm, G is 64 mm from A along the diagonal.

Exercise 24

Find the position of the CG of each of the following laminae, referring
to the axes AX and AY: all measurements are in m.

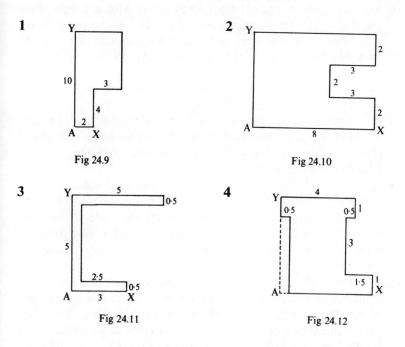

Fig 24.9   Fig 24.10

Fig 24.11   Fig 24.12

5 If the shapes in figs 24.9, 24.10, 24.11, 24.12 are suspended freely
from the corner Y what angle will AY make with the vertical in
each case?

6 ABC is a uniform rod of mass 10 kg and length 5 m. ABC is a
right angle and AB = 2 m. If suspended freely from A, what hcri-
zontal force will keep it in equilibrium with BC vertical? What
is the least force to keep the rod in equilibrium in this position?

7 If figs 24.9 and 24.10 represent shapes made from thin uniform
wire, find the position of the CG referred to the axes AX and AY.

8 Two circular cylinders, of the same length 50 cm but with radii
20 cm and 40 cm, are glued together to have a common axis of
symmetry. Find the position of the CG from the common base.

9 A circular disc of radius 50 cm has a circular hole in it of radius
20 cm. The centre of the hole is 25 cm from the centre of the disc.
Find the distance of the CG of the remainder from the centre of
the disc.

10 A rectangle 2 m by 3 m is cut out from one corner of a square of side 5 m. Find the distance of the CG of the remainder from the other two sides of the square.

11 ABC is a uniform rod of length 40 cm where ABC is a right angle and AB = BC. Find the position of the CG. If the rod is freely suspended from A what angle does AB make with the vertical?

12 Two solid spheres, of the same material and radii 20 cm and 10 cm, are connected by a light thin rod of length 18 cm, the line of which passes through both centres. Find the position of the CG of the complete solid from the centre of the larger sphere.

13 A hemisphere radius 10 cm is glued centrally to one face of a cube (made of the same material) of side 20 cm. Find the position of the CG of the combined solid from the common face. (See example 3 for CG of a hemisphere).

14 A solid circular cone of height 10 cm and base radius 5 cm is joined to a circular cylinder of the same radius and length 15 cm to have the same central axis. Both solids are made of the same material. Find the distance of the CG of the joint solid from the common base. (Volume of cone = $\frac{1}{3}\pi r^2 h$ and CG is $\frac{1}{4}$ way up central axis from the base).

15 A straight pillar is built up of two parts. One part is made of iron of diameter 20 cm and of uniform cross-section and of length 40 cm. The iron has a mass of 7 g per cm³ of volume. The other part is made of wood of square cross-section of side 10 cm and length 2 m, attached to the iron part so that they have the same central axis of symmetry. The wood used has a mass of 0·8 g per cm³. Find the position of the CG from the common surface.

16 An open tin can is taken as a cylinder with one circular end removed. Its radius is 5 cm and its length 12 cm. Find the position of the CG of the can on the central axis from the bottom of the can. If the can is suspended freely from a point on its rim, what angle will the central axis make with the vertical?

17 A square piece of cardboard of uniform thickness and of side 20 cm has two circular holes cut (each radius 1 cm) in it as shown in fig 24.13. Find the position of the CG of the remainder measured from AB and AC.

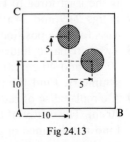

Fig 24.13

18 A circular cone of radius 10 cm and height 20 cm is glued to a hemisphere of equal radius, made from the same wood, so that they have a common central axis. Find the position of the CG of the combined solid from the common surface. (See Example 3 for the CG of a hemisphere and Question 14 for the CG of cone.) If the solid is suspended freely from a point on the edge of the common surface, what angle would the central axis make with the vertical?

19 Two hemispheres of equal radius 8 cm are glued together to make a sphere but the density of one is twice that of the other. Find the position of the CG as measured from the common face.

20 Fig 24.14 shows a cylinder of length 50 cm and diameter 20 cm into one end of which has been drilled a hole of circular cross-section (diameter 10 cm) and depth 20 cm along the axis of the cylinder. Find the position of the CG of the remainder from the other end of the cylinder.

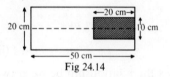

Fig 24.14

25 Momentum and Impulse

Suppose a body, mass m kg, moving in a straight line with velocity in m/s, is acted on by a force F N for t s (fig 25.1).

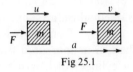

Fig 25.1

Then the acceleration of the body is given by $F = ma$ ... (i) and the velocity v reached after t s is $v = u + at$ ... (ii).

From (ii)
$$a = \frac{v - u}{t}$$

and (i) becomes
$$F = m\frac{v - u}{t}.$$

or
$$Ft = mv - mu \text{ ... (iii)}$$

This equation shows the relationship between two important quantities. **Ft** is the **time-effect** of the force and is called the **impulse** of the force. It is the product of the force (assumed constant) and the time for which it acts. Unit Force (1 N) acting for 1 s gives the unit of impulse as 1 N s (Newton second). If a force of 50 N acts on a body for 0·1 s, the impulse is $50 \times 0·1 = 5$ N s.

The right hand side of (iii) measures the change in a quantity called the **momentum** of the body, defined as **mass × velocity**. The initial momentum of the body was mu and its final momentum mv. Hence $mv - mu$ is the change of momentum (in this case an increase).

Equation (iii) shows that

   Impulse of force on a body = Change of momentum of body.

Since the unit of impulse is the N s, the unit of momentum must also be N s. So a body of mass 3 kg travelling at 5 m/s has momentum $3 \times 5 = 15$ N s. In finding momentum, mass must be expressed in kg, velocity in m/s. It must be noted that as momentum is the product of a scalar (mass) and a vector (velocity) it is itself a vector quantity. Hence care must be taken with its direction.

If the initial and final velocities are in the same line as the impulsive force, measure the momentum in the direction of the force and use the equation, impulse = final momentum − initial momentum, the arrows showing that all are measured in the same direction.

If the two momenta are in different lines, the change in momentum must be found by vector subtraction. Such cases are not dealt with at this stage.

Example 1

A force of 20 N *acts on a body of mass* 2 kg *travelling at* 4 m/s *for* 0·3 s *in the direction of its motion. What is the final velocity of the body?*
The impulse of the force = 20 × 0·3
= 6 N s.
Then measuring in the direction of the force (fig 25.2). 6 = 2v − 2 × 4
(final momentum − initial momentum)
giving v = 7 m/s.

Fig 25.2

Example 2

A body of mass 0·5 kg *travelling at* 3 m/s *encounters a constant frictional force of* 2·5 N. *What is the speed of the body after* 2 s?
The impulse of the force = 2·5 × 2
= 5 N s.
Measuring momentum in the direction of the force (fig 25.3),
5 = (− 0·5v) − (−0·5 × 3)
    final   initial
    momentum   momentum
  = −0·5v + 1·5
∴ v = 0·2 m/s.

Fig 25.3

Example 3

A ball of mass 0·2 kg *is dropped from a height of* 5 m *onto a concrete floor and rebounds to a height of* 2 m. *Find the impulse of the floor on the ball. If the contact lasts* 0·05 s *find the average force on the ball.*
(g = 10 m s^{-2}).

Let the velocities on arrival and lift-off be v_1 and v_2 respectively, and the impulsive force be P N. (Fig 25.4)
To find v_1 and v_2 use the equation
$v^2 = u^2 + 2as$ where $a = g = 10$.
Downwards, $v_1{}^2 = 2g5 = 100$ and
$v_1 = 10$ m/s.
Upwards, $0 = v_2{}^2 − 2g2$,

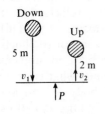

Fig 25.4

and $v_2{}^2 = 40$

giving $v_2 = 2\sqrt{10}$ m/s.

Measuring momentum in the direction of the force P,

$$\text{Impulse} = (+ 0.2 \times 2\sqrt{10}) - (-0.2 \times 10)$$
$$= 1.26 + 2 = 3.26 \text{ N s}$$
$$\text{Impulse} = Pt \text{ and } t = 0.05 \text{ s}$$
$$\therefore \quad P \times 0.05 = 3.26$$
$$\text{or} \quad P = 65.2 \text{ N}$$

Example 4.

A hose (cross-section 2 cm^2) *delivers a jet of water with a speed of* 20 m/s. *With what average force does the water hit a man?*

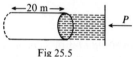

Fig 25.5

The momentum of the water is destroyed by the impulsive force exerted by the man on the water, which equals the force exerted by the water on him (fig 25.5).

Consider the momentum destroyed in 1 s.

In 1 s 20 m of water of area 2×10^{-4} m^2 issues from the pipe, a volume of $20 \times 2 \times 10^{-4} = 4 \times 10^{-3}$ m^3.

1 m^3 of water has a mass of 10^3 kg.

Hence the total momentum of the water $= 4 \times 10^{-3} \times 10^3 \times 20$ N s
$$= 80 \text{ N s}$$

The impulse lasts for 1 s and therefore

$$P \times 1 = 80 \text{ or } P = 80 \text{ N}.$$

Exercise 25.1

1 Find the momentum of the body in each of the following: **(a)** mass 2 kg, velocity 5 m/s; **(b)** mass 0.25 kg, velocity 2 m/s; **(c)** mass 80 g, velocity 1.5 m/s; **(d)** mass 300 kg, velocity 20 km h^{-1}.

2 A force of 10 N acts on a body of mass 2 kg for 0.5 s. What is the increase in momentum? If the body was originally travelling at 5 m/s what is its final speed? How far will it travel in this time?

3 A body of mass 1.5 kg travelling at 3 m/s is acted on by a force P N for 0.75 s. If its velocity at the end of that time is 5 m/s, find the value of P.

4 A ball of putty, mass 0·75 kg, moving at 3 m/s hits a wall at right angles and stops dead. Find the impulse on the ball.

5 A hammer of mass 5 kg, travelling at 4 m/s, hits a nail directly and does not rebound. What is the impulse on the hammer? If the contact effectively lasts 0·5 s what is the average force between the two?

6 A mass of 1·2 kg travelling at 2 m/s strikes a wall at right angles and rebounds (also at right angles) with a velocity of 1·5 m/s. If the contact lasted 0·25 s find the force on the mass.

7 A ball of mass 0·25 kg falls freely onto a concrete floor from a height of 20 m and rebounds to a vertical height of 5 m. If the ball was in contact with the floor for 0·8 s find the average force exerted on the ball. (Take $g = 10 \text{ m s}^{-2}$)

8 A truck of mass 50 kg has its speed reduced from 4 m/s to 1·5 m/s in 30 s. Find the braking force (assumed constant). After what further time will the truck come to rest under this force?

9 A tennis ball of mass 30 g travelling horizontally at 20 m/s is hit straight back at 30 m/s. If the impact lasted 0·04 s find the average force on the ball.

10 A box of mass 10 kg is dragged across a rough floor (coefficient of friction 0·5) by a force P N. If the speed of the box is increased from 0·5 m/s to 1·9 m/s in 10 s, find P. ($g = 9·8 \text{ m s}^{-2}$).

11 A hose (cross-section 4 cm^2) delivers water horizontally with a speed of 25 m/s. What is the impulse of the water on a vertical wall (assuming no rebound)? What average force acts on the wall?

12 A horizontal jet of water is emitted from a circular pipe of radius 1 cm at a speed of 12 m/s. Find the mass of water emitted per s and the average force exerted on a vertical wall.

Conservation of Momentum

The importance of momentum lies in the fact that momentum is not lost during a collision between two bodies.

If two bodies, A and B, collide, each will exert an impulse on the other (fig 25.6). By Newton's Third Law, A will exert a force P on B and B will exert an equal but opposite force P on A. The time of contact is naturally the same for both. Hence the impulse of A on B is Pt and impulse of B on A is Pt. Since the impulse on a body equals

Fig 25.6

the change of momentum produced, the two changes of momentum are equal but opposite in direction. So the *total* momentum of the system is unaltered when measured in the same direction. One body gains in momentum, the other loses an equal amount. This gives the principle of the **conservation of momentum**:

In any collision between two bodies, the total momentum in any direction is unchanged, provided no external force acts in that direction.

In examples 3 and 4, momentum was not conserved as the forces were external forces. Gravity is not however an external force in this context. We can also use the principle in the form: momentum before = momentum after collision (in the same direction). As an example of the principle, consider a gun being fired resting against the shoulder of a man. The explosion gives the bullet forward momentum and so the gun must acquire an equal amount of momentum backwards, thus producing the recoil of the gun against the man's shoulder. A more sophisticated example is how a spaceship can change direction in space. As there is no resistance to motion in empty space, any object will move continually in a straight line. It is impossible to produce an external force to change direction or to slow down, as there is no atmosphere or friction to "push against". If however a small rocket is fired, momentum in that direction is created and an equal but opposite amount of momentum affects the object thus changing its direction of motion or its speed. So to increase the speed of a spacecraft, a retro-rocket is fired backwards, giving the craft additional momentum (and hence increased speed) forwards. To slow it down, a forward-facing rocket is fired, so reducing the momentum of the spacecraft.

In the examples which follow, changes of momentum between two bodies moving in the same line are considered.

Example 5

A gun of mass 8 kg fires a bullet of mass 80 g at a speed of 300 m/s. With what initial speed does the gun recoil?

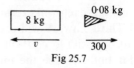

Fig 25.7

The total momentum, which initially was zero, is unchanged by the explosion. Therefore taking the direction of the bullet as positive (fig 25.7).

$$0.08 \times 300 + (-8v) = 0$$

or $\qquad 24 = 8v$ giving $v = 3$ m/s.

Example 6

Two trucks, masses 80 and 50 kg, are travelling on the same track with speeds of 4 and 2 m/s respectively in the same direction. They collide and hook together. With what speed do they combine?

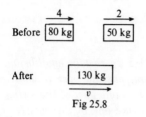

Fig 25.8

Momentum is conserved in the direction of motion as there is no external force (fig 25.8).

Momentum before collision $= 4 \times 80 + 2 \times 50 = 420$ N s.

$\qquad$,, $\qquad$ after $\qquad = 130v$ N s.

$\qquad \therefore \quad 130v = 420$

$\qquad$ or $v = 3.2$ m/s

Example 7

Two balls, masses 0·8 and 0·5 kg, roll towards each other in the same line at speeds of 2 and 3·4 m/s respectively. After the collision the first ball is observed to have a speed of 1·5 m/s in the opposite direction. What is the speed of the second ball after the collision? What was the impulse between them?

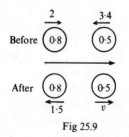

Fig 25.9

Measuring in the direction indicated, (fig 25.9)

Momentum before $= 2 \times 0.8 - 3.4 \times 0.5 = -0.1$ N s.

Momentum after $= -1.5 \times 0.8 + 0.5v$

$\therefore \quad -0\cdot1 = -1\cdot5 \times 0\cdot8 + 0\cdot5v$ giving $v = 2\cdot2$ m/s in the direction shown.

The change of momentum for the $0\cdot8$ kg ball is
$$(-0\cdot8 \times 1\cdot5) - (0\cdot8 \times 2) = -2\cdot8 \text{ N s}.$$
The change of momentum for the $0\cdot5$ kg ball is
$$(0\cdot5 \times 2\cdot2) - (-0\cdot5 \times 3\cdot4) = +2\cdot8 \text{ N s}.$$
The second ball has gained $2\cdot8$ units of momentum and the other had lost an equal amount.

The impulse = change of momentum of either ball. Hence the impulse = $2\cdot8$ N s numerically.

Example 8

Two equal masses of $1\cdot5$ kg rest in equilibrium suspended at the ends of an inextensible string passing over a small pulley. A small ring of mass $0\cdot4$ kg is threaded on one part of the string and falls down through a distance of 4 m to strike one of the masses. If it remains in contact with this mass, with what velocity does the system begin to move? ($g = 10 \text{ m s}^{-2}$).

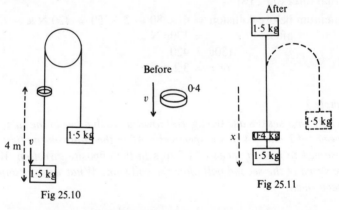

Fig 25.10   Fig 25.11

We first find the speed of the ring just before the impact (fig 25.10).
We use
$$v^2 = u^2 + 2as$$
$$v^2 = 2 \times g \times 4 = 80$$
or
$$v = 8\cdot9 \text{ m/s}.$$
Now take the downward vertical as the + direction of momentum, and let the speed of all three bodies after the collision be x m/s. Since the three masses are connected by a taut string, they are in effect one body and their momenta are all positive (fig 25.11).

Momentum before collision $= 0\cdot4 \times 8\cdot9$ N s

„   after   „ $= (1\cdot5 + 1\cdot5 + 0\cdot4)x$ N s

$\therefore \quad 0\cdot4 \times 8\cdot9 = 3\cdot4x$ giving $x = 1\cdot05$ m/s.

Exercise 25.2

1 Two trucks, masses 30 kg and 20 kg, travelling at 4 and 1 m/s respectively in the same direction, collide and continue together. Find their common speed after the collision.

2 A toy railway truck, mass 0·3 kg, travelling at 2 m/s, collides with another stationary truck, mass 0·25 kg, and they couple together. Find the common speed after impact and the impulse between them.

3 A gun of mass 450 kg fires a shell of mass 2 kg horizontally at a speed of 300 m/s. Find the initial recoil velocity of the gun.
If the gun comes to rest (moving horizontally) in 10 s find the average resisting force.

4 Two masses 3 and 2 kg move towards each other at speeds of 1·5 and 2 m/s. After the collision they move together. Find their common velocity.

5 A mass of 0·1 kg travelling at 10 m/s overtakes and collides with a mass of 0·5 kg moving at 2 m/s. They move on together. Find their common velocity.

6 Two billiard balls of equal mass (0·8 kg) are moving in opposite directions (in the same line) with speeds of 12 m/s and 5 m/s. They collide and the slower ball is now seen moving at 8 m/s in the opposite direction. Find the new speed of the other ball and the impulse between them.

7 A body of mass 10 kg is moving horizontally with a speed of 20 m/s. It explodes and splits into two parts of masses 6 kg and 4 kg. The 4 kg part continues to move in the original direction but with a speed of 30 m/s. Find the speed of the 6 kg part.

8 A spacecraft of mass 450 kg is moving in space with a speed of 3×10^3 m/s. A rocket is fired straight ahead, emitting 1·5 kg of gas at a speed of 2×10^4 m/s. Ignoring the slight reduction in mass of the spacecraft, find the change in its speed.

9 Two masses of 4 kg and 3 kg are connected by a light string over a smooth pulley. After moving for 5 s the 3 kg mass picks up a third mass of 1 kg instantaneously. Find the speed of the 3 masses after the pickup. ($g = 9·8$ m s^{-2})

10 Two masses of 5 kg and 2 kg are connected by a light string over a smooth pulley. The 5 kg mass is at rest on a horizontal table (below the pulley) and the 2 kg mass is released from rest. After it falls freely for 2 s the string is tightened and the 5 kg mass is jerked off the table. Find the velocity with which the masses now continue ($g = 9·8$ m s^{-2}). Also find their common acceleration.

11 Two masses 5 kg and 2 kg are connected by a light string over

a smooth pulley. The system is released from rest. Find the common acceleration (g = 9·8). After falling for 2 s the 5 kg mass hits a horizontal table and does not rebound. Find the velocity of the 2 kg mass at this time and find how much higher it will continue to rise before coming to rest. Find also the common velocity when the string tightens again.

12 A particle A, of mass 4 kg, is travelling in a straight line due north with a speed of 3 m/s; another particle B, of mass 3 kg, is travelling in the same straight line towards A with a speed of 5 m/s. After the collision A is moving south with a speed of 2 m/s. Calculate (i) the velocity of B after the collision, (ii) the impulse between the particles.

13 A particle of mass m hangs at rest by a string from a fixed point. A second particle of mass M, moving horizontally with speed u hits the first particle and sticks to it. Find the initial speed of the combined particle after the collision and the impulse between them.

14 Two balls of the same mass are moving in the same direction along a straight line at speeds of 4 m/s and 3 m/s. After colliding, the speed of one of them is 0·5 m/s more than the speed of the other. Find the speed of each ball.

15 A ball A moving with velocity u, collides with a similar ball B (of different mass) which is at rest. After the collision B moves with a velocity $\frac{1}{2}u$ and A moves with a velocity $\frac{1}{4}u$ in the opposite direction. Find the ratio of the masses of A and B.

26 Work, Energy and Power

Work

When a force acts on a body and causes it to move, we say the force does **work** on the body. The measurement of this work will involve the distance through which the body is moved by the force and the size of the force (or its component in the direction in which the body moves).

Fig 26.1

(1) If P moves the body through a distance s (fig 26.1) in the direction of the force, the work done $= Ps$.

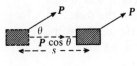

Fig 26.2

(2) In fig 26.2 P acts at an angle θ to the direction in which the body moves. The component of P along the line of motion is $P \cos \theta$. Hence the work done in this case $= P \cos \theta \times s$.

Work done by a force is also defined as the product of the force and the distance its point of application moves *in the direction of the force*.

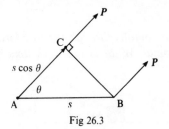

Fig 26.3

Thus in fig 26.3 the work done by $P = P \times$ AC (AC is the component of AB in the direction of the force and is the distance the point of application moves in that direction)

$$= P \times AB \times \cos \theta = P \times s \cos \theta$$

which is the same result as before.

The other component of the force ($P \sin \theta$) does no work as its point of application does not move in the direction of the force.

As work is the product of a force and a distance, the unit of work will
be 1 N × 1 m = 1 N m (Newton-metre) which is given the name
Joule (J). It is named in honour of the scientist Joule of the 19th century
who did fundamental work on the conservation of energy.

So if a force P N moves its point of application through s m in the
direction of the force, the work done = Ps J

Just as Impulse (Pt) was regarded as the time-effect of a force so Work
(Ps) may be regarded as **distance-effect** of a force.

Example 1

*I lift a mass of 4 kg at a steady speed through a vertical distance of
1·5 m. What is the work done by me?*

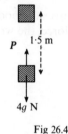

The force I exert (P N) will equal the
weight of the mass ($4g$ N) and the
distance lifted is 1·5 m in the direction
of the force (fig 26.4). Hence the work
done = $4g \times 1·5 = 6 \times 9·8 = 58·8$ J.

Fig 26.4

Note: If I just hold the 4 kg mass in my hand without moving it, I
am doing no work, according to our definition. This is more restricted
in Applied Mathematics than in ordinary life!

Example 2

*A trolley is pulled horizontally through 4 m by a force of 60 N at an angle
of 30° to the horizontal. What is the work done?*

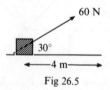

Fig 26.5

The component of the force in the direction of motion is
$$60 \times \cos 30° = 30\sqrt{3} \text{ N} \quad \text{(fig 26.5)}$$

Hence the work done = $30\sqrt{3} \times 4$
$$= 120\sqrt{3} \text{ J}.$$

Kinetic Energy (KE, E_k)

Suppose a body mass m kg is travelling at u m s^{-1} horizontally in a straight line. A force P N now acts on it in the direction of motion and gives it an acceleration a m s^{-2}. The force ceases to act when the velocity reached is v m s^{-1}.

If the distance travelled is s m,

then $$v^2 = u^2 + 2as \text{ or } as = \tfrac{1}{2}v^2 - \tfrac{1}{2}u^2 \dots\dots \text{ (i)}$$

Now the acceleration equation is $P = ma$.

The work done by the force is $Ps = mas$

But from (i) $mas = \tfrac{1}{2}mv^2 - \tfrac{1}{2}mu^2$

$\tfrac{1}{2}mv^2$ and $\tfrac{1}{2}mu^2$ are quantities of the same kind typified as

$$\tfrac{1}{2}(\text{mass}) \times (\text{velocity})^2$$

and this quantity is called the **Kinetic Energy (KE)** of a moving body. So $\tfrac{1}{2}mu^2$ = the **initial KE** of the body and $\tfrac{1}{2}mv^2$ = the **final KE**.

Our result shows that

Work done by a force = Final KE − Initial KE = Increase in KE.

The distance-effect of a force (the work done) is a change in the KE of the body acted on. The unit of KE is therefore the same as the unit of work, 1 J. In using the formula, KE $= \tfrac{1}{2}(\text{mass}) \times (\text{velocity})^2$, standard SI units must be used – kg for mass, m s^{-1} for velocity. For example, the KE of a body mass 2 kg moving at 4 m s^{-1} $= \tfrac{1}{2}(2)(4)^2 = 16$ J.

The work done by the force in raising the velocity of the body from u to v is converted into the increased KE of the body. Conversely, some or all of the KE possessed by a body can be converted into work. Hence loss of KE = work done *against* a force.

The quantity $\tfrac{1}{2}mv^2$ is always positive and is *not* a vector quantity. The KEs of two equal bodies moving in any two different directions with the same speed are equal. Again as work can be converted into KE and conversely, work is also a scalar quantity. Contrast this with momentum and impulse which are vector quantities.

Example 3

A body of mass 2 kg is brought from rest to a speed of 3 m s^{-1} over a distance of 1·2 m. What force was acting?

The initial KE = 0; the final KE $= \tfrac{1}{2}(2)(3)^2 = 9$ J

Therefore the work done by the force = force × 1·2

$$= \text{gain of KE} = 9$$

and the force $= \dfrac{9}{1\cdot2} = 7\cdot5$ N.

If more than one force acts on a body then the total work done = gain in KE of the body. The total work done is the algebraic sum of all the separate amounts of work done by or against the separate forces.

Example 4

A particle of mass 4·5 kg is projected up an incline of angle θ where sin θ = 0·2 *with speed* 2 m s⁻¹. *How far will it travel up the incline* (i) *if the surface is smooth,* (ii) *if the coefficient of friction is* 0·6? (g = 9·8 m s⁻²)

(i) The only resisting force in the direction of motion is the downward component of the weight (4·5g N) (fig 26.6).

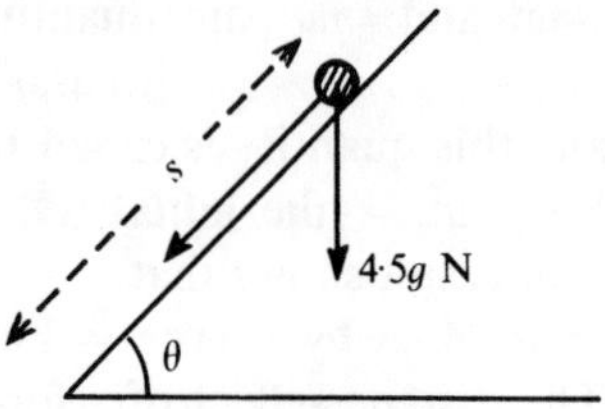

This component $= 4·5g \times \sin θ$
$$= 4·5g \times 0·2 \text{ N}$$

The work done against this resistance
$$= \text{loss of KE}$$
Initial KE $= \frac{1}{2}(4·5)(2)^2 = 9$ J

Then $4·5g \times 0·2 \times s = 9$
giving $s = 1·02$ m (2 sig figs).

Fig 26.6

(ii) There is now an additional resistance F due to friction (fig 26.7).
Resolving perpendicular to the surface,
$$R = 4·5g \cos θ$$
If sin θ = 0·2, then cos θ = 0·98
$$(\cos θ = \sqrt{1 - \sin^2 θ})$$
and $F = \mu R = 0·6 \times 4·5g \times 0·98$.

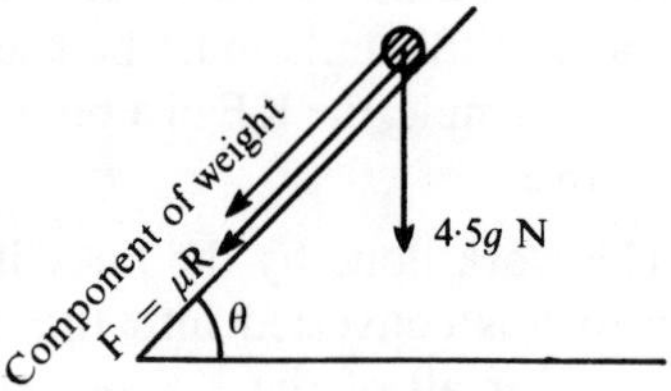

Fig 26.7

The work done against weight-component and friction
$$= \text{loss of KE}$$
∴ $(4·5g \times 0·2 + 0·6 \times 4·5g \times 0·98) \times s = 9$

or $s = \dfrac{9}{4·5g(0·2 + 0·6 \times 0·98)} = \dfrac{9}{4·5 \times 9·8 \times 0·79} = 0·26$ m.

Example 5

A gun of mass 50 kg *fires a shell of mass* 0·5 kg *with a velocity of* 250 m s⁻¹ *horizontally. In recoiling horizontally the gun encounters a resistance of* 4 N *per* kg *of mass. How far will it travel before coming to rest?*

Using the momentum principle,
$$\text{momentum of gun} = \text{momentum of shell}$$

$\therefore$ $\qquad\qquad$ $50v = 0.5 \times 250$ (where v = recoil velocity of gun)
giving $\qquad\qquad$ $v = 2.5 \text{ m s}^{-1}$

This is the initial recoil velocity of the gun. Its KE is $\frac{1}{2}(50) (2.5)^2$ J.
The resistance $= 4 \times 50 = 200$ N and the work done in recoiling
through a distance s m $= 200s$ J

$\therefore$ $\qquad\qquad$ $200s = \frac{1}{2}(50) (2.5)^2$
giving $\qquad\qquad$ $s = 0.78$ m.

Example 6

*A hammer of mass 4.5 kg falls through a vertical height of 1 m and
hits a nail of mass 50 g directly without rebounding. If the nail is then
driven into a piece of wood to a depth of 2 cm what is the average resistance of the wood?*

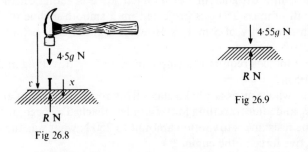

Fig 26.8   Fig 26.9

First find the speed of the hammer just before the impact (fig 26.8).
From $v^2 = u^2 + 2as$, $v^2 = 2 \times 9.8 \times 1 = 19.6$ giving $v = 4.4 \text{ m s}^{-1}$.
Now using the momentum principle find the initial common velocity
of the hammer and nail (x).

$\qquad$ Initial momentum $= 4.5 \times 4.4$ N s
$\qquad$ Final momentum $= (4.5 + 0.05) \times x$ N s

Therefore $4.5 \times 4.4 = 4.55x$ giving $x = 4.35 \text{ m s}^{-1}$ (the nail having
a small mass in comparison with the hammer, makes little difference
to the velocities before and after).

The hammer and nail now have KE $= \frac{1}{2}(4.55) (4.35)^2 = 43$ J.

If the resistance of the wood is R N, the net opposing force is
$(R - 4.55g)$ N (fig 26.9).

The work done against this force equals the loss of K.E.

Therefore $\qquad\qquad$ $(R - 4.55g) \times 0.02 = 43$
which gives $\qquad\qquad\qquad$ $R = 220$ N (to 2 sig figs).

Exercise 26.1

Take $g = 9.8$ m s^{-2}.

1 Find the work done when a load of 50 kg is lifted vertically through 10 m.

2 A block is pulled horizontally through 4 m at a steady speed by a force of 20 N, inclined at an angle of 60° to the line of motion. Find the work done.

3 A mass of 20 kg is pulled across a rough horizontal floor (coefficient of friction 0·4) through 2 m at a steady speed by a horizontal force. Find the work done.

4 If a mass of 10 kg at rest acquires a velocity of 2 m s^{-1} after being pulled through 1·5 m what force is acting in the direction of motion?

5 A body of mass 1 kg travelling at 2·5 m s^{-1} on a horizontal surface meets a rough patch and comes to rest in 2 m. What is the resisting force? Also find the coefficient of friction of the rough surface.

6 The velocity of a body of mass 0·5 kg is reduced from 3 to 1·5 m s^{-1} in a distance of 1·5 m. What force is acting on the body?

7 A ball of mass 250 g is projected up a smooth incline of $\frac{1}{70}$ (sine) with a velocity of 5 m s^{-1}. How far will it travel before coming to rest?

8 What force is required to stop a mass of 5 kg travelling at 2 m s^{-1} in 1·5 m?

9 A car whose mass is 500 kg starts from rest at the foot of an incline of $\frac{1}{70}$ and after travelling for 0·5 km has reached a speed of 5 m s^{-1}. If the resistances to motion amount to 250 N what was the average tractive force of the engine?

10 A car of mass 400 kg travelling at 9 m s^{-1} comes to rest in 200 m. What was the resistance?

11 A train of mass 100 t travelling at 0·5 m s^{-1} hits the buffers at a station and comes to rest in a distance of 30 cm. What is the average resistance of the buffers?

12 A ship of mass 5000 t moving at 0·01 m s^{-1} hits a quayside and continues to move for 15 cm before coming to rest. What average force does the quay exert on the ship?

13 A particle of mass 1·5 kg is projected up an incline of $\frac{1}{7}$ with an initial speed of 1 m s^{-1}. How far will it travel up the incline if (i) the surface is smooth, (ii) the coefficient of friction is 0·5?

14 A gun of mass 45 kg fires a shell of mass 0·9 kg at a speed of 100 m s^{-1} horizontally. Find the initial recoil velocity of the gun. If this recoil is opposed by a constant force of 250 N how far does the gun recoil?

15 A horizontal force of 10 N is applied to a body A of mass 2 kg,

initially at rest on a smooth surface, for 5 s. What velocity is gained by A? A now collides with another body B of mass 4 kg at rest and the two continue together. Find the common velocity.

16 Find the average force required to stop a 5 t lorry travelling at 10 m s^{-1} on a level road in a distance of 15 m.

17 What is the braking force operating if a lorry of mass 3·5 t can be stopped in 20 m if travelling at 20 m s^{-1}?

18 A mass of 10 kg falls vertically through a height of 5 m onto a wooden stake of mass 2 kg and does not rebound. The stake is driven into the ground through a vertical distance of 10 cm. Find the average resistance of the ground. (Take $g = 10$)

19 A shell of mass 5 kg is travelling horizontally at 200 m s^{-1} when it explodes into two parts. One part (of mass 3 kg) continues in the same direction at a speed of 400 m s^{-1}. What will be the velocity of the other part? What are the KEs before and after the explosion?

20 A bullet of mass 50 g strikes a fixed piece of wood 10 cm thick with a velocity of 300 m s^{-1} and emerges with a velocity of 100 m s^{-1}. Find the average resistance of the wood.

21 A particle of mass 4 kg is projected up an incline of $\frac{1}{70}$ with a speed of 4 m s^{-1}. The coefficient of friction is 0·4. How far will the particle have travelled when its speed is halved?

22 A nail of mass 20 g is driven horizontally into wood by a hammer of mass 3 kg. Just before the impact the hammer is moving horizontally at 4 m s^{-1}. Find the common velocity of hammer and nail after the impact, if they move together, and if the nail penetrates the wood to a depth of 5 cm, find the average resistance of the wood.

Potential Energy (PE, E_p)

Suppose I lift a mass of 4 kg vertically through a height of 2 m. The lifting force = the weight of the body = $4g$ N. Hence the work I do $= 4g \times 2 = 78 \cdot 4$ J and this work has been done against gravity. At this point the body is now at rest and has no KE, but if I let go, the body will acquire KE in falling and can do work on the downward path. Hence in its state of rest at a height of 2 m the body has a potential for doing work and we say it possesses **potential energy (PE)**. PE is the ability to do work because of the position of the body, in the sense that if released, the body will move to a lower position and its PE will be converted into KE. The PE of a body has no absolute value, but is relative to some datum level, say the surface of the earth or some other level above which the body is raised and to which it can fall.

Suppose a body of mass m kg is raised through a height of h m from a floor (fig 26.10). The work done against gravity $= mgh$ J and hence the body now posses-ses $PE = mgh$ J. If the body now falls it will acquire a velocity v on reaching its original level, where $v^2 = 2gh$. Its KE is now $\frac{1}{2}mv^2 = \frac{1}{2}m \times 2gh = mgh$. Hence all the PE has been converted into KE.

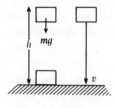

Fig 26.10

Note that this result is true if the particle descends by any route (provided it is smooth) through a vertical drop of h m (fig 26.11). The distance travelled by the point of application of the weight (the CG) in the direction of the weight is always h. Hence the work done by

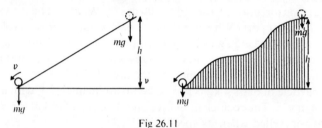

Fig 26.11

gravity is mgh, which is converted into KE. Similarly if a body of mass m kg is raised through a vertical height (h m) by whatever path (provided smooth), the work done against gravity $= mgh$ and this is the value of the PE of the body.

If the body now strikes the floor some KE will be lost, i.e., converted into another form of energy, heat, light, sound for example. This is a simple illustration of the principle of the **conservation of energy**, which states that the total energy in a closed system is constant. This principle is true provided *all* forms of energy, mechanical and non-mechanical are taken into account, such as heat, sound, light, chemical, electrical energy, etc. It also illustrates the fact that energy can be converted from one form to another. For example in a hydroelectrical scheme, the water in a reservoir possesses PE. This can be converted into KE by allowing the water to fall through a sluice gate. It then strikes turbine wheels, being converted into another form of kinetic energy, kinetic energy of rotation. This in turn, is converted into electrical energy, which is used in factories and homes to be converted into light, heat and kinetic energy again.

From a mechanical point of view, energy dissipated through friction, heat, sound etc, is energy lost and wasted. If there were no such losses it would be possible to achieve perpetual motion mechanically. In applied mathematics, we deal only with KE and PE. The principle of conservation will then appear in the form:

KE + PE = constant

Hence, the (KE + PE) of a body at any time =
original KE + PE + any work done by a force on the body.

In certain problems, particularly involving collisions, we shall find that KE is lost. Actually this means that it has been converted to a non-mechanical form of energy.

Example 7

Two trucks, of masses 1 t and 0·8 t, travel towards each other with speeds of 6 and 5 m s^{-1} respectively. They collide and couple together. Find their common speed after the collision and the loss of energy.

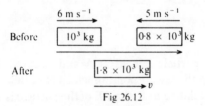

Fig 26.12

Let v be the common final speed (fig 26.12). No momentum is lost in the collision.

Momentum before = $6 \times 10^3 - 5 \times 0\cdot8 \times 10^3$ N s

Momentum after = $1\cdot8 \times 10^3 \times v$ N s

∴ $6 \times 10^3 - 5 \times 0\cdot8 \times 10^3 = 1\cdot8 \times 10^3 \times v$

or $v = \dfrac{2}{1\cdot8} = 1\cdot1$ m s^{-1} in the direction of the force.

KE before = $\frac{1}{2}(10^3)(6)^2 + \frac{1}{2}(0\cdot8)(10^3)(5)^2$

= $18 \times 10^3 + 10 \times 10^3 = 28 \times 10^3$ J

KE after = $\frac{1}{2}(1\cdot8 \times 10^3)(1\cdot1)^2 = 1\cdot1 \times 10^3$ J

∴ Loss of energy = $26\cdot9 \times 10^3$ J

Note that the percentage loss of energy

$= \dfrac{\text{Loss}}{\text{original KE}} \times 100 = \dfrac{26\cdot9 \times 10^3}{28 \times 10^3} \times 100 = 96\%.$

96% of the original energy is lost in the collision, which will give some idea of the great wastage of energy in such collisions.

This loss of energy would mostly appear as sound and heat in the couplings.

Example 8

A particle of mass 2 kg is suspended at the end of a light string length 40 cm which is fixed at its other end. The particle is held with the string taut at an angle of 30° to the downward vertical and let go. What is its maximum velocity in the ensuing pendulum motion?

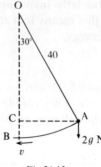

Fig 26.13

In fig 26.13 the particle is held at A and released. It then describes an arc of a circle, the lowest point of which is B. BC = $40 - 40 \cos 30°$ = $40 - 20\sqrt{3}$. Relative to B, the PE of the particle is $2g(40 - 20\sqrt{3})$ J. As the particle descends this PE is gradually converted to KE which reaches a maximum at B (zero PE). Hence the maximum speed of the particle (v) will occur at B, with the particle travelling horizontally.

Then $\frac{1}{2}(2) v^2 = 2g(40 - 20\sqrt{3})$

or $v^2 = 2 \times 9·8 \times 5·36$

which gives $v = 10·25 \text{ m s}^{-1}$.

It is worth considering this problem theoretically. If the particle is of mass m and is released from the position where the string makes an angle θ with the vertical and the string is of length l, then the

$$PE = mg(l - l \cos \theta).$$

The KE at the lowest position = $\frac{1}{2}mv^2$

Hence $\frac{1}{2} mv^2 = mg(l - l \cos \theta)$

or $v^2 = 2gl(1 - \cos \theta) \dots \text{(i)}$

Note: The result is independent of the mass of the particle. Some of our calculation above was thus unnecessary.

The converse problem is similarly solved. Equation (i) above will give the angle which the string will finally make when the particle is projected with a horizontal velocity v (provided the particle does not rise above the horizontal through O).

Example 9

A truck of mass 100 kg *is travelling in a straight line on level ground with a speed of* 20 m/s. *After a distance of* 30 m *the ground slopes upwards at an angle of* 30° *to the horizontal. The frictional resistance of the ground is* 5 N *per kg. Find how far up the slope the truck will travel before coming to rest.* $(g = 9.8 \text{ m s}^{-2})$

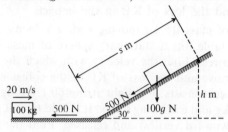

Fig 26.14

The forces acting and the distance travelled are shown in fig 26.14. The ground resistance is 500 N.

Initially the truck has KE $= \frac{1}{2}(100)(20)^2 = 20\ 000$ J

Let the truck travel s m up the slope equivalent to a vertical rise of h m, where $h = s \sin 30°$.
Then the total work done against ground resistance

$$= (500 \times 30 + s \times 500) \text{ J}$$
$$= (15\ 000 + 500s) \text{ J}$$

In climbing the slope the truck acquires PE $= 100gh$
$$= 100gs \sin 30°$$
$$= 490s \text{ J}$$

Now, the initial energy $=$ final energy $+$ work expended
$$20\ 000 = (15\ 000 + 500s) + 490s$$

or $5000 = 990s$
giving $s = 5.05$ m.

Exercise 26.2

If required take $g = 9.8 \text{ m s}^{-2}$.

1 Two trucks, masses 30 kg and 20 kg, travelling at 6 and 2 m s^{-1} respectively in the same direction, collide and continue together. Find the loss of KE due to the collision, and the percentage loss of energy.

2 Two masses, of 3 and 2 kg, move towards each other at speeds 2 and 1 m s^{-1} respectively. After colliding they move together. Find the percentage loss of energy in the collision.

3 A force of 2 N is applied for 5 s to a mass of 2 kg resting on a smooth horizontal surface. The mass now collides with a second mass of 3 kg at rest, and they continue together. Find the common velocity and the loss of KE in the impact.

4 A sphere of mass 0·5 kg moving with a velocity of 4 m s^{-1} on a smooth table hits a stationary sphere of mass 0·75 kg and is brought to rest. Find the velocity with which the second sphere starts to move and the loss of KE in the collision.

5 A pendulum consists of a light string 60 cm long attached to a mass of 5 kg and can swing freely. It is held taut at an angle of 60° to the downward vertical and released. Find the velocity of the mass at its lowest point.

6 Two masses of 5 and 3 kg move directly towards each other and collide. Their speeds before impact are 4 m s^{-1} and 3 m s^{-1} respectively. After the collision the 3 kg mass reverses at a speed of 2 m s^{-1}. Find the velocity of the 5 kg mass after the collision. Also find the percentage loss of KE.

7 A mass of 4 kg suspended by a light string 2 m long and at rest is projected horizontally with a velocity of 1·5 m s^{-1}. Find the angle made by the string when the mass comes to momentary rest.

8 A mass A travelling at 6 m s^{-1} collides with a mass of 2·5 kg travelling in the opposite direction at 2 m s^{-1}. After the collision both masses continue in the original direction of motion of mass A with speeds of 3 and 4 m s^{-1} respectively. Find the mass of A and the loss of KE in the collision.

9 A mass of 10 kg slides down a slope of 30° from rest. The coefficient of friction is 0·5. If the length of the slope is 5 m find the velocity of the mass at the foot of the slope.

10 A ball of mass 0·5 kg is dropped from a height of 5 m onto a table and rebounds to a height of 1 m. Find the percentage loss of KE due to the impact.

11 A boy of mass 45 kg dives from the stern of a rowing boat of mass 90 kg. The boat is motionless in the water but can move freely. If the boy gives himself a horizontal velocity of 3 m s^{-1} relative to the boat, find the horizontal velocity given to the boat. Find also the KE generated, assuming that neither boy nor boat is given a vertical velocity.   (C)

12 Two particles, masses 2 kg and 4 kg, are suspended by equal strings of length 2 m from the same point. They are held taut at angles

of 10° and 8° respectively to the downward vertical and on opposite sides. Then they are released simultaneously. If the particles stick together on impact, what is their common velocity immediately afterwards? What is the percentage loss of energy due to the impact?

13 Two pendulums with light strings each 1 m long carry masses of 1 and 2 kg and are suspended side by side from the same point. The larger mass is raised until its string is horizontal and taut and is then released. If the two masses stick together on impact, find the vertical height to which they rise after the impact.

14 Masses of 10 kg and 4 kg are connected by a light string passing over a smooth pulley. After the 10 kg mass descended from rest for a time of 2 s, find (**i**) the velocity of each mass; (**ii**) the potential energy lost by the system.

15 A pendulum consists of a mass of 8 kg attached to a fixed point by a light string 50 cm long. The mass is at rest when it is struck a blow lasting 0·05 s by a force of 200 N, acting horizontally. Find the angle made by the string with the downward vertical when the mass first comes to rest.

Power

One machine does a certain amount of work in 1 s, a second machine does the same amount of work in 2 s. Hence the first machine is more **powerful** than the first. Its **rate of doing work** is greater (twice as great in fact). This gives us a concept of power or rate of doing work.

Power = rate of doing work = number of J/s

The unit of power is 1 J/s = 1 **Watt (W)**
For very powerful machines we use a unit of 1 **kW** (kilowatt = 10^3 W) or 1 **MW** (Megawatt = 10^6 W).
A well known unit in the British system of units was the horsepower (HP) which is approximately 746 W. This was the original unit of power, established by James Watt in the 18th century, when working on the development of steam engines and the new unit has been named in his honour.

Example 10

My mass is 76 kg *and I run up a flight of stairs to a vertical height of* 10 m *in* 4 s. *At what rate am I working?*

Work done against gravity = $76g \times 10$ J and this is done in 4 s.

Hence the power developed $= \dfrac{76g \times 10}{4} = 186$ W.

Compare this with the power consumed in an electric light bulb of say 150 W or developed by a small electric motor of 250 W. It would of course be possible for a man to develop much higher rates of working but only for short periods.

Example 11

A cyclist is travelling at a steady speed of 4 m s^{-1}. *The total mass of the cyclist and machine is* 80 kg. *All resistances* (*wind, friction, etc*) *amount to* 1·2 N *per* kg *of mass. At what rate is he working?*

At a steady speed, the driving force exerted by the cyclist equals the resistance. The total resistance $= 96$N

In 1 s the cyclist moves 4 m.

Hence the work done per s $= 96 \times 4$ J $= 384$ J

and so the power $= 384$ W.

Example 12

A train of mass 200 t *is travelling on a level track at a steady speed of* 50 km h^{-1} *and working at* 75 kW. (**i**) *What is the resistance per* t *to motion?* (**ii**) *If the engine continues working at the same rate and the resistance is unchanged, what is the maximum speed with which it can climb a slope of* $\frac{1}{400}$? (**iii**) *If it travels up the same slope at a speed of* 20 km h^{-1} *what acceleration will it have?* (*Power and resistance as before.*)

(**i**) Consider the train on the level (fig 26.15). Let the resistance be P N/t.

The total resistance is then 200P N.

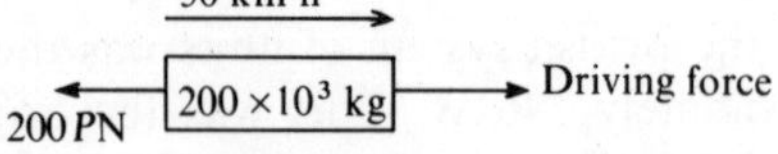

50 km h^{-1}

200 PN ← $\boxed{200 \times 10^3 \text{ kg}}$ → Driving force

Fig 26.15

The distance travelled per s $= \dfrac{50}{3600}$ km

$$= \dfrac{50 \times 10^3}{3600} \text{ m}$$

$$= \frac{5}{36} \times 10^2 \text{ m.}$$

At a steady speed the driving force = resistance = $200P$ N

∴ The work done per s = $200P \times \frac{5}{36} \times 10^2$ J which equals the stated power 75×10^3 W.

Hence $\qquad 200P \times \frac{5}{36} \times 10^2 = 75 \times 10^3$

which gives $\qquad\qquad\qquad P = 27$ N.
(The total resistance is then $27 \times 200 = 54 \times 10^2$ N)

(ii) Consider the train going up the slope (fig 26.16).

$$\sin \theta = \tfrac{1}{400}.$$

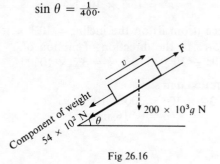

Fig 26.16

In addition to the resistance already found there is the component of weight down the slope.
Hence the total resistance to motion
$$\begin{aligned} &= 54 \times 10^2 + 200 \times 10^3 \times g \times \sin \theta \\ &= 54 \times 10^2 + 200 \times 10^3 \times 9{\cdot}8 \times \tfrac{1}{400} \\ &= 5{\cdot}4 \times 10^3 + 4{\cdot}9 \times 10^3 \\ &= 10{\cdot}3 \times 10^3 \text{ N} \end{aligned}$$

When travelling at its *maximum* speed (v m s^{-1}) the train has no acceleration. Hence the driving force F = the total resistance. In 1 s the driving force moves v m, so the work done per s = Fv J, i.e., a power of Fv W which equals the stated power 75×10^3 W.

∴ $\qquad\qquad 10{\cdot}3 \times 10^3 \times v = 75 \times 10^3$

giving $\qquad\qquad\qquad v = 7{\cdot}3$ m s^{-1}.

This is equivalent to $\frac{7{\cdot}3}{10^3} \times 60 \times 60 = 26$ km h^{-1} which is the maximum speed up this incline using the given power.

(iii) 20 km h^{-1} is equivalent to $\dfrac{20 \times 10^3}{3600} = \dfrac{50}{9}$ m s^{-1}.

The train is not travelling at its maximum speed (found in **ii**) but the power developed remains the same. Hence there is spare power available to accelerate the train.
Let the driving force at this speed ($\dfrac{50}{9}$ m s^{-1}) be F' N.

The work done per s will be $F' \times \dfrac{50}{9}$ J and this will equal the given power 75 $\times$ 10^3 W.

Then $F' \times \dfrac{50}{9} = 75 \times 10^3$

i.e. $F' = 13 \cdot 5 \times 10^3$ N.

But the total resistance (from **ii**) on the incline = $10 \cdot 3 \times 10^3$ N. So there is a resultant force in the direction of motion of
$$(13 \cdot 5 \times 10^3 - 10 \cdot 3 \times 10^3)\, \text{N} = 3 \cdot 2 \times 10^3\, \text{N}$$

which gives an acceleration a m s^{-2}.
Using $F = ma$, $3 \cdot 2 \times 10^3 = 200 \times 10^3 \times a$
giving $a = 0 \cdot 016$ m s^{-2}.

Example 13

A pump, working at a steady rate, lifts water at rest, from a well of vertical depth 8 m and then discharges it at a speed of 8 m s^{-1} through a pipe of cross-section 10 cm^2. At what rate is the pump working, if it is only 60% efficient? (1 m^3 of water has a mass of 10^3 kg).
The work done by the pump consists of (**a**) lifting the water through 8 m against gravity and (**b**) giving it KE (speed 8 m s^{-1}).
We find the amount of work done per s. This will give the rate of working of the pump, i.e. the power.
First, we must find the volume and then the mass of water carried per s. In 1 s a length of 8 m of water issues from the pipe (area 10 cm^2).
So the volume of water carried per s = 8 $\times$ $\dfrac{10}{10^4}$ m^3 and this has a

mass of 8 $\times$ $\dfrac{10}{10^4}$ $\times$ 10^3 = 8 kg or a weight of 8g N.

The work done in lifting this weight of water through a height of 8 m = 8g $\times$ 8 = 64g J and this is done in 1 s.
The KE of the water = $\frac{1}{2}$(8) $\times$ (8)^2 = 256 J and this is given in 1 s.
Hence the total amount of work done in 1 s = 64g + 256 = 883 J.

As the pump is 60% efficient, this output from the pump is 60% of the actual work done by the pump (P J).
Hence 60% of $P = 883$ J

and $$P = 883 \times \frac{100}{60} \text{J} = 1472 \text{ J}.$$

As this is done in 1 s, the actual power of the pump is 1472 W. (Note that 40% of this is wasted, against friction mostly.)

Exercise 26.3

($g = 9{\cdot}8$ m s^{-2})

1 If a car travels at a steady speed of 15 m s^{-1} against resistances of 200 N what power is being exerted by the engine?

2 A boy of mass 44 kg runs up a flight of stairs of vertical height 4 m in 5 s. What power is he sustaining?

3 A man runs 100 m in a time of 5 s. If the resistances to motion are estimated at 50 N what power does he use?

4 The power of the engine of a car is 7 kW. What would be the maximum speed of the car on the level against resistances of 250 N?

5 A pump raises water through a height of 15 m at the rate of 0·05 m^3 per s. What is the power of the pump?

6 A train of total mass 300 t travels at a constant speed of 20 m s^{-1} on the level, the resistances being 100 N per t of mass. What is the power of the engine?

7 A fire hose delivers water horizontally at a speed of 20 m s^{-1} through a nozzle of cross-sectional area 10 cm^2. Find the power of the pump if it is only 70% efficient.

8 A car of mass 800 kg is travelling at a steady speed of 20 m s^{-1} on the level. The engine is developing a power of 8 kW. Find the resistance to motion.

9 A pump delivers water from a depth of 15 m and delivers it at a rate of 0·1 m^3/s at a speed of 10 m/s. Find the power of the pump.

10 On the level a car develops a power of 15 kW. If the resistance to motion is 300 N what is the maximum speed of the car? Working at the same power and with the same resistance operating, what would be the maximum speed possible up an incline of $\frac{1}{70}$ if the mass of the car is 500 kg?
What is the acceleration at the time when the car is moving up this incline at 30 m s^{-1}?

11 A car of mass 800 kg working at 12 kW can climb a slope of $\frac{1}{140}$ at a steady speed of 30 m s^{-1}. What is the resistance due to motion? If the resistance and the power are unchanged, what would be the maximum speed of the car on level ground?

12 A diesel electric engine has a power rating of 3000 kW. If it travels at a steady speed of 120 km h^{-1} on the level find the resistance to motion.

If the total mass of the same engine and its train is 450 t and the same power is used, what is the acceleration on the level if the speed is 100 km h^{-1}?

13 If a car, of mass 900 kg, can travel at a maximum speed of 40 m s^{-1} on the level and at a maximum speed of 30 m s^{-1} up an incline of $\frac{1}{70}$, find the resistance to motion (assumed the same in both cases) and the power of the engine.

14 A car is rated at 30 HP. Taking 1 HP $\approx$ 750 W, find the maximum speed of the car up an incline of $\frac{1}{280}$ if the resistance to motion is 700 N and the mass of the car is 800 kg.

15 A car of mass 800 kg freewheels at a steady speed of 20 m s^{-1} down a slope of $\frac{1}{70}$. Find the resistance to motion. Assuming that the resistance to motion is proportional to the square of the speed, find the resistance at a speed of 30 m s^{-1}. Now find the power required to drive this car up the same incline at a steady speed of 30 m s^{-1}.

16 A train of mass 400 t is moving up an incline of $\frac{1}{280}$ with an acceleration 0·2 m s^{-2}. If the resistance to motion is equivalent to 10 N/t find the power being developed by the engine when the speed of the train is 10 m s^{-1}. (Find the total resistance to motion; also find (using $F = ma$) the force required to accelerate the train. The total driving force is then known and now find the work done at the given speed.)

27 Machines

A machine is a device producing a force which can move or lift a **load** when another force, the **effort**, is applied to the machine. The simplest machine is a **lever**.

Levers

Any rigid body supported at a point, called the **fulcrum**, can be used as a lever. A lever is used **(a)** to apply a force at a point other than where the effort is applied, **(b)** to move a heavy load by a smaller effort. Fig 27.1 shows a lever, supported at its fulcrum F, being used to

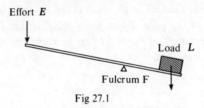

Fig 27.1

lift a load L by an effort E (directed vertically downwards). (The force produced by the lever is equal and opposite to the weight of the load). Fig 27.2 shows the forces acting on the lever.

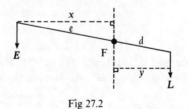

Fig 27.2

When the lever is in balance about F, we can take moments.
Taking moments about F, $E \times x = L \times y$ (neglecting the weight of the lever itself at present).

But, by similar triangles, $\dfrac{x}{e} = \dfrac{y}{d}$ and therefore

$$E \times e = L \times d$$

$$\text{or } \frac{L}{E} = \frac{e}{d}.$$

The ratio $\dfrac{L}{E} = \dfrac{\text{Load}}{\text{Effort}}$ is called the **Mechanical Advantage (MA)** of the machine.

$$\boxed{\text{MA} = \frac{\text{Load}}{\text{Effort}}}$$

In the above simple case, the MA $= \dfrac{e}{d}$.

Example 1

A uniform wooden pole rests on a stone, which acts as a fulcrum. The pole is then used to lift a slab of concrete of mass 50 kg. The distance of the fulcrum from the CG of the slab is 0·8 m and the effort is applied at the other end of the pole, 1·4 m from the fulcrum. If the mass of the pole is 5 kg, find the effort required just to move the slab and calculate the MA. ($g = 9\cdot 8 \text{ m s}^{-2}$)

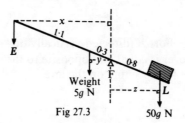

Fig 27.3

The length of the pole $= 1\cdot 4 + 0\cdot 8 = 2\cdot 2$ m.
The CG of the pole is at its midpoint, 0·3 m from the fulcrum (fig 27.3).
The force exerted by the load $= 50g$ N and the weight of the pole $= 5g$ N.
Taking moments about F (the fulcrum),
$$E \times x + 5g \times y = 50g \times z.$$
But, by similar triangles, $\dfrac{x}{1\cdot 4} = \dfrac{y}{0\cdot 3} = \dfrac{z}{0\cdot 8}$
Hence,
$$E \times 1\cdot 4 + 5g \times 0\cdot 3 = 50g \times 0\cdot 8$$
$\therefore$ $1\cdot 4E = g(50 \times 0\cdot 8 - 5 \times 0\cdot 3) = 9\cdot 8 \times 38\cdot 5$
Hence $E = 270$ N
The MA $= \dfrac{\text{Load}}{\text{Effort}} = \dfrac{50g}{270} = \dfrac{490}{270} = 1\cdot 8$ (no units as it is a ratio).

Types of Levers

These depend on the different relative positions of load, effort and fulcrum.

(a) The fulcrum lies between the load and effort, as in fig 27.1;

(b) the load is placed between fulcrum and effort, as in fig 27.4;

(c) the effort is applied between the load and the fulcrum, as in fig 27.5.

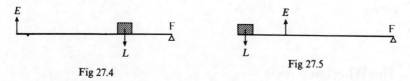

Fig 27.4

Fig 27.5

Common examples of the three types are:

(a) see-saw; a pair of scissors; a simple balance; a steelyard; a pair of pliers; a crowbar.

(b) a wheelbarrow; nutcrackers; a trap-door; an oar; a crowbar.

(c) a pair of tongs; the human forearm.

Example 2

A uniform steel crowbar, length 1 m, is placed under a packing case, so that the floor acts as fulcrum. The weight of the packing case is equivalent to a force of 10 000 N acting at a point 15 cm along the crowbar from the fulcrum. If the mass of the crowbar is 10 kg and the effort is applied vertically at the other end of the bar, calculate (i) the effort required just to raise the case, (ii) the mechanical advantage and (iii) the vertical reaction of the floor on the bar. ($g = 9 \cdot 8$ m s^{-2})

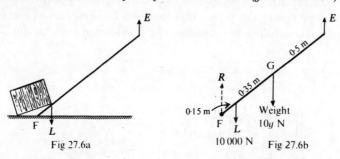

Fig 27.6a

Fig 27.6b

The problem is illustrated in fig 27.6a and the forces acting on the bar are shown in fig 27.6b.

  (i) Taking moments about F (the fulcrum) and using similar triangles as above:

$$10\,000 \times 0{\cdot}15 + 10g \times 0{\cdot}5 = E \times 1$$

Hence, $E = 1500 + 49 = 1549 \text{ N}$

(ii) The MA $= \dfrac{\text{load}}{\text{effort}} = \dfrac{10\,000}{1549} \approx 6{\cdot}5$

(iii) Taking moments about G and using similar triangles:

$$R \times 0{\cdot}5 - 10\,000 \times 0{\cdot}35 = 1549 \times 0{\cdot}5$$

Hence, $R \times 0{\cdot}5 = 3500 + 775 = 4275$

giving $R = 8550 \text{ N}.$

The Wheel and Axle

The basic form of this machine is shown in fig 27.7.

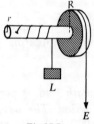

Fig 27.7

A large wheel, radius R, is fixed to an axle of radius r. A rope wound round the wheel is pulled by an effort E, turning the wheel and axle together. Another rope wound round the axle in the opposite direction to the rope round the wheel, lifts the load L. Examples in everyday use of this machine are a windlass (fig 27.8) and a capstan (fig 27.9).

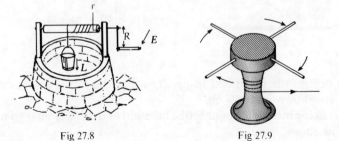

Fig 27.8   Fig 27.9

The Principle of Work

The work put into a machine = the effort × the distance moved by the effort. This is called the **input**. The **output** from a machine = the load × the distance travelled by the load. In an ideal machine, where no work is consumed or wasted by friction etc., the input = the output. In all practical machines however, work is wasted in overcoming friction or in moving parts of the machine itself, and so we have the **principle of work: input = output + wasted work.**

The Velocity Ratio (VR)
The velocity ratio of a machine is defined as:

$$\boxed{\text{VR} = \frac{\textbf{distance moved by the effort}}{\textbf{distance moved by the load}}}$$

For the wheel and axle, consider the wheel making one revolution.
Then the distance moved by the effort = $2\pi R$
and the distance moved by the load = $2\pi r$.
Hence the VR $= \dfrac{2\pi R}{2\pi r} = \dfrac{R}{r}$.

The efficiency of a machine (η)*

The efficiency of a machine is defined as $\dfrac{\text{output of work}}{\text{input of work}}$.

Hence $\eta = \dfrac{\text{load} \times \text{distance moved by load}}{\text{effort} \times \text{distance moved by effort}}$

$= \dfrac{\text{load}}{\text{effort}} \times \dfrac{\text{distance moved by load}}{\text{distance moved by effort}}$

$= \dfrac{\text{L}}{\text{E}} \times \dfrac{1}{\text{VR}} = \dfrac{\text{MA}}{\text{VR}}$

Therefore
$$\boxed{\eta = \frac{\textbf{MA}}{\textbf{VR}}}$$

In an ideal machine, where no work is wasted, and output = input, $\eta = 1$.
Hence in an ideal machine MA = VR.

---

*Greek letter, "eeta".

The efficiency of a machine is usually expressed as a percentage, i.e. by multiplying the fraction obtained from this formula by 100.

Example 3

In a wheel and axle machine, the radius of the wheel is 45 cm and the radius of the axle is 7·5 *cm. A force of* 160 N *is needed to raise a load of mass* 90 kg. *Find* **(i)** *the velocity ratio and* **(ii)** *the efficiency of the machine expressed as a percentage. (Take g = 10)*

$$\text{VR} = \frac{\text{distance effort moves}}{\text{distance load moves}}$$

$$= \frac{2\pi \times 45 \text{ cm}}{2\pi \times 7\cdot5 \text{ cm}}$$

$$= \frac{45}{7\cdot5} = 6$$

$$\text{MA} = \frac{\text{load}}{\text{effort}}$$

$$= \frac{90g \text{ N}}{160 \text{ N}} = \frac{900 \text{ N}}{160 \text{ N}} = \frac{45}{8}$$

$$\eta = \frac{\text{MA}}{\text{VR}}$$

$$= \frac{45}{8} \div 6 = \frac{45}{48}$$

$$= \frac{15}{16} \times 100\% = 93\cdot75\%$$

Exercise 27.1

Take $g = 10 \text{ m s}^{-2}$

1 A crowbar, 250 cm long, rests on a fixed fulcrum at one end. A load effectively placed at 10 cm from the fulcrum is raised by a force of 500 N. Calculate the load, in kg, and the mechanical advantage of the lever.

2 A crowbar, 250 cm long, rests on a fixed fulcrum 10 cm from one end. A load of mass 1200 kg is placed at the end near the fulcrum. Calculate the effort, applied at the other end, needed to raise the load.

3 A lever, 3 m long, is used to lift a load of mass 35 kg by applying an effort of 100 N. If the fulcrum is between the load and the effort, find the distance of the fulcrum from the load.

4 In a wheel and axle machine, the radius of the wheel is 80 cm and the radius of the axle is 10 cm. A 100 kg load is raised by a force of 200 N. Find **(i)** the velocity ratio and **(ii)** the efficiency of the machine.

5 The efficiency of a wheel and axle is 95% when a load of mass 80 kg is raised. Find the effort, in N, required if the radius of the wheel is 50 cm and of the axle is 10 cm.

6 A capstan consists of an upright drum, 1 metre in diameter, with two poles placed opposite each other in holes in the drum so that a form of wheel and axle is obtained. Two men operate the capstan, one at the end of each pole; the distance between the men is 8 m. The capstan is used to haul a boat out of the water and up a beach, by means of a rope wound round the drum. Draw a diagram of the capstan and find its velocity ratio. If the efficiency of the capstan is 65%, and each man exerts a force of 250 N, calculate the force pulling a boat out of the water.

Pulleys

The simplest application of a pulley is shown in fig 27.10. A string, or rope, passes round the pulley, and an effort, E, is applied to the rope. The effort, E, moves a distance of $2x$ upwards to raise the pulley a distance x. The velocity ratio is thus $\dfrac{2x}{x}$ i.e. 2. If the pulley is frictionless, the tension in the rope is the same on both sides of the pulley, and is equal to the effort. (Friction in the pulley reduces the tension in the rope on the side of the pulley opposite to the effort.)

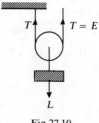

Fig 27.10

Hence, in an ideal machine, with pulleys of negligible weight,
$$L = 2T = 2E \text{ (since } T = E).$$
The MA $= \dfrac{L}{E} = \dfrac{2E}{E} = 2$, i.e. the MA $=$ the VR in an ideal machine.

In practical machines, the efficiency is less than 100% because **(a)** friction in the pulley absorbs work and **(b)** the pulleys are not of negligible weight. However this method of finding the VR can be applied to other pulley systems.

Single-string pulley systems
This system of pulleys is the most common in use. Practical machines

using this system are called **block and tackle** machines. The simplest example, using two pulleys, is sketched in fig 27.11.

To find the VR, assume that the load is to be raised through a distance x. The two parts, P_1 and P_2, of the rope have each to be shortened by x, so the whole rope must shorten by $2x$, i.e. the effort moves down a distance $2x$. Hence

the VR $= \dfrac{2x}{x} = 2$.

Fig 27.11

With three pulleys (fig 27.12), three sections of the rope connect the upper and lower blocks of pulleys and hence each section has to be shortened by x to raise the load through a distance x. The effort must move therefore through a distance $3x$ to take in the whole

rope. Hence the VR $= \dfrac{3x}{x} = 3$.

Fig 27.12

A five pulley system is shown in fig 27.13. Note that five sections of the rope connect the two pulley blocks. By similar reasoning to that above, the VR is found to be 5. From these cases, it is seen that the VR of a single-string pulley system equals the number of pulleys in the system.

Fig 27.13

Block and tackle

A common form of a block and tackle is shown in fig 27.14. Such machines are chiefly used on ships for raising and lowering sails or moving cargo. Blocks with varying numbers of pulleys are available so that any desired VR can be obtained. The weight of the lower block of pulleys combined with friction in the bearings, reduces the efficiency of the system.

E

Fig 27.14

Example 4

A block and tackle has two pulleys in each block. The lower block has a mass of 7 kg. An effort of 80 N raises a load of mass 21 kg. Calculate (**i**) *the efficiency of the system,* (**ii**) *the amount of work wasted in over-coming friction when the load is raised through* 1 m. ($g = 10$ m s^{-2})

(**i**) Referring to fig 27.14, the VR of the system is 4 (4 pulleys, 4 rope sections).

$$\text{The MA} = \frac{L}{E}$$

$$= \frac{21g}{80} = \frac{210}{80} = \frac{21}{8}$$

Hence, $$\eta = \frac{\text{MA}}{\text{VR}}$$

$$= \frac{21}{8} \div 4 = \frac{21}{32} = 66\% \text{ approx.}$$

(**ii**) When the load is raised through 1 m, the output = 210 N × 1 m = 210 J.
As the VR = 4, the effort moves through 4 m.
Hence the input = 80 N × 4 m = 320 J.
The work wasted = 320 − 210 = 110 J.
The work done in raising the block of mass 7 kg through 1 m
$$= 7g \times 1 \text{ m} = 70 \text{ J}.$$
Hence the work wasted against friction = 110 − 70 = 40 J.

The Weston differential pulley

This pulley system consists of an upper block containing two concentric pulleys of different radii r_1 and r_2 ($r_1 > r_2$) rotating together and a lower block containing one pulley (fig 27.15). The two blocks are connected by a continuous chain running over the pulleys. The load is attached to the lower block, and the effort is applied by pulling down the outer section of the chain. When the effort moves down $2\pi r_1$, both pulleys in the upper block rotate through one revolution. The part C_1 of the chain moves upward through a distance $2\pi r_1$ and the part C_2 moves downward $2\pi r_2$.

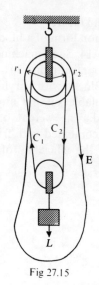

Fig 27.15

The net shortening of the chain is therefore $2\pi r_1 - 2\pi r_2$. The lower block moves upward through *half* this distance, i.e. $\pi(r_1 - r_2)$.

Hence the VR $= \dfrac{\text{distance moved by effort}}{\text{distance moved by load}} = \dfrac{2\pi r_1}{\pi(r_1 - r_2)} = \dfrac{2r_1}{r_1 - r_2}$.

By making r_1 nearly equal to r_2 a very large VR can be obtained. Due to friction, the efficiency is very low but nevertheless the MA is still quite high, and so this system is widely used for lifting heavy loads. It has practical advantages in being very robust and easily transportable.

Multiple string pulley systems

Fig 27.16 shows such a system. These systems are best considered by assuming the machine to be an ideal one. The VR of the system in fig 27.16 can be found at follows:

(1) Let the effort $= E$. Then $E = T_1$.

(2) Pulley P_3 has no effect on the VR as it only changes the direction of the effort. Pulley P_2 is supported by 2 sections of a string (assumed both vertical) in which the tension is T_1.
Hence $T_2 = 2T_1$.

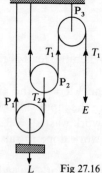

(3) Pulley P_1 is supported by a force of $2T_2 = 4T_1$.
Therefore, $L = 4T_1$.

Fig 27.16

The MA $= \dfrac{4T_1}{T_1} = 4$ and this equals the VR for an ideal machine.

As the VR is the same whether the machine is ideal or not, the VR of the system $= 4$.

A second system of pulleys is shown in fig 27.17. Once again we calculate the VR from the MA of an ideal machine.

(1) $T_2 = 2T_1$ (considering pulley P_1)

(2) $T_3 = 2T_2 = 4T_1$ (considering pulley P_2)

(3) The load is therefore raised by a force
$= T_3 + T_2 + T_1 = 7T_1$

Hence the MA $= \dfrac{7T_1}{T_1} = 7$ and this is the VR of the system.

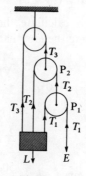

Fig 27.17

A third system is shown in fig 27.18. Using the same method as above,

(1) $T_2 = 2T_1$

(2) The load is therefore raised by a force
$= 2T_1 + T_2 = 4T_1$

Hence the MA $= \dfrac{4T_1}{T_1} = 4$ and this is the VR of the system.

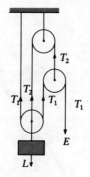

Fig 27.18

Example 5

In the system of pulleys shown in fig 27.19, when a load of mass 50 kg is raised, the efficiency of the system is 80%. Calculate the effort required.

(**1**) Assuming the machine is an ideal one, we first find the VR from the theoretical MA. From the diagram $T_2 = 2T_1$. Hence the force raising the load $= T_2 + T_1 = 3T_1$.

Hence the MA $= \dfrac{3T_1}{T_1} = 3$

and so the VR $= 3$.

(**2**) Now calculate the effort from the real MA using the fact that the efficiency $= 80\%$.

$$\eta = \frac{MA}{VR} = 0.8$$

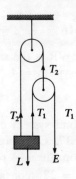

Fig 27.19

but   VR $= 3$.

Hence   MA $= 0.8 \times 3 = 2.4$.

But   MA $= \dfrac{L}{E} = \dfrac{50g}{E} = 2.4$

which gives $E = \dfrac{50g}{2.4} \approx 208$ N, taking $g = 10$ m s^{-2}.

The Screw

A wheel or crank handle is used to turn a screw which advances through a fixed nut, as shown in fig 27.20a.

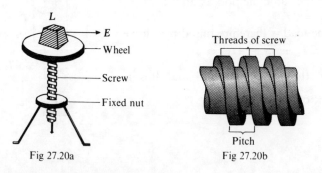

Fig 27.20a   Fig 27.20b

One revolution of the wheel produces one revolution of the screw, which advances by a distance equal to the distance between two consecutive threads on the screw (fig 27.20b). This distance is called the **pitch** of the screw. Common examples of the screw are the car-jack (fig 27.21) and screw presses.

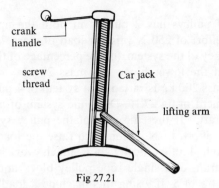

crank handle

screw thread

Car jack

lifting arm

Fig 27.21

Velocity ratio of the screw

In fig 27.20a, an effort E is applied to the wheel of radius r, and the rotation of the wheel advances a screw, of pitch d, and raises a load, L. Let the wheel make 1 revolution.

Distance moved by effort $= 2\pi r$

Distance moved by load $= d$

$$VR = \frac{2\pi r}{d}$$

Example 6

A car-jack has a crank handle of radius 14 cm and the pitch of the screw is 1 cm. If an effort of 20 N raises a load of mass 44 kg, what is the efficiency of the jack? Why is it so low? (Take $\pi = \frac{22}{7}$ and $g = 10$).

$$VR = \frac{2\pi r}{d} = 2 \times \frac{22}{7} \times \frac{14 \text{ cm}}{1 \text{ cm}} = 88$$

$$MA = \frac{\text{load}}{\text{effort}} = \frac{44g}{20} = \frac{440}{20} = 22$$

$$\eta = \frac{MA}{VR} = \frac{22}{88} = \frac{1}{4} = 25\%.$$

The efficiency is low because the friction between the moving parts is very high.

Exercise 27.2

Take $g = 10 \text{m s}^{-2}$ and $\pi = \frac{22}{7}$

1 In a block and tackle, the lower block has 2 pulleys and the upper block has 3 pulleys. A rope passes round all the pulleys with one end attached to the block, and an effort of 150 N is applied to the free end. If the efficiency of the system is 60%, calculate the load, in kg, that can be raised by the effort.

2 A system of pulleys has 2 pulleys in both the upper and lower blocks. An effort of 250 N raises a load of mass 70 kg. Calculate (i) the efficiency of the system, (ii) the percentage of the input wasted in friction, if the lower block has mass 20 kg.

3 A load of mass 200 kg is raised by a system of 4 pulleys similar to the arrangement in fig 27.16. Draw the system of pulleys and find its velocity ratio. If the efficiency of the pulley system is 62·5%, calculate the effort, in N, required to raise the load. If the load is raised through 4 m vertically, how much work is wasted?

4 Draw a diagram of a single rope pulley block and tackle having a velocity ratio of 5. If, using this machine a load of mass 50 kg is raised a vertical distance of 2 m by an effort of 160 N, find (i) its efficiency, (ii) the work wasted in the machine.

5 A car-jack has a screw with two threads to a length of 3 cm and the crank handle describes a circle of 28 cm radius. If the efficiency of the jack is 30%, calculate the force applied to the handle to raise a mass of 0·5 t.

6 In a Weston differential wheel and axle, the radii of the wheel and the two parts of the axle are 14 cm, 4 cm, and 3 cm respectively. If the efficiency of the machine is 80%, calculate the effort needed to raise a load of mass 100 kg. If the load is raised a vertical distance of 1·5 m, calculate the amount of work, in J, wasted in the machine.

7 A screw jack advances 2 cm for every complete turn of the screw. The effort is applied at the end of an arm of radius 21 cm. If the efficiency of the machine is 65%, calculate (i) the velocity ratio of the machine, (ii) the mechanical advantage of the machine, and (iii) the greatest load in t, correct to 2 significant figures, that could be raised by an effort of 750 N.   (L)

8 In a block and tackle system, the upper block has 10 pulleys, and the lower block has 9. Calculate the velocity ratio of the system. If the efficiency is $33\frac{1}{3}$%, what load will be raised by an effort of 240 N? If the lower block has mass 50 kg, what percentage of the input is wasted on overcoming friction?

Linear Law for a Machine

Work is wasted in machines due to two causes: (**a**) friction between moving parts and (**b**) in raising part of the machinery (e.g. as for the lower pulley block in a block and tackle). Irrespective of the type of machine used, the efficiency (η) depends on the ratio of MA $\div$ VR, and hence calculations for any type of machine can be made, without knowing its mechanism. The efficiency of most machines varies with

the load and the load and effort are often connected by a linear law,
such as $E = aL + b$, where E, L are in N and a, b are constants.
For example, using a block and tackle machine with 2 pulleys in both
blocks, it was found that if the load was of mass 5·5 kg the effort was
30 N and if the load was of mass 31 kg the effort was 122 N.
Hence assuming a linear law, of the above type:

$$30 = a(5\cdot5g) + b$$
$$122 = a(31g) + b$$

giving $\qquad\qquad a \approx 0\cdot36, b \approx 10$ (taking $g = 10$)

Hence $\qquad\qquad E \approx 0\cdot36L + 10$

This relationship is approximately true over a range of loads. It can
be used to calculate the effort required for a load in that range.

The term, b, represents the effort for *no load*, i.e., the effort required
to operate the machine's moving parts by themselves. The factor,
a, gives a measure of the friction present.

Example 7

A machine produced the following set of results:

L (kg)	0	10	20	30	50	70
E (N)	32	58	86	112	160	209

Using a graph, derive a formula giving E in terms of L.

First plot a graph of effort against load, with the loads expressed

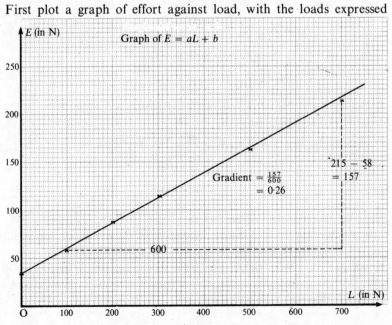

Fig 27.22

in N (taking $g = 10$). Draw a line of best fit through the points. (fig 27.22) From this line, $b = 32$ and a (= the gradient of the line) = 0·26. Hence $E = 0·26L + 32$ is the required formula.

Power of a Machine

The power consumed by a machine is measured from the input. The power developed by a machine for useful work is measured from the output. In each case, we recall that power (measured in W or kW) is a *rate* of using or doing work.

Example 8

An electric motor operates a machine with a VR of 12, and an efficiency of 25%. The machine raises a load of mass 240 kg through 5 m vertically at a constant speed in 50 s. Find (i) the effort applied to the machine, (ii) the power developed by the machine, (iii) the power supplied by the electric motor. ($g = 10$ m s^{-2})

(i) First calculate the MA of the machine from its efficiency.

$$\eta = \frac{\text{MA}}{\text{VR}} = \frac{\text{MA}}{12} = 25\%.$$

Hence $\text{MA} = 12 \times \dfrac{1}{4} = 3.$

Now calculate the effort of the machine from the MA:

$$\text{MA} = \frac{L}{E} = \frac{240g}{E} = 3.$$

Hence $E = 80g = 800$ N.

(ii) The power output of the machine.
The load ($240g$ N) is moved vertically through 5 m in 50 s.
Therefore the work done in 50 s = $240g \times 5$ J

Hence the power developed = $\dfrac{240g \times 5}{50}$ J s^{-1}.

$$= 240 \text{ W}$$

$$= 0·24 \text{ kW}.$$

(iii) The power output of the electric motor.
The machine is 25% efficient. Therefore 25% of the power input (supplied by the electric motor) appears as output.
Hence 25% × power input = 0·24 kW giving the power input as 0·96 kW.
So the electric motor supplies 0·96 kW.

Exercise 27.3

Take $g = 10 \text{ m s}^{-2}$.

1 In a certain type of lifting tackle with a VR of 8, the effort E N required to lift a load of W N is given by $E = aW + b$, where a and b are constants. Experiment shows that an effort of 100 N will just lift a load of mass 40 kg, and an effort of 120 N will just lift a load of 60 kg. Find the values of a and b and calculate the efficiency of the tackle when a load of mass 115 kg is being lifted.   (L)

2 With a pulley system whose VR = 8 a boy finds that he can lift an object of mass 60 kg with an effort of 110 N. Calculate (i) the MA of the system, (ii) the efficiency of the system as a percentage, (iii) the work done by the boy in raising the load vertically through 6 m.

3 A fallen tree of mass 2 t and length 10 cm rests on horizontal ground, its CG being 4 m from one end. Its other end is lifted from the ground by a pulley system, of efficiency 45% and VR = 6, which exerts a vertical force on the tree. Calculate the effort required just to raise the end of the tree from the ground.   (L)

4 Fig 27.23 shows a machine which is being used to lift a load of W kg. The rope along which the effort P N is exerted passes over a a fixed pulley A and is fixed at the end B. The movable pulleys C and D each have mass 0·5 kg. The ropes are vertical except where they pass round a pulley. Find the velocity ratio of the system.

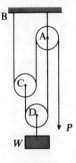

Fig 27.23

If frictional resistances and the weight of the rope are neglible, write down an equation giving P in terms of W and find the efficiency of the machine when the load has a mass 55 kg.   (L)

5 A screw of pitch 2 mm is turned by a lever arm of length 50 cm and raises a load of mass 500 kg by means of an effort of 20 N applied at right angles to the lever. Find the mechanical advantage and efficiency of the device.   (O)

6 In a sheaved pulley system with a velocity ratio 4, the lower block has mass 7 kg and a load of mass 21 kg is raised by an effort from a mass of 8 kg hanging freely. Find (a) the efficiency, (b) the proportion of work that is expended in overcoming friction.   (O)

7 Draw a sketch of a pulley system with velocity ratio 4. If the efficiency of such a system is 60% calculate (a) the effort required just to raise a load of mass 100 kg, (b) the work that must be done by the effort in raising this load through a vertical distance of 6 m.

8 A windlass is used to raise a cage of mass (including contents) 5 t through a vertical height of 200 m in 400 s. The diameter of the axle is 15 cm and that of the driving wheel is 1·2 m. If the efficiency of the windlass is 75%, calculate the power input.

Revision Papers

(Take $g = 9.8 \text{ m/s}^2$)

I

1 Find x if $(0.7)^x = (1.2)^{2x-1}$

2 If one root of the equation $x^2 - ax + b = 0$ is the square of the other, find the roots in terms of a and b.

3 The nth term of a series is given by the formula $an^2 + bn + c$ where a, b and c are constants. If the second, third and fifth terms are 8, 16 and 38 respectively, find the fourth term.

4 Evaluate **(a)** $\displaystyle\int_{-3}^{3} (x - x^2)\, dx$.

   (b) $\displaystyle\int_{0}^{\pi/6} \sin \frac{3x}{2}\, dx$.

5 If $2x^3 + ax^2 + bx + 6$ has factors $(2x-1)$ and $(x+2)$, find the values of a and b. What then is the third factor?

6 Sketch the curve $y = 1 + x^2$. Find the equation of the tangent to this curve at the point A(2, 5), and the coordinates of the point B where this tangent cuts the y-axis. Calculate the area enclosed by the line BA, the curve and the y-axis.

7 Find the first three terms of the expansion of $(8 - 3x)^{\frac{1}{3}}$. Hence find an approximate value for the cube root of 7.97, correct to five significant figures.

8 A train of total mass 150 t is travelling on level ground and the engine is working at the rate of 90 kW. If the maximum possible speed is 54 km h^{-1}, what is the resistance per t to motion? If these resistances remain the same and the engine works at the same rate, what is the maximum speed (in km h^{-1}) up a slope of angle $\theta°$, where $\sin \theta° = \frac{1}{300}$?

9 At 13.00 h a ship B is 40 km east of another ship A. A is sailing due north at 20 km/h and B is sailing in the direction N 60° W at 30 km/h. Find the distance between A and B when they are closest together and the time when this happens. What is the actual position of A at this time and how far is B north and east of A's original position then?

10 Five horizontal wires are attached to a bolt fixed in the ground. The tension in each wire is 50 N. The directions of the wires are N, E, NW, N 30° E and SW. Find the resultant force on the bolt in magnitude and direction.

II

1 (i) If $\dfrac{dy}{dx} = 9x^2 - 2x + 2$ and $y = 23$ when $x = 2$, find the value
   of y when $x = 3$.

   (ii) Find $\dfrac{dy}{dx}$ if $y = \dfrac{x^2 - 2x}{x + 2}$.

2 For the circle whose equation is $x^2 + y^2 - 4x - 6y + 4 = 0$ (i) find
   the coordinates of the centre and the length of the radius; (ii) prove
   that the x-axis is a tangent; (iii) find the equation of the other
   tangent from the origin to the circle; (iv) find the angle between
   these two tangents.

3 Solve the following equations in the range $0°$ to $360°$:
   (i) $\sin 2x = \sin x$;   (ii) $2 \sin^2 x = 3 \cos x + 3$;
   (iii) $\sin 3x + \sin 5x = \cos x$.

4 (i) Prove that $3x^2 - x + 2$ is positive for all real values of x.
   What is the minimum value of this expression?

   (ii) If $y = 6 - x - x^2$, find the range of values of x for which y is
   positive. Find the maximum value of y and the value of x at which
   this occurs.

5 Expand $\sqrt{\dfrac{1 + x}{1 - x}}$ as a series of ascending powers of x as far as the

   term in x^3. (Write the expression as $(1 + x)^{\frac{1}{2}} (1 - x)^{-\frac{1}{2}}$, expand
   each part as far as its x^3 term and then multiply.)
   By substituting $x = \frac{1}{10}$ find the value of $\sqrt{11}$ correct to four significant figures.

6 Find the equation of the locus of the point $A(x, y)$ in each of the
   following:
   (i)   A is equidistant from the points $(1, 2)$ and $(-2, 1)$;
   (ii) the distance of A from $(1, 2)$ is twice its distance from $(-2, 1)$;
   (iii) A is equidistant from the point $(-2, 1)$ and the line $y = x$.

7 (i) Find the value of $\displaystyle\int_{1}^{4} \dfrac{x^2 + 1}{\sqrt{x}} dx$.

   (ii) Sketch the curve $y = (x - 1)^2$. The area bounded by the curve
   and the coordinate axes is rotated about the y-axis through $360°$.
   Find the volume generated.

   (iii) Differentiate $x \cos x$ wrt x. Hence find $\int x \sin x \, dx$.

8 A body of mass 5 kg lies on a rough inclined plane which makes
   an angle θ with the horizontal where $\sin \theta = \frac{3}{5}$. A force of 10 N
   applied to the body up the plane along a line of greatest slope is
   just sufficient to prevent it slipping down the plane. Find the

coefficient of friction.

9 Forces of 2, 4, 6, 8 N act along the sides AB, BC, CD and DA respectively of a square of side a in the direction of the letters. Find where their resultant cuts the sides AB and AD of the square, measured from A.

10 A body travels in a straight line but its speed at any time does not exceed 5 m/s. If it accelerates and decelerates at 2 m/s², find the shortest time needed to cover a distance of 30 m from rest to rest.

III

1 A straight line of variable gradient m passes through the fixed point $(1, 1)$ and cuts the x-axis at A and the y-axis at B. The rectangle AOBT is then drawn. Using m as a parameter, find the coordinates of T in terms of m and hence find the Cartesian equation of the locus of T.

2 (i) Functions f and g are defined as $f : x \to (x+3)$ and $g : x \to (2x-1)$ where x is a real number. Define in a similar way the functions fg and gf, simplifying your answers.

(ii) Show, using an arrow diagram, the relation "has no common factor > 1 with" among the set of numbers $\{3, 5, 6, 7, 9, 10\}$.

3 The values of x and y in the table below are believed to fit approximately the equation $y = ax^n$. By drawing a suitable graph find the values of a and n to two significant figures as accurately as you can.

x	1·3	1·5	1·9	2·1	2·4
y	4·7	5·6	7·6	8·3	9·9

4 (i) Find the values of $\dfrac{\mathrm{d}^2 y}{\mathrm{d}x^2}$ on the curve $y = 3x(x - 2)^2$ where $\dfrac{\mathrm{d}y}{\mathrm{d}x} = 0$.

(ii) Find the coefficient of x^3 in the expansion of $(2x - \dfrac{1}{x^2})^{12}$.

(Leave the answer in terms of factorials.)

5 A, B, C and D are the points whose coordinates are $(-5, 2)$, $(2, 3)$, $(-1, -6)$ and $(-5, -4)$ respectively. Find **(i)** the coordinates of the point M where AD and BC intersect; **(ii)** the equation of the line through M perpendicular to AB.

6 (i) If $\log_e a^3 b = u$ and $\log_e ab^2 = v$, find a in terms of e, u and v.
(ii) Solve the equation $x^4 = 5$.

7 (i) Prove that $\dfrac{\sin 5\theta - \sin 3\theta}{\cos 5\theta + \cos 3\theta} = \tan \theta$. For this result to be valid, explain why $\cos 4\theta$ must not equal zero.

(ii) Solve the equation $\sin 2\theta + 1 = \cos^2 \theta$ where $0° \leqslant \theta < 360°$.

8 A man of mass 60 kg stands on the floor of a lift. Calculate the reaction of the floor on his feet when **(i)** the lift is moving upwards at a steady speed of $1\cdot5$ m/s; **(ii)** when the lift accelerates upwards at $1\cdot2$ m/s^2; **(iii)** when the lift accelerates downwards at $1\cdot8$ m/s^2.

9 OABCDE is a uniform lamina. O is the origin of coordinates and the coordinates of A, B, C, D and E are (0, 3), (4, 3), (4, 1), (6, 1) and (6, 0) respectively, and all the edges of the lamina are straight lines. Find (i) the coordinates of the CG and (ii) the angle (to the nearest degree) OA makes with the vertical when the lamina is freely suspended from A.

10 A car of mass 400 kg is travelling up a slope which makes an angle α with the horizontal, where $\sin \alpha = 0\cdot05$, at a constant speed of 54 km/h. The resistance to motion from friction etc is 94 N. Find the power of the engine.

IV

1 (i) If $f : x \rightarrow 2x - 1$ and $g : x \rightarrow x^2 + 1$, find the values of $f^{-1}g(-3)$ and $gf^{-1}(-2)$.

(ii) A linear function f is such that $f(2) = 2$ and $f(-1) = -7$. Find $f(3)$.

2 (i) If a curve is given by the parametric equations

$$x = \frac{2 + t^2}{\sqrt{1 + t^2}}, \; y = \frac{t^2}{\sqrt{1 + t^2}}, \quad \text{find its Cartesian equation.}$$

(ii) How many numbers greater than 4000 can be made up from the digits 2, 3, 4, 5, 6 if no digit can be used more than once in any number?

3 Find $\dfrac{dy}{dx}$ if $y^2 = 4x$. Hence prove that the equation of the normal to the curve $y^2 = 4x$ at the point $P(t^2, 2t)$ where t is a parameter, is $y + tx = t^3 + 2t$. If this normal cuts the x-axis at G and M is the midpoint of PG, find the coordinates of M in terms of t. Hence find the Cartesian equation of the locus of M as P varies.

4 (i) Differentiate **(a)** $\dfrac{2x - 1}{2x + 1}$ and **(b)** $x^2 \tan 2x$, wrt x.

(ii) For a certain curve $\dfrac{dy}{dx} = x^2 - x$. If the curve passes through the point (1, 1), find its equation.

(iii) Sketch the curve $y = x(2 - x)$. Find the area lying between the curve and the line $2y = x$.

5 (i) If the roots of the equation $2x^2 - x - 4 = 0$ are α and β, form the equation whose roots are $\dfrac{1}{\alpha}$ and $\dfrac{1}{\beta}$.

(ii) For what range of values of x is $2x^2 + 3 \leqslant 7x$?

6 (i) A pyramid VABCD stands on a square base ABCD of side 8 cm and the slant edges VA, VB, VC and VD are each 13 cm long. Calculate the height of the pyramid and the size of the angle VAB.

(ii) Prove that $1 + \cos 2A = \sin 2A \cot A$.

(iii) If A and B are acute angles and $\tan A = \frac{4}{3}$ and $\tan B = \frac{1}{9}$ show, without using tables, that $A + B = 45°$.

7 A uniform rod ABCD of length 2 m rests horizontally over two supports B and C where AB = 0·5 m and CD = 0·2 m. The mass of the rod is 5 kg. A downward force of 35 N is applied at a point P of the rod so that the reactions of the supports are equalized. Find the distance AP.

8 Ball A (mass 3 kg) is travelling due east in a straight line with speed 5 m/s when it collides with ball B (mass 4 kg) which was at rest. After the collision both balls move east with relative speed 2·5 m/s. Find their separate speeds.

9 A pump lifts water from rest through a vertical height of 10 m and then forces it through a pipe of cross-sectional area 20 cm^2 at a speed of 4 m/s. At what rate is the pump working? [1 m^3 of water has a mass of 10^3 kg].

10 To a man in a helicopter travelling horizontally in a direction 030° at 40 km/h, a light aeroplane appears to be moving due north. If the speed of this aeroplane is known to be 120 km/h, in what direction is it moving?

V

1 (i) If $r = a(\cos \theta + 1)$ where a is a constant, prove that $r\dfrac{d\theta}{dr} = -\cot \dfrac{\theta}{2}$.

(ii) If $y(x^2 - 2) = 3$, find the value of $\dfrac{dy}{dx}$ when $x = -1$.

(iii) If $x = \sec t$ and $y = \tan t$ are the parametric equations of a certain curve, find its Cartesian equation and the gradient at the point whose parameter is t.

2 (i) The sum of the first ten terms of an AP is 145 and the sum of the *next* ten terms is 445. Find the first three terms of the AP.

(ii)   The polynomial $ax^3 + bx^2 - 5x + 2$ has $(2x - 1)$ and $(x - 1)$ as two of its factors. Find a and b and the third factor.

3 ABCD is a trapezium where AB is parallel to CD, AB > CD and CD = 10 cm. The angles at A and B are each equal to $\theta°$ and AD = BC = 8 cm.

(i)   Prove that the area of the trapezium is $64 \sin \theta \cos \theta + 80 \sin \theta$.

(ii)   By differentiating this expression wrt θ and then solving a quadratic equation, find the value of θ which will make the area a maximum.

4 A circle touches the y-axis at the point $(0, 3)$ and passes through the point $(4, -1)$. Find the equation of the circle.

5 There are 9 boys going to spend the morning in 2 groups, one group to go swimming and the other to go for athletics practice. In how many ways can the group for swimming be selected if there must be at least 3 boys in each group?

6 (i)   Prove that
$$13 \cos 2A + 5 = 2(3 \cos A + 2 \sin A)(3 \cos A - 2 \sin A).$$

(ii)   Solve the equation $2 \sin x = \cos 40°$ where $0° \leqslant x \leqslant 360°$.

7 A particle moves in a straight line with constant acceleration and passes in succession three points A, B and C. The speed of the particle at B is three times its speed at A. The particle takes 20 s to go from B to C. The distance AB = 90 m and the distance BC = 260 m. Find (i) the speed of the particle at A, (ii) its speed at C and (iii) its acceleration.

8 A block and tackle consists of a fixed block with two pulleys vertically above a movable block which has two pulleys. The effort is applied at the end of a rope which passes through the four pulleys with the other end fastened to the movable block. The rope is vertical except where it passes round a pulley. The effort is applied vertically upwards. Make a sketch of the block and tackle and state its velocity ratio. The machine is used to raise loads vertically. If the mass of the movable block and its pulleys is 5 kg, find the mechanical advantage and the efficiency of the machine when it raises a load of 60 kg. (Answer to 3 significant figures).

9 A helicopter flies from a point A to a point B due north of A. The airspeed of the helicopter is 200 km/h and a steady wind of 60 km/h is blowing from the NE. By drawing or calculation, find the course taken by the helicopter and its resultant speed. The helicopter now flies back from B to A with the same wind blowing. Find the resultant speed.

10 A cyclist and his machine have a total mass of 80 kg. He reaches

the top of a hill AB with a speed of 3 m/s. The hill is 400 m long
and has a slope of $\theta°$ where $\sin \theta = 0.09$. The cyclist now freewheels
down the hill against resistances of 70 N. What is his speed when
he reaches B, the bottom of the hill?

VI

1 Prove that the curve $y = (x - 4)^2$ touches the x-axis at the point A
(4, 0). The curve cuts the y-axis at B. Find the equation of the
tangent to the curve at B and show that this tangent passes through
the midpoint of AO, where O is the origin.
The area bounded by OA, OB and the arc AB of the curve is now
rotated about the x-axis through four right angles. Calculate the
volume formed (leave your answer as a multiple of π).

2 (i) Find the equation of the tangent to the curve $y^2 = 12x$ which
is parallel to the line $y - 3x + 2 = 0$. Find the coordinates of the
point of contact of the tangent with the curve.
(ii) Two tangents are drawn to the curve $y^2 = 12x$ from the point
$(-3, -8)$. Prove that they are perpendicular.

3 (i) If $\log_p 4 = x$ and $\log_p 5 = y$, find in terms of x and y (a) $\log_p 100$;
(b) $\log_p 2.5$; (c) $\log_p \dfrac{25p^2}{8}$.

(ii) Given that $\cos \theta = x$ and $\cos 2\theta = y$, find x in terms of y.

4 (i) Prove the identities
(a) $(\sin A - 2 \cos A)^2 + (2 \sin A + \cos A)^2 = 5$.
(b) $2(\sin A - 2 \cos A)^2 = 3 \cos 2A - 4 \sin 2A + 5$.
(ii) Solve the equation $\cos x + \cos 2x = 0$, in radians, where
$0 \leqslant x \leqslant 2\pi$.

5 (i) If α and β are the roots of the equation $3x^2 - 3x - 1 = 0$,
form the equation whose roots are $\alpha - \dfrac{1}{\alpha}$, $\beta - \dfrac{1}{\beta}$.

(ii) The volume of a spherical balloon is growing at the rate of
0.1 cm^3/s when its radius is 12 cm. Find the rate of change of its
surface area at this time. (Volume $= \frac{4}{3}\pi r^3$, area $= 4\pi r^2$.)

6 (i) If $y = \sqrt{3x + 2}$, prove that $y\dfrac{d^2y}{dx^2} + (\dfrac{dy}{dx})^2 = 0$.

(ii) If $y = 2u + \dfrac{1}{2u}$ and $u = x + \dfrac{1}{x}$, find the value of $\dfrac{dy}{dx}$ when $x = 1$.

7 A school committee of 4 members is to be formed from 5 teachers,
2 students and the headmaster. In how many ways can the committee be formed if (i) it *must* include the headmaster; (ii) the head-

master is *not* to be included; (iii) exactly three teachers must be included; (iv) not more than three teachers are included?

8 A steel ball of mass 0·2 kg is dropped from rest from a height of 5 m above horizontal ground. After striking the ground, it rebounds to a height of 3 m. Find (i) the speed of the ball just before it strikes the ground; (ii) the speed of the ball as it leaves the ground; (iii) the magnitude of the impulse which the ball receives from the ground; and (iv) the loss of kinetic energy during the impact.

9 A uniform straight rod ABCD of length 8 m and mass 30 kg rests horizontally on two supports at B and C where AB = 3 m and CD = 2 m. A body of mass 40 kg is placed on the rod at a point P where AP = x m. Find x when the rod is on the point of turning about (i) B; (ii) C.

10 A car of mass 1000 kg is ascending a slope of angle $\theta°$ to the horizontal where $\sin \theta = \frac{1}{20}$. The engine is working at a rate of 20 kW. (i) If the car is travelling at a steady speed of 72 km/h, find the resistance to motion; (ii) if the resistance to motion is reduced to 310 N, find the acceleration of the car up the slope when the speed is 72 km/h.

VII

1 (i) If the polynomials $x^3 + x^2 - 4x + 5$ and $x^3 + 3x - 7$ leave the same remainder when they are each divided by $(x - a)$, find the possible values of a.

(ii) If the roots of the equation $x^2 - 5x + 1 = 0$ are α and β, form an equation with roots $\alpha + 3\beta$, $3\alpha + \beta$.

2 A vertical radio mast AB, stands with its foot A on horizontal ground. P is a point on the ground due south of A and from P the angle of elevation of the top of the mast, B, is 30°. Q is a point due east of A and the bearing of Q from P is 050°. The distance PQ = 400 m. Find the angle of elevation of B from Q.

3 (i) Solve the equation $3^x \times 4^x = 9^{x+2}$, giving the answer correct to three significant figures.

(ii) If $y = b \sin 2x$, where b is a constant, satisfies the equation $\frac{d^2y}{dx^2} + 8y = 4 \sin 2x$, find the value of b.

4 Write down the binomial expansions of $(3 + x)^4$ and $(3 - x)^4$. Hence express $(3 + x)^4 - (3 - x)^4$ as a series of powers of x. Use this result to calculate the exact value of $(3·1)^4 - (2·9)^4$.

5 (i) Show that the line $y = 4x$ is a tangent to the curve $y = \dfrac{3 - 4x}{2 - 3x}$
at the point $(\frac{1}{2}, 2)$.

(ii) Find the coordinates of the point on the curve $y = 3x^2 - x + 4$
where the gradient of the normal is $-\frac{1}{5}$.

6 P is a variable point whose coordinates are (t, t) where t is a parameter. A is a fixed point, coordinates $(1, 0)$. Write down the
equation of the line PA and find the coordinates (in terms of t)
of the point Q where this line cuts the y-axis. Hence find the coordinates of M, the midpoint of PQ and deduce the Cartesian
equation of the locus of M.

7 (i) If the three functions f, g, h are given as $f : x \to 2x$, $g : x \to x - 1$
and $h : x \to x^2$, write in a similar way the composite functions
(a) fgh and **(b)** $f^{-1}g^{-1}$.
Express the function $x \to (2x - 1)^2$ in terms of f, g and h.

(ii) A quadratic function f is given by $f : x \to x^2 + ax + b$. If
$f(2) = 7$ and $f(3) = 14$, find the values of a and b.

8 The centre of gravity of a uniform solid circular cone of height h
is at a distance $\dfrac{3h}{4}$ from the vertex along the axis of the cone.

A solid cone of this type is 56 cm high and has a symmetrical
conical cavity of similar shape of height 28 cm measured from its
base. Calculate the distance of the centre of gravity of the remaining solid from the vertex of the original cone.

9 A cyclist and his machine together have a mass of 75 kg. To maintain a steady speed of 10 m/s along a straight level road the cyclist
must work at a rate of 0·15 kW. The cyclist now rides up a hill
which has a constant slope. If the resistance to motion remains the
same and his speed drops to 5 m/s, find the sine of the angle of the
slope.

10 A car is travelling along a road in the direction NE. A lorry is
travelling on a bearing of 275° along a second road which intersects
the first road. The lorry is at the crossroads when the car is approaching the crossroads and 3 km distant from it. If the lorry travels
at a speed of 70 km/h and the car at 90 km/h, find the least distance
between them.

VIII

1 (i) The roots of the equation $2x^2 - 3x + 5 = 0$ are α and β and
the roots of the equation $px^2 + x + q = 0$ are $\alpha - 1$ and $\beta - 1$.
Find the values of p and q.

(ii) Factorize the polynomial $x^3 + 7x^2 + 8x - 16$.

2 Values of y corresponding to certain values of x were found experimentally as follows:

x	2	4	6	8	10	12
y	6·4	17·5	33·7	54·2	80	110·5

It is known that these values satisfy an equation $y = ax^2 + bx$ where a and b are constants. Plot values of $\dfrac{y}{x}$ against corresponding values of x and from your graph estimate the values of a and b.

3 A closed rectangular box has a square base of side m cm and a height of 10 cm. When the volume of box is 1000 cm³ the volume is increasing at the rate of 20 cm³/min. Calculate the rate of change of the surface area at this time.

4 (i) How many numbers of two digits can be formed by using in each number two *different* integers chosen from 1, 2, 3, 4, 5, 6, 7, 8?
(ii) How many of these numbers are less than 50?
(iii) If the two integers are chosen at random, what is the probability that the number selected will be less than 50?
(iv) Find the probability that the sum of the digits in the number selected will be less than 6.

5 The coordinates of A and B are $(2, -4)$ and $(0, 3)$ respectively. C is a variable point such that the area of the triangle ABC is 10 square units. Find the equation of the locus of C.

6 Write down the expansion of $(1 + 3x)^6$, simplifying each term. Hence obtain the value of $(1·03)^6$ correct to six significant figures.

7 (i) Solve the simultaneous equations $\dfrac{x + 6}{8} + \dfrac{y - 1}{2} = 2$, $(x + 2)(y - 1) = 8$.
(ii) Find in its simplest form a quadratic function in x which has a maximum value of 25 when $x = 2$ and a value of 9 when $x = 0$.

8 Two particles X and Y of masses 0·2 kg and m ($>0·2$) kg respectively are joined by a light inextensible string which passes over a smooth pulley. The particles are held at rest and then released. They move freely under gravity with an acceleration of 4·2 m s⁻². Find the value of m. After moving for 2 s Y is suddenly brought to rest. Calculate the total distance travelled by X from this time to the instant when it is next momentarily at rest.

9 A locomotive of mass 16 000 kg pulls a truck of mass 4000 kg along a horizontal track. The resistance to motion for both the locomotive and the truck is 0·1 N/kg. If the locomotive is working at 180 kW, find (i) the acceleration when both are travelling at

15 m s^{-1}; (ii) the tension in the coupling at this time.

10 A particle A of mass 500 g is moving in a straight line at 9 m s^{-1}
It collides with a particle B of mass 250 g which is moving at 3 m s^{-1}
along the same line in the opposite direction. As a result of the
collision, half the momentum of A is transferred to B. Find the
velocities of A and B after the collision. If the collision takes 0·01 s,
find the average force acting on A and B during the collision.

IX

1 The curve $y = 3ax(b - 2x)$ passes through the point $(1, 3)$ and
$b > 2$. (i) Find a in terms of b; (ii) find the area enclosed between
the curve and the x-axis in terms of b; (iii) show that the minimum
value of this area, as b varies, is $\frac{27}{8}$ square units.

2 (i) If $x = t^2 + \dfrac{1}{t^2}$, $y = t - \dfrac{1}{t}$, find (a) $\dfrac{dy}{dx}$ in its simplest form;

(b) the Cartesian equation of the curve.

(ii) Show, in an arrow diagram, the relation "is the father of"
on the set of men A, B, C, D, E, F where A is the father of C and
the grandfather of F and B is the father of D and E.

3 (i) If $y = 2x - \tan 2x$, prove that $\dfrac{dy}{dx} = -2 \tan^2 2x$.

(ii) If $\dfrac{d^2y}{dx^2} = \dfrac{18}{x^3}$ and $y = 0$ and $\dfrac{dy}{dx} = 0$ when $x = 1$, find y

in terms of x.

4 Find the equation of the locus of the point P in each of the following:
(i) P is equidistant from the point $(2, -3)$ and the y-axis.

(ii) A straight line $y = (\tan \dfrac{2\pi}{3})x$ is drawn through the origin. A

circle centre P is drawn to touch this line and to touch the positive
x-axis.

(iii) The distance of P from the point $(4, 2)$ is twice its distance
from the point $(1, 2)$. Show that this locus is a circle and find the
coordinates of its centre.

5 (i) If $(m\sqrt{2} + n)^2 = 34 - 24\sqrt{2}$, find m and n.

(ii) Factorize $2x^3 - 3x^2 - 8x - 3$.

(iii) Find the number of different ways in which the letters of the
word ADDITIONAL can be arranged.

6 (i) Three points A, B, C have coordinates $(2, -1)$, $(-3, 2)$,
$(4, 3)$ respectively. Find the equation of the line which passes
through A and is perpendicular to BC.

(ii) If $f : x \to 2x$ and $g : x \to \sin x$, sketch, using the same axes and scales, the graphs of the functions fg and gf, taking as domain $0 \leqslant x \leqslant 2\pi$. How many solutions will there be of the equation $fg(x) = gf(x)$?

7 (i) Find the equation of the normal to the curve $x^2 y = x - 1$ at the point where $x = 2$.

(ii) Find the value of c if $c \log_9 N = \log_3 N$. Hence find the value of N if $\log_3 N + \log_9 N = 6$.

8 A sphere of mass 10 kg and a sphere of mass 4 kg are travelling towards each other in a smooth linear groove. The speed of the heavier sphere is 4 times that of the lighter sphere. After colliding, the heavier sphere has a speed of 4 m/s and the lighter sphere a speed of 8 m/s. Calculate (i) the speeds of the spheres before the collision; (ii) the loss in energy caused by the collision; (iii) the average force between the spheres if contact lasted 0·01 s.

9 A small aeroplane flies from A to B where B is 200 km due east of A. The speed of the aeroplane relative to the air is 200 km/h but there is a wind of 40 km/h blowing *from* the direction N 60° E. Find the time taken on the flight.

10 A fixed wedge has a cross section in the shape of an isosceles triangle ABC with a right angle at B. The base AC is horizontal. At B is a frictionless pulley over which passes an inextensible string. One end of the string is attached to a block of mass 1·8 kg and the other end to a block of mass m kg. The string on each side of B is parallel to a line of greatest slope. The coefficient of friction between each block and the wedge is 0·4. Find limits for the value of m if the blocks are not to slide on the slopes. If $m = 7·2$ kg, find the acceleration of the blocks.

X

1 (i) If $P = 6t^{\frac{3}{2}}$, find the approximate increase in the value of P if t is increased from 0·25 to 0·27.

(ii) A bag contains 3 red balls and some white balls, all of identical size. Two balls are drawn at random, one after the other, the first one not being replaced after drawing. If the probability of drawing 2 red balls is $\frac{3}{28}$, how many white balls were in the bag at the start?

2 (i) Find the derivatives wrt x of (a) $(1 - x + x^2)(1 + x - x^2)$;

(b) $\dfrac{x}{1 - x + x^2}$; (c) $x \cos 5x$.

(ii) Find the area between the curves $y = \sqrt{x}$ and $y = x^3$. If this area is rotated about the x-axis through $360°$, find the volume generated.

3 P is a variable point on the line $x + y = 2$. Taking t as a parameter, write the coordinates of P if its x-coordinate is t. The line through P perpendicular to $x + y = 2$ meets the y-axis at Q and M is the midpoint of PQ. Find the coordinates of M in terms of t and hence find the Cartesian equation of the locus of M.

4 (i) The gradient of the tangent to a curve is given by $\dfrac{dy}{dx} = \dfrac{1}{(4x+1)^3}$. Find the equation of the curve, if it passes through the point $(0, -\frac{1}{8})$.

(ii) Find the equations of the tangent and normal at the point $(1, 1)$ on the curve given by the parametric equations $x = \dfrac{1}{2t-1}$, $y = (2t - 1)^2$.

5 (i) Find the range of values of x for which $\dfrac{x^2 - 4x + 2}{x - 4} < 1$.

(ii) Write down the binomial expansion of $(1 - x)^{10}$ in ascending powers of x as far as the term in x^5. Hence find the value of $(0.99)^{10}$ correct to six significant figures.

6 With the usual axes on horizontal ground, A and B are the points with coordinates $(2, 3)$ and $(14, 6)$ respectively. C is a point 4 units vertically above A and D is 9 units vertically above B. Calculate (i) the length of CD, (ii) the angle CD makes with the horizontal.

7 A small metal sphere is projected with a speed of 21 m/s along a line of greatest slope up a plane inclined at an angle θ to the horizontal where $\tan \theta = \frac{4}{3}$. The coefficient of friction between the sphere and plane is 0.4. Calculate (i) the retardation of the sphere as it moves up the plane; (ii) the acceleration of the sphere as it returns down the plane; (iii) the speed of the sphere when it returns to the point of projection.

8 Two flat bodies are connected by a light string and placed on a plank AB of length 45 cm. The body nearer A is smooth and of mass 5 kg. The other body is rough and of mass 25 kg. The end A of the plank rests on horizontal ground and the end B is raised. The string tightens and both bodies commence to slide down the plank when B is 27 cm above the horizontal. Calculate the coefficient of friction between the rough body and the plank.

9 A uniform rod AB of mass 18 kg and length 15 cm, is suspended from a point C by two strings CA and CB, of lengths 9 cm and 12 cm respectively. A horizontal force P N is applied at the end A of the rod, and it is sufficient to keep the rod in equilibrium in a

horizontal position. Find P and the tensions in the strings.

10 A uniform straight rod ABC is bent at B so that the two parts AB and BC, of lengths 0·2 m and 0·8 m respectively, are at right angles. Find the distances of the centre of gravity of the rod from AB and BC. If the rod is freely suspended from A, find the angle AB makes with the vertical.

XI

1 (i)   The nth term of a series is $n^2 - n - 2$. Find the first four terms and the term which is equal to 208.

(ii)   Find the square roots of $28 - 16\sqrt{3}$.

2 A field ABC is triangular in shape. AB = 200 m, AC = 150 m. The bearings of C and B from A are 050° and 075° respectively. Find the area of the field, correct to three significant figures.

3 (i)   Given that x is acute and $\cot x = b$, prove that
$$b \sin 2x - \cos 2x = 1.$$

(ii)   If $y = \dfrac{1 + \sec \theta}{\tan \theta}$, prove that $\dfrac{1}{y} = \dfrac{\sec \theta - 1}{\tan \theta}$.

4 A curve is given by the equation $y = ax^4 + bx^2$. The gradient of the tangent to this curve at the point $(1, 1)$ is -2. Find the values of a and b. Find the coordinates of the points where the tangent is parallel to the x-axis.

5 (i)   Prove that $3x^2 - 7x + 8$ is positive for all real values of x.

(ii)   If $y = 6 + x - x^2$, find the range of values of x for which y is positive and find the maximum value of y.

6 (i)   Differentiate wrt x: (a) $\dfrac{3x^4 - x^2 + 1}{2x^2}$; (b) $\sqrt{3x^2 - 1}$; (c) $\sin^3 2x$.

(ii)   Find the values of (a) $\displaystyle\int_{-1}^{2} (2x^2 - x + 1)dx$; (b) $\displaystyle\int_{1}^{4} x^{\frac{5}{2}} dx$.

7 ABC is a triangle where the coordinates of A, B, C are $(5, 6)$, $(2, -1)$ $(6, 3)$ respectively. If M is the midpoint of BC, prove that $AB^2 + BC^2 = 2BM^2 + 2AM^2$.

8 On a straight level road, car A starts from rest and accelerates at 1 m/s². 4 s after starting, A passes another car B which is at rest. B immediately starts to accelerate at 2 m/s². Draw a v/t graph to illustrate these statements. When will B overtake A and what is the speed of B at this time?

9 (i)   Three forces act at the origin O and are represented by the vectors OA, OB and OC where the coordinates of A, B and C

are (6, 3), (− 2, 4) and (− 4, − 7) respectively. Show that the forces are in equilibrium.

(ii) Forces of 10, 8, 6 and 2 N act along the sides AB, BC, CD and DA respectively of a square ABCD of side 5 cm. Find where their resultant cuts the sides AB and AD, measured from A.

10 A 1 tonne railway wagon moving at a speed of 3 m/s on a straight track hits a second wagon of mass 2 tonne which is at rest. Find the speed of the second wagon if the first wagon has a speed of 0·5 m/s in the opposite direction after the collision, and the loss of kinetic energy due to the collision.

XII

1 (i) Express in the form $a + b\sqrt{3}$: $\dfrac{2\sqrt{3} - 1}{1 + 3\sqrt{3}}$.

(ii) Solve the simultaneous equations: $x + 3y = 1$, $xy = y^2 - 3$.

(iii) Find x if $x^{2\cdot2} = 4\cdot4$.

2 (i) If α, β are the roots of the equation $2x^2 - 3x - 4 = 0$, find the equation whose roots are $\dfrac{1}{\alpha} + \beta$, $\dfrac{1}{\beta} + \alpha$.

(ii) For what values of x is $(x-4)(x-3) < 2$?

3 In a triangle ABC, AB = 7 cm, BC = 5 cm and AC = 10 cm. Find the angles of the triangle and its area.

4 Expand $(1 - 2x)^9$ as far as the term in x^4. Deduce an approximation for $(0\cdot98)^9$ correct to six significant figures.

5 The area of a sector of a circle is 8 cm² and the perimeter of the sector is 12 cm. Find the possible lengths of the radius of the circle and the angle of the sector in each case to the nearest degree.

6 (i) For a certain linear function f, $f(2) = 1$, $f(-2) = -7$. Find $f^{-1}(3)$.

(ii) A is the set $\{4, 8, 10, 15, 20, 21, 23, 25\}$ and B is the set of all prime numbers. Draw an arrow diagram between the sets A and B to show the relation "x is mapped onto the nearest prime number not larger than x" where $x \in A$.

(iii) If $f: x \rightarrow x^2 - 2$ and $g: x \rightarrow \dfrac{x - 1}{3}$, find $fg(4)$ and $gf(4)$.

7 A motor working at 0·2 kW is pulling a load of 15 kg (including the motor) horizontally at a speed of 0·8 m/s. Find the resistance to motion. If the motor now pulls the same load, with the resistance unchanged, up a slope of $\theta°$ to the horizontal, where $\sin \theta = \frac{2}{7}$, what is the greatest speed possible?

8 A string passes over a smooth frictionless pulley and carries masses

of 2 kg and 1 kg at its ends. The system is held at rest and then released. After 3 s the 2 kg mass is stopped. How much higher does the 1 kg mass travel and in what time? When the 1 kg mass returns to its original position and the 2 kg mass is released simultaneously, with what common speed do the masses begin to move?

9 Motorboat A is travelling on the calm surface of a lake with speed 12 km/h in direction 050°. At a certain time motorboat B is 4 km due east of A and travelling at 10 km/h in direction 330°. After what time are the two boats closest together? (Answer to the nearest minute).

10 A box of mass 10 kg is being dragged at a steady speed across a rough horizontal floor by a force of 20 N inclined at an angle of 60° to the floor. Find the coefficient of friction between the box and the floor.

XIII

1 (i) If $F = \dfrac{1}{\sqrt{2x-1}}$, find the approximate percentage change in F if x is decreased by 0.5% when $x = 5$.

(ii) The lengths of the sides of a cube are each increasing at the rate of 0.25 m/s. At what rate is the volume increasing when it is 125 cm^3?

2 (i) Differentiate wrt x: (a) $\sqrt{x^2 - x - 1}$; (b) $\sin 2x \cos x$.

(ii) Find the area between the curve $y = 3 + 2x - x^2$, the line $y = x + 1$ and the y-axis.

(iii) Sketch the curve $y = 3x - x^2$. The portion above the x-axis is rotated completely about the x-axis. Find the volume created.

3 (i) O is the origin and the points A, C have coordinates (4, 3), (2, 6) respectively. If OABC is a parallelogram find (a) the coordinates of B; (b) the length of OA; (c) the distance of C from OA; (d) the area of the parallelogram.

(ii) A curve is given by the parametric equations $x = t^2 - \dfrac{1}{t}$, $y = t^2 + \dfrac{1}{t}$. Find the equations of the tangent and normal at the point where $t = 2$. Show that the Cartesian equation of the curve is $(x-y)^2 (x+y) = 8$.

4 (i) Without using tables find the value of $\dfrac{\sin 105° + \sin 15°}{\cos 345° + \cos 105°}$.

(ii) Find the solutions of the equation $\sin 2x = 0.75$ in the range $0° \leqslant x \leqslant 360°$.

(iii) Solve the equation $2 \sin (x° + 60°) = \cos (x° - 30°)$ in the range $0 \leqslant x \leqslant 180$.

5 A particle moves along a straight line so that its velocity, t s after passing a fixed point O on the line, is $(3t^2 - 4t + 1)$ m/s. What is its distance from O 5 s after passing O? At what time and where is its acceleration zero?

6 **(i)** The length of the perimeter of a sector of a circle is 20 cm. Write down an expression for the area of the sector in terms of the radius of the circle (r) and hence find the maximum area of the sector.

(ii) If $\log_3 (x^2 - x + 1) = 1$, find the possible values of x.

7 A lorry of mass 20 t draws a trailer of mass 5 t. The resistance to motion for each is 200 N/t. The engine of the lorry works at a rate of 75 kW. What is the maximum speed on the level and what is the tension in the tow-bar?

If the two travel at a speed of 10 m/s on the level, and the resistances and rate of working are unchanged, what is the acceleration at this speed and the tension in the tow-bar?

8 A heavy plank ABCD of length 5 m rests on two supports at B and C where AB = 2 m and CD = 1 m. A mass of 10 kg placed at D just tips the plank about C, while a mass of 8 kg placed at A just tips the plank about B. Find the mass of the plank and the distance of its centre of gravity from A.

9 A uniform rod AB of length 2 m and mass 3 kg is hinged at A and rests against a smooth peg C where AC = 1·5 m. The rod makes an angle of 60° with the horizontal. Find the reaction at C and the reaction at the hinge.

10 To a man walking along a horizontal road at 1·5 m/s the rain is coming towards him and appears to be falling at 3·5 m/s at an angle of 30° to the vertical. Find, by calculation or drawing, the true speed of the rain and the angle this makes with the vertical.

XIV

1 **(i)** How many different arrangements are there of the letters of the word CALCULUS?

If one of these arrangements is chosen at random, what is the probability that it will begin with C?

(ii) A bag contains 7 balls identical in size of which three are black and four are green. A ball is drawn at random and not replaced. A second ball is then drawn. What is the probability that they will be of different colours?

2 (i) The fourth term of an AP is 14 and the eleventh term is 35. What is the eighth term?

(ii) In a GP the product of the second and fourth terms is double the fifth term. The sum of the third and fourth terms is twelve times the second term. Find the GP if all terms are positive.

3 (i) Solve the equations $2x + 3y = 4$, $x^2 + xy + 1 = 0$.

(ii) Find the coordinates of the turning points on the curve $y = x^3 - 3x^2 - 9x + 4$ and give the nature of each one.

4 OABC is a parallelogram drawn in the first quadrant and O is the origin. The equation of the line through O and A is $2y = x$ and the equation of the line through O and C is $y = 2x$. The co-ordinates of B are (6, 5). Find the coordinates of A and C and the equation of AC.

5 The angle of elevation of the top of a vertical mast due North of an observer is 25°. From a point 100 m due West of the observer the angle of elevation is 17°. Find the height of the tower.

6 (i) If $\sin 2A = \frac{3}{5}$ and A is acute, without using tables find the values of: **(a)** $\cos 2A$; **(b)** $\sin A$; **(c)** $\sin 3A$.

(ii) Express $(2 - 3\sqrt{2})^3$ in the form $a + b\sqrt{2}$.

(iii) Using the remainder theorem factorize $x^3 - x^2 - 8x + 12$.

7 (i) Differentiate wrt x: **(a)** $\dfrac{2 + x}{1 - 2x}$; **(b)** $\sec 3x$; **(c)** $(x^2 - x + 5)^2$.

(ii) A circular cone has an angle of 60° at its apex. It is held, apex downwards and liquid is poured in at a rate of 40 cm^3/s. Find the rate at which the level of the liquid is rising when its depth is 5 cm.

8 A uniform ladder AB, 8 m long and mass 30 kg, rests with end A on smooth horizontal ground and end B against a smooth vertical wall at an angle of 30° to the wall. One end of a rope is tied to the ladder at a point C and the other end to the point O where the wall meets the ground, so that OC is at right angles to AB. Find the tension in the rope.

9 Two small balls, one of mass 1 kg and the other of mass 2 kg, are suspended side by side by equal strings of length 5 m from the same point. The 2 kg mass is then held with the string taut and inclined at 60° to the vertical and released so that it strikes the other ball. They stick together at the impact. Find the loss of energy due to the collision and the height the balls rise after the collision.

10 A balloon starts from rest on the ground and rises vertically with a constant acceleration 0·01 m/s^2. When it is 200 m from the ground a stone is released from the rising balloon. Find the time taken by the stone to reach the ground.

XV

1 (i) Simplify without using tables: $2 \log(1 \cdot 5) - \log(0 \cdot 75) + \log(3\frac{1}{3})$.
(ii) Solve the equation $2^{2x+2} - 5 \times 2^x = -1$.

2 In an acute-angled triangle ABC, BC $= a$ and the altitude of the triangle is h. A rectangle PQRS is drawn inside the triangle so that P and Q are on BC, R is on CA and S is on AB. If PS $= x$, derive an expression for the area of this rectangle in terms of a, h, x and hence find its maximum area.

3 (i) If f: $x \to \cos x$ and g: $x \to 2x$, sketch in the same graph the functions fg and gf for the domain $0 \leqslant x \leqslant 2\pi$. How many solutions will there be of the equation $\mathrm{fg}(x) = \mathrm{gf}(x)$ for this domain? Solve the equation $\mathrm{fg}(x) = \mathrm{gf}(x)$ giving your solutions in degrees and minutes.
(ii) A line of variable gradient m passes through the fixed point $(3, 2)$ and cuts the y-axis at A. The line through $(3, 2)$ perpendicular to the first line cuts the x-axis at B. Find the coordinates of M, the midpoint of AB, in terms of m, and hence find the Cartesian equation of the locus of M.

4 (i) Differentiate wrt x: (a) $(x - 1)(x^2 + 1)$; (b) $x\sqrt{x^2 + 1}$ (simplify your answer).
(ii) Find the values of: (a) $\displaystyle\int_{-2}^{-1} \frac{x-1}{x^3}\,\mathrm{d}x$; (b) $\displaystyle\int_{0}^{3} (x-3)(2x+1)\,\mathrm{d}x$.
(iii) A particle moves in a straight line so that its velocity at time t s after starting from rest at a point O on the line is given by $v = 2t - t^2$. Find its distance from O after 5 s and its acceleration at that time.

5 (i) If α, β are the roots of the equation $2x^2 - x = 5$, find the equation whose roots are $\alpha + 2\beta$, $\beta + 2\alpha$.
(ii) Solve the equation $\sqrt{2x - 3} + \sqrt{x + 10} = 7$.
(iii) If the roots of the equation $t^2x^2 - (t + 2)x + 1 = 0$ are equal, find the values of t.

6 (i) Three boys and two girls sit at random on five chairs in a row. What is the probability that the two girls will be sitting on the end chairs?
(ii) Give the first four terms of the expansion of $(2-x)^{-1}$ and state for what range of values of x the expansion is valid.
Hence calculate the value of $\dfrac{1}{1 \cdot 99}$ correct to 7 significant figures.

7 An aeroplane travels from A to B, where B is 120 km on a bearing of $060°$ from A. The airspeed of the plane is 200 km/h and there is a wind of 40 km/h blowing from the north-east. By drawing or

calculation, find (i) the course taken by the pilot; (ii) the time taken for the journey.

8 ABCD is a square of side 0·5 m. The following forces act in the directions given: 10 N along AB, 6 N along BC, 6 N along CA and 2 N along AD. Find their resultant and where its line of action cuts AB and AD (measured from A).

9 A pump raises 50 litres of water per minute from a depth of 12 m and sends it through a pipe at a speed of 5 m/s. Find the rate at which the pump is working.

10 A plastic sheet ABCD of uniform thickness is in the shape of a rectangle where AB = 35 cm and BC = 50 cm. A square of side 20 cm is removed from the sheet where two sides of the square lie along BC and CD. Find the position of the centre of gravity of the remainder measured from AB and AD.

If the sheet is now suspended freely from the corner D, what angle does AD make with the vertical?

Answers

Exercise 1.1 (*p.* 5)

1 1 **2** $\frac{1}{5}$ **3** $\frac{1}{1000}$ **4** 2

5 10 **6** 4 **7** $\frac{1}{9}$ **8** -2

9 1 **10** $\frac{1}{9}$ **11** $\frac{1}{5}$ **12** $\frac{1}{x^2}$

13 $-\frac{1}{7}$ **14** $\frac{1}{100\,000}$ **15** $9\cdot2 \times 10^7$ **16** $5\cdot3 \times 10^2$

17 $3\cdot4 \times 10^{-6}$ **18** $2\cdot3 \times 10^{-10}$ **19** $2\cdot15 \times 10^5$ **20** $1\cdot01 \times 10^{-2}$
21 $1\cdot02 \times 10^3$ **22** $3\cdot46 \times 10^9$ **23** 230 000 **24** 0·0023
25 0·000 004 05 **26** 130 000 000

Exercise 1.2 (*p.* 6)
1 $x = 5$ **2** $x = 4$ **3** $x = -\frac{1}{2}$ **4** $x = -3$
5 $x = -4$ **6** $x = -\frac{2}{5}$ **7** $x = 2$ or 0 **8** $x = 2$ or 1
9 $x = -1$ or 4 **10** $x = -1$ **11** $x = 3$ **12** $x = -2$ or 3

Exercise 1.3 (*p.* 7)
(i) 2·41; **(ii)** 3·74; **(iii)** $\cdot x = 1\cdot46$

Exercise 1.4 (*p.* 7)
1 $2 = \log_7 49$ **2** $5 = \log_2 32$ **3** $0 = \log_{10} 1$ **4** $3 = \log_{10} 1000$

5 $-2 = \log_5 \left(\frac{1}{25}\right)$ **6** $2^x = 8, x = 3$ **7** $3^x = 27, x = 3$

8 $4^x = 64, x = 3$ **9** $5^x = 125, x = 3$ **10** $2^x = \frac{1}{8}, x = -3$

11 $10^x = 0\cdot001, x = -3$ **12** 1 **13** 1 **14** 2
15 0 **16** -1 **17** -2 **18** 5 **19** 5 **20** -2
21 x **22** x

Exercise 1.5 (*p.* 10)
1 $\frac{1}{2}\log_5 3$ **2** $\frac{1}{3}\log_2 5$ **3** $3\log_4 3$ **4** $3\log_a x$
5 $3\log_r 10$ **6** $3\log_2 5$ **7** $4\log_{10} 10 = 4$ **8** $2\log_7 7 = 2$
9 $3\log_3 \frac{1}{2} = -3\log_3 2$ **10** $2\log_6 11$ **11** $4\log_x x = 4$
12 $x\log_4 4 = x$ **13** $\log_5 120$ **14** 0 **15** $2\log_3 6$
16 2 **17** $\log_5 9\sqrt{6}$ **18** $\log_a \frac{7}{6}$ **19** 1·113
20 0·251 **21** 2·046 **22** $-0\cdot251$ **23** 1·293
24 1·431 **25** 0·341 **26** 1·544 **27** 1·615
28 $-0\cdot862$ **29** $-0\cdot318$ **30** 2·862

Exercise 1.6 (*p.* 11)
1 1·465 **2** 2·807 **3** 1·861 **4** 1·826 **5** 1·107
6 $-1\cdot849$ **7** 1·829 **8** 2·039 **9** $-1\cdot288$ **10** 2·322

11 1·465	**12** 1·691	**13** 1·431	**14** −0·228	**15** −0·7563
16 7·454	**17** 1·5	**18** −0·605	**19** 5·849	**20** −1·449

Exercise 1.7 (*p.* 12)

1 (a) $\frac{1}{243}$; (b) $\frac{1}{1\cdot2}$; (c) $\frac{64}{125}$; (d) 4; (e) $\frac{1}{1000}$ **2** 1000

3 (i) −2; (ii) −3; (iii) $1\frac{1}{2}$; (iv) $-\frac{1}{2}$; (v) $-\frac{2}{3}$; (vi) −2
4 2·771; 129·6   **5** $\frac{3}{2}$ **6** 0·725   **7** (i) 1   (ii) 49

8 (i) −1·322; (ii) 6, −12   **9** $x = \frac{y^2}{1000}$

10 (a) 0; (b) 0·064   **11** 5·294   **12** 0
13 $x = 20$, $y = \frac{1}{5}$ **14** 28·13 yr.   **15** 14·9 yr.
16 $t = 1\cdot428$ **21** 3·495

Exercise 2.1 (*p.* 17)

1 $6\sqrt{2}$ **2** $5\sqrt{2}$ **3** $3\sqrt{2}$ **4** $2\sqrt{5}$ **5** $10\sqrt{10}$ **6** 2   **7** $\frac{1}{2\sqrt{2}}$

8 2   **9** $12\sqrt{6}$ **10** $3\sqrt{5}$ **11** $7\sqrt{2}$ **12** $11\sqrt{3}$ **13** 4   **14** $5\sqrt{3}$
15 $9\sqrt{7}$ **16** 24   **17** 4   **18** $7 + 4\sqrt{3}$ **19** $4 - 2\sqrt{3}$
20 $9 - 4\sqrt{2}$ **21** $3\sqrt{2}$ **22** $34 + 24\sqrt{2}$ **23** $2\sqrt{2}$
24 $9 - 7\sqrt{2}$ **25** $31 - 8\sqrt{2}$ **26** 0·707   **27** 1·15   **28** 3·54
29 2·83   **30** 0·354   **31** 5·20   **32** 0·289   **33** 3·15
34 $\pm(1 + 3\sqrt{2})$ **35** $\pm(2 - 3\sqrt{7})$ **36** $\pm(\sqrt{5} - 2\sqrt{3})$
37 $\pm(2\sqrt{2} - 3\sqrt{3})$ **38** $\pm(3 + 4\sqrt{5})$

39 $\sqrt{2} + 1$ **40** $\frac{1}{2}(\sqrt{5} + 1)$ **41** $\frac{(2\sqrt{3} + 1)}{11}$ **42** $3 + 2\sqrt{2}$ **43** $5 - 2\sqrt{6}$

44 $\frac{(6\sqrt{6} + 3\sqrt{2} + 2\sqrt{3} + 1)}{17}$

45 $\frac{36 - 11\sqrt{10}}{2}$ **46** $5 - 2\sqrt{6}$ **47** $\sqrt{5} - \sqrt{6}$

48 $h + \frac{2}{3}k$; $x = \sqrt[3]{\frac{16}{7}}$ **49** $\cos 45° = \sin 45° = \frac{1}{\sqrt{2}}$

50 $\sin 60° = \frac{\sqrt{3}}{2} = \cos 30°$; $\tan 60° = \sqrt{3}$; $\tan 30° = \frac{1}{\sqrt{3}}$

Exercise 2.2 (*p.* 19)

1 $x = \frac{9}{4}$ **2** $x = 0$ or 16   **3** $x = 0$ **4** $x = 5$ **5** $x = 7$ **6** $x = 3$
7 0, 4   **8** 3   **9** 2

Exercise 3.1 (*p.* 21)

1 $x = 1$, $y = -1$, $z = 2$ **2** $x = -2$, $y = 3$, $z = 1$

3 $x = -2$, $y = 3$, $z = 4$ **4** $x = \frac{1}{2}$, $y = 2$, $z = 1\frac{1}{2}$

5 $x = 1$, $y = -1\frac{1}{2}$, $z = \frac{1}{2}$ **6** $x = -3$, $y = 0$, $z = 5$

7 $x = -\frac{1}{2}$, $y = \frac{2}{3}$, $z = -\frac{4}{5}$ **8** $a = 2$, $b = -3$, $z = 4$

9 $p = 3$, $q = -1$, $r = 2$

Exercise 3.2 (*p.* 23)

1 $x = 2$, $y = 3$ or $x = 3$, $y = 2$

2 $x = 5$, $y = 2$ or $x = -\frac{23}{5}$, $y = -\frac{14}{5}$

3 $x = 3$, $y = -1$ or $x = -\frac{4}{3}$, $y = \frac{23}{3}$

4 $x = 3$, $y = 2$ or $x = 2$, $y = -1$

5 $x = 2$, $y = -1$ or $x = \frac{43}{10}$, $y = -\frac{25}{2}$

6 $x = 3$, $y = 2$ or $x = \frac{5}{7}$, $y = \frac{38}{7}$

7 $x = \frac{27}{4}$, $y = \frac{49}{6}$ or $x = -4$, $y = 1$

8 $x = 3$, $y = 1$ or $x = -\frac{74}{27}$, $y = -\frac{35}{27}$

9 $x = 5$, $y = -1$ or $x = -\frac{41}{11}$, $y = \frac{53}{11}$

10 $x = -2$, $y = 1$ or $x = -\frac{3}{11}$, $y = -\frac{62}{33}$

Exercise 3.3 (*p.* 24)

1 $x = 4$, $y = -1$ **2** $x = 4$, $y = 1$

3 $x = 3$, $y = -2$ or $x = \frac{7}{3}$, $y = -\frac{10}{3}$ **4** $x = 3$, $y = -1$

5 $x = 3$, $y = -1$ **6** $x = 5$, $y = 1$ or $x = \frac{7}{2}$, $y = \frac{13}{4}$

Exercise 3.6 (*p.* 31)

1 1, 2, 3 but not 4 (6 has 2 arrows) or 5 (as e.g. 4 is the square of ± 2).

2 (**b**) and (**c**) are not functions.   **4** 3, 5, 7, 9, 11

5 No   **7** Yes

8 (**a**) f: $x \to x + 2$; 5, -3;   (**b**) g: $x \to 3 - x$; -1, 5;   (**c**) h: $x \to x^2 + 1$; 10, 17

9 $-3 \leqslant y \leqslant 13$ **10** $3 \leqslant y \leqslant 19$ **11** 1·8; 1·8, -1, $-0·6$, 3·8

14 -2, 1, 1   **15** -11, 54

Exercise 3.7 (*p.* 33)

1 fg: $x \to 2x - 1$; gf: $x \to 2(x - 1)$; 3, 2, -3, -4

2 $x \to (x + 1)^2$, $x \to x^2 + 1$; 1, 9, 5, 5

3 $x \to x + 2$; 1, 2, 5   **4** ab; ba   **5** fg; gf

6 20, sin 10°, 2 sin 10°, sin 20°, 2 sin 15°, sin 30°

7 $2(\frac{x}{2} + 1) = x + 2$; $x + 1$; $x + 1$; $\frac{1}{2}(2x + 1) = x + \frac{1}{2}$

8 1, 2, 3

10 $\frac{1}{2}$, 2, 2, $\frac{1}{2}$, 3, x; ff is the identity function; $\dfrac{1}{x}$ is undefined for $x = 0$

11 13   **12** 1 to 16   **13** $x^2 + 6x + 11$, $x^2 + 5$

14 g: $x \to x - 1$

Exercise 3.8 (*p.* 36)

2 $x \to \dfrac{x}{2}$ **3** $x \to x + 2$ **4** $x \to \dfrac{x}{5}$

5 $x \to 2x$ **6** $x \to x - 4$ **7** $x \to \dfrac{x - 1}{2}$

8 $x \to \dfrac{x + 1}{3}$ **9** $x \to 5 - x$ **10** $x \to \dfrac{1}{x}$

11 $x \to 2x - 1$ **12** $x \to \dfrac{4x + 1}{3}$ **13.** $x \to \dfrac{5 - x}{2}$

14 $x \to \dfrac{12}{x}$ **15** 9, 10, 14

16 $x \to 3x + 6$; $x \to \dfrac{x - 6}{3}$; $x \to \dfrac{x}{3}$; $x \to x - 2$; $x \to \dfrac{x}{3} - 2 \; (= \dfrac{x - 6}{3})$

17 $x \to \dfrac{x}{3} - 2$; $x \to 3x + 6$; $x \to 3x$; $x \to x + 2$; $x \to 3(x + 2) = 3x + 6$

18 $x \to \dfrac{1}{3} \sin x$; $x \to \sin \dfrac{x}{3}$; **19** $-1, 2, 0, 0, 4$

Exercise 3.9 (*p.* 38)
1 (a) -5; (b) 7 **2** (a) 4; (b) 16 **3** (a) $-4\frac{1}{4}$; (b) 22
4 (a) $8\frac{3}{4}$; (b) 17 **5** (a) 20; (b) $-\frac{5}{8}$ **6** (a) $-4\frac{26}{27}$; (b) $-13\frac{7}{8}$
7 (a) -47; (b) $13\frac{1}{2}$ **8** $0, 0; (x - 2)(x - 3)$
9 0 in each case; $x + 1, x - 1, x - 2$ are factors

Exercise 3.10 (*p.* 40)
1 $(x - 1)(x + 2)(x - 3)$ **2** $(x - 1)(x - 1)(x - 2)$
3 $(x - 1)(x + 1)(x + 2)$ **4** $(x - 1)(x^2 + x + 1)$
5 $(x + 1)(x + 2)(x + 3)$ **6** $(x - 1)(2x - 1)(x + 2)$
7 $(x + 1)(x + 1)(2x + 3)$ **8** $(x - 1)(3x + 2)(2x - 1)$
9 $(x - 2)(4x - 3)(x + 3)$ **10** $(x - 1)(x + 1)(x - 3)(x - 2)$
11 -3 **12** 2
13 $a = -2, b = -11; (x - 4)$ **14** $a = -1, b = -5; (2x + 1), (x + 1)$
15 $a = 3, b = 2, c = -19; (3x - 1)$ **16** $x = 1$ or 5 or -1
17 $x = 2$ or 3 or 4 **18** $x = 2$ or $\frac{1}{2}$ or -3
19 $x = 1$ or -1 or -2 or -3

Exercise 3.11 (*p.* 40)
1 $a = 2, b = -6; (x + 3)$ **2** $x = 9, y = -1$ or $x = -3, y = 2$
3 $a = 3, b = -3$ **4** $a = -\frac{11}{30}, b = \frac{71}{30}$
5 $a = 3, b = -1, c = 4$ **6** $x = 4, y = 2$ or $x = -2, y = 4$

7 $x = \frac{18}{13}, y = -\frac{5}{13}, z = \frac{19}{13}$ **8** $(x - 1)(x^2 + 2x - 4)$

9 $p = -4; (x - 4), (x + 2)$ **10** 0

11 $x = 1, y = -2$ **12** $a = \frac{8}{5}, b = \frac{3}{5}, c = \frac{9}{5}$

13 $c = \dfrac{f(a) - f(b)}{a - b}$, $d = \dfrac{a\,f(b) - b\,f(a)}{a - b}$, remainder $= \dfrac{(x - b)\,f(a) - (x - a)\,f(b)}{a - b}$

14 $a = 12, b = -19; 6x^2 + 6x - 7$ **15** $x = \frac{41}{10}, y = \frac{16}{3}$ or $x = 1, y = -5$

17 (i) $x \to x - 1$; (ii) $x \to \dfrac{x + 3}{2}$; (iii) $x \to 3 - x$; (iv) $x \to \dfrac{1}{x} - 1$

18 $x \to x - 2, \ x \to 3 - x, \ x \to 5 - x, \ x \to 1 - x, \ x \to 1 - x, \ x \to 5 - x$

20 $\frac{1}{3}, 0, 7, 4$ **22** $8, 3, 3, 0, 1$ **23** $2, -3$ **25** -6

26 (i) 3 to 6; (ii) $\frac{1}{4}$ to 1; (iii) 1 to $\frac{1}{4}$; (iv) 0 to 1

27 3 **28** 2 **29** $x = 1, 2, -1$ **30** -2 to 1

31 $x \to 3x; \ x \to 3(x - 1)$ **32** $x \to x + 1$

Exercise 4.1 (p. 44)

1 1, 3, 5, 7, 9, 11, 13, 15 **2** 8, 11, 14, 17 **3** 1, 8, 27, 81

4 $1, 1\frac{1}{2}, 2, 2\frac{1}{2}$ **5** 2, 4, 8, 16 **6** 1, 3, 9, 27

7 7, 11, 15, 19 **8** $3, 1, -1, -3$ **9** 4, 9, 16, 25

10 3, 9, 19, 33 **11** 3, 5, 7, 9 **12** $-1, 1, -1, 1$

13 $1, \frac{1}{2}, \frac{1}{3}, \frac{1}{4}$ **14** $-1, 2, -3, 4$ **15** 3, 6, 12, 24

16 0, log 2, log 3, log 4 **17** $\frac{1}{2}, \frac{2}{3}, \frac{3}{4}, \frac{4}{5}$ **18** $2n$

19 $3n + 2$ **20** $11 - 2n$ **21** 2^{n-1} **22** $4n + 3$

23 n^3 **24** $3 \times (-2)^{n-1}$

Exercise 4.2 (p. 48)

	(a)	(b)	(c)	(d)	(e)	(f)
1	20	110	30	240	$2n$	$n(n + 1)$
2	7	80	27	195	$2n - 3$	$n(n - 2)$
3	39	210	59	465	$4n - 1$	$n(2n + 1)$
4	76	400	116	900	$8n - 4$	$4n^2$
5	-9	$+45$	-24	-45	$-3n + 21$	$\frac{3n}{2}(13 - n)$
6	36	135	61	390	$5n - 14$	$\frac{n}{2}(5n - 23)$
7	$15\frac{1}{2}$	$87\frac{1}{2}$	23	$187\frac{1}{2}$	$\dfrac{3n + 1}{2}$	$\dfrac{n(3n - 5)}{4}$

8 15th term **9** 10 **10** 4, 7, 10, ...; 175

11 13, 467 **12** $-7\frac{1}{2}, -5\frac{1}{2}, -3\frac{1}{2}, ...$ **13** $a = 7\frac{1}{2}, b = 11$

14 $-\frac{1}{3}$

Exercise 4.3 (p. 51)

1 (a) $\frac{1}{2}$ (b) $\dfrac{1}{32}$ (c) $\dfrac{2^5 - 1}{2} = 15\frac{1}{2}$ (d) $\dfrac{2^9 - 1}{2^5}$ (e) 2^{4-n} (f) $2^4\left(1 - \dfrac{1}{2^n}\right)$

2 (a) -3×4^4 (b) -3×4^8 (c) $-\frac{3}{5}(4^5 + 1)$ (d) $-\frac{3}{5}(4^9 + 1)$

    (e) $(-1)^n 3 \times 4^{n-1}$ (f) $\frac{3}{5}[(-4)^n - 1]$

3 (a) 2   (b) $\frac{1}{8}$ (c) $\frac{2}{3}(2^5 + 1) = 22$ (d) $\dfrac{2^9 + 1}{3 \times 2^3}$ (e) $(-1)^{n-1}2^{6-n}$

   (f) $\dfrac{2^6}{3}\left[1 - (-\frac{1}{2})^n\right]$

4 (a) 3   (b) $\dfrac{3^5}{2^4}$ (c) $\dfrac{2^5}{3^3}\left[(\frac{3}{2})^5 - 1\right]$ (d) $\dfrac{2^5}{3^3}\left[(\frac{3}{2})^9 - 1\right]$ (e) $\dfrac{3^{n-4}}{2^{n-5}}$

   (f) $\dfrac{2^5}{3^3}\left[\dfrac{3^n}{2^n} - 1\right]$

5 $6, 4, \frac{8}{3}, \dots$ **6** $12, 3$ **7** $8, 4, 2 \dots$ **8** $p = 2, r = 2; 5 \times 2^6$ **9** $\frac{3}{2}, 1, \frac{2}{3}$

Exercise 4.4 (p. 52)

1 $\frac{16}{3}$ **2** $1165\frac{1}{2}$ **4** $328; 40 \times 2^{n-1} + 17 - 3(n - 1)$

5 $3; 41$ **6** 1240 **7** $2, 1\frac{1}{2}; 19$ **9** 7

10 $2a + d; 4d$ **12** $0{\cdot}4, 3{\cdot}8, 7{\cdot}2, \dots$ **13** 2 **14** $3n$

15 $1, 5, 9, 13$ **16** 5 **17** $\frac{1}{9}, 3, \frac{3280}{9}$ **18** $19{:}5$

19 $3, 5; 1 + 2n + 5 \times 2^{n-1}$ **20** (i) $9\frac{3}{8}$;   (ii) $\frac{5}{8}$

21 (a) $\dfrac{n(n + 5)}{12}$;   (b) $\dfrac{3}{2}\left[(\frac{4}{3})^n - 1\right]$ **22** $-\frac{11}{4}$ or 2

23 $149, 7500$ **24** $25, -15, 9, \dots$ or $4, 6, 9, \dots$

25 $2, 4, 7, 12; 2^{n-1} + n; 1078$ **26** 3244

27 1108 **28** $3; 72$ **29** 31

Exercise 5.1 (p. 58)

1 $\frac{1}{3}, -\frac{1}{3}$ **2** $-4, -1$ **3** $\frac{1}{2}, \frac{5}{2}$ **4** $0, -\frac{7}{3}$

5 $\dfrac{q}{p}, -\dfrac{r}{p}$ **6** $\dfrac{a + 3}{2}; \dfrac{a^2}{2}$

7 (i) $3x^2 - 2x - 4 = 0$;   (ii) $9x^2 - 7x + 1 = 0$;
   (iii) $x^2 + x - 3 = 0$;   (iv) $27x^2 - 10x - 1 = 0$;
   (v) $3x^2 - 7x + 3 = 0$

8 (i) $x^2 + 3x - 8 = 0$;   (ii) $4x^2 - 25x + 16 = 0$;
   (iii) $4x^2 - 3x - 2 = 0$;   (iv) $8x^2 + 99x - 64 = 0$;
   (v) $2x^2 - x - 5 = 0$

9 (i) $3x^2 - 2x - 36 = 0$;   (ii) $9x^2 - 55x + 81 = 0$;
   (iii) $9x^2 + x - 3 = 0$;   (iv) $27x^2 - 82x - 729 = 0$;
   (v) $3x^2 - 7x - 5 = 0$

10 (i) $x^2 + 2x - 32 = 0$;   (ii) $x^2 - 17x + 64 = 0$;
   (iii) $8x^2 - x - 1 = 0$;   (iv) $x^2 + 25x - 512 = 0$;
   (v) $x^2 - x - 8 = 0$

11 (i) $ax^2 + 2bx + 4c = 0$;   (ii) $a^2x^2 - (b^2 - 2ac)x + c^2 = 0$;
   (iii) $cx^2 + bx + a = 0$;   (iv) $a^3x^2 + b(b^2 - 3ac)x + c^3 = 0$;
   (v) $ax^2 - (2a - b)x + (a + c - b) = 0$

12 $15x^2 - 13x - 169 = 0$

13 (i) -3;   (ii) -5;   (iii) 19;   (iv) -72;   (v) 311

14 $ax^2 - (2a - b)x + (a - b + c) = 0$

15 $x^2 + 4x + 1 = 0$ **16** $\frac{1}{9}$ **17** 3 **18** ± 6

19 (a) $-\alpha, -\beta$; (b) $\dfrac{1}{\alpha}, \dfrac{1}{\beta}$; $a^2 x^2 + a(b - c)x - bc = 0$

20 $16x^2 - 40x + 9 = 0$

Exercise 5.2 (*p.* 62)

1 2 rat. roots   **2** 2 real roots   **3** Complex roots
4 Equal roots   **5** 2 real roots   **6** Equal roots
7 2 real roots   **8** Min $-\frac{9}{8}$, $x = \frac{1}{4}$ **9** Max 6, $x = -2$
10 Min $-6\frac{1}{12}$, $x = \frac{1}{6}$ **11** Max $5\frac{1}{8}$, $x = \frac{1}{4}$ **12** ± 6
13 $-\frac{1}{3}$. 1

Exercise 5.3 (*p.* 64)

1 $x \leqslant -4$. $x \geqslant 2$ **2** $x < -2$, $x > \frac{1}{3}$ **3** $-2 < x < \frac{1}{3}$

4 $x \leqslant -4$, $x \geqslant \frac{1}{2}$ **5** $x < \dfrac{-1 - \sqrt{21}}{2}$, $x > \dfrac{-1 + \sqrt{21}}{2}$

6 $1 < x < 2$ **7** $x < -\frac{1}{2}$, $x > 3$

8 $-3 < x < -2$ and $1 < x < 2$ **9** $-\frac{1}{3} < x < -2$

10 $k > \frac{1}{8}$ **12** $a = 1$ **13** $-2 \leqslant p \leqslant 6$

15 (i) $\dfrac{2(a - c)}{a + b + c}$; (ii) $-3 < x < -1$ **16** $-6, 2$; (i) 1; (ii) 3

17 (a) $\frac{17}{4}$; (b) $-\frac{1}{4}$; $x^2 - 4x + 4 = 0$ **18** ± 10
19 $9b = 2a^2$ **20** $x \leqslant -4$ or $x \geqslant \frac{1}{2}$

Exercise 6.1 (*p.* 68)

1 $100°$ **2** 70   **3** $231°$ **4** $144° \, 32'$ **5** $43° \, 02'$ **6** 2nd
7 4th   **8** 4th   **9** 2nd   **10** 1st   **11** 4th   **12** 1st
13 (a) $0 \cdot 3649$: (b) $0 \cdot 3644$; (c) $0 \cdot 7393$: (d) $0 \cdot 9061$; (e) $0 \cdot 0561$
14 (a) $0 \cdot 9311$: (b) $0 \cdot 9313$; (c) $0 \cdot 6734$; (d) $0 \cdot 4231$; (e) $0 \cdot 9984$;
15 (a) $0 \cdot 6445$: (b) $1 \cdot 4957$; (c) $0 \cdot 3259$; (d) $5 \cdot 0976$

Exercise 6.2 (*p.* 72)

1 $+0 \cdot 6820$ **2** $-0 \cdot 9272$ **3** $-8 \cdot 144$ **4** $-0 \cdot 7660$ **5** $-0 \cdot 8391$
6 $-0 \cdot 9816$ **7** $-0 \cdot 2946$ **8** $-0 \cdot 5823$ **9** $-0 \cdot 2746$ **10** $+6 \cdot 3138$
11 $+0 \cdot 5878$ **12** $+0 \cdot 8425$ **13** $+0 \cdot 0250$ **14** $+1 \cdot 2985$ **15** $-0 \cdot 1831$
16 (a) $45° \, 18'$; (b) $74° \, 46'$; (c) $26° \, 02'$
17 (a) $41° \, 24'$; (b) $62° \, 51'$; (c) $63° \, 44'$
18 $65°$; $115°$ **19** $53° \, 30'$; $306° \, 30'$ **20** $153° \, 30'$; $333° \, 30'$
21 $216° \, 50'$; $323° \, 10'$ **22** $63° \, 26'$; $243° \, 26'$ **23** $132° \, 58'$; $227° \, 02'$
24 $117° \, 30'$ **25** $559° \, 18'$ **26** $574° \, 13'$
27 $672° \, 45'$ **28** $567° \, 20'$

Exercise 6.3 (*p.* 80)

1 (a) $\sqrt{3}$; (b) $\frac{1}{2}$; (c) $\dfrac{1}{\sqrt{2}}$; (d) -1; (e) $-\sqrt{3}$

2 $\frac{1}{2}$ 3 1   4 $\frac{3}{2}$ 5 1
6 63°, 297°; 49°, 131°, 229°, 311°   7 130°; 310°

Exercise 6.5 (p. 88)
1 0, 60°, 180°   2 10°, 110°, 130°, 230°, 250°, 350°
3 38°; 142°   4 210°; 330°
5 Max. $y = 5$, $x = 90°$; min. $y = 1$, $x = 270°$. The equation has no roots
    as the x-axis does not intersect the graph
6 81°   7 205°; 352°   8 13·5°, 163·5°, 253·5°
9 0·95; 5·85   10 0·25; 3·65

Exercise 6.6 (p. 93)
1 1·0625   2 0·7701   3 −4·0765   4 −3·5978
5 −1·2489   6 −1·9278   7 0·653 radians   8 3·49 radians
9 0·670 radians   10 80° 14′   11 263° 37′   12 572° 56′
13 (i) 0·8 radians; (ii) 45° 50′;   14 6·76 cm.   15 14·4 cm^2
16 45·8 cm^2 17 0, 1·9   18 0·74   19 0·97   20 0·64

Exercise 6.7 (p. 96)
1 11° 32′; 90°; 168° 28′; 270°   2 0°; 180°
3 41° 49′; 90°; 138° 11′; 270°   4 30°, 150°; 210°; 330°
5 30°; 150°; 210°; 330°   6 270°
7 120°; 180°; 240°   8 90°; 210°; 330°
9 26° 34′; 135°; 206° 34′; 315°   10 23° 35′; 156° 25′
11 270°   12 45°; 63° 26′; 225°; 243° 26′
13 30°; 41° 49′; 138° 11′; 150°   14 56° 09′; 90°; 236° 09′; 270°
15 45°; 63° 26′; 225°; 243° 26′   16 30°; 150°
17 45°; 135°; 225°, 315°
18 54° 44′; 125° 16′; 234° 44′; 305° 16′
19 0°; 180°   20 20° 54′; 69° 6′; 200° 54′; 249° 6′
21 48° 11′; 90°; 270°; 311° 49′   22 211° 40′, 328° 20′
23 45°; 116° 34′; 225°; 296° 34′   24 63° 26′; 243° 26′
25 63° 26′; 153° 26′; 243° 26′; 333° 26′

Exercise 7.1 (p. 103)
1 49·7 cm   2 64° 25′   3 76° 56′   4 2·07x; 41° 37′
5 4·5 cm; 39° 49′; 66° 3′   6 86° 11′   7 38° 13′
8 36° 20′; 62° 43′; 80° 57′   9 25° 51′; 110° 55′; 43° 14′

Exercise 7.2 (p. 107)
1 4·83 cm   2 F = 63° 44′; EF = 2572 m; DF = 1800 m
3 A = 41° 54′; a = 6·3 cm; b = 7·93 cm
4 A = 38° 50′; b = 23·2 m; c = 42 m
5 (a) Ambiguous case; $b < c$;
    (b) Not the ambiguous case; $c > a$;
    (c) Not the ambiguous case; Q is obtuse;
    (d) Not the ambiguous case; $y > x$;
    (e) Ambiguous case; $p < q$

Exercise 7.2 (*continued*)

6   A = 53° 49'; C = 98° 05·5'; c = 0·736 cm
     A = 126° 11'; C = 25° 43·5'; c = 0·323 cm
7   A = 74°;   C = 63° 44'; a = 2·57 cm
     A = 21° 28'; C = 116° 16'; a = 0·979 cm
8   B = 21° 36'; C = 30° 07'; b = 11·4(5) cm
9   A = 58° 08'; C = 52° 38'; a = 9·08 cm
10   A = 75° 45'; C = 39° 07'; c = 20·3 m
     A = 104° 15'; C = 10° 37'; c = 5·93 m
11   A = 69° 39'; C = 65° 14'; c = 86·4 m
     A = 110° 21'; C = 24° 32'; c = 39·5 m
12 A = 33° 34'; B = 50° 42'; b = 70 cm
13 B = 15° 34', C = 139° 1', c = 12·23 mm
14 a = 6·52 cm, C = 26° 20', B = 13° 40'

Exercise 7.3 (*p.* 110)

1   7 cm^2 2   12·2 cm^2 3   9·80 cm^2 4   268 cm^2
5   120° 41'   6   3·94 cm   7   6·61 cm^2 8 p = 5
9   A = 30°; B = 90°   10 a = 4 cm   11 x = $\frac{1}{2}$
12   154 cm^2

Exercise 8.1 (*p.* 113)

2   (0, −1)   3   (3, 5) and (−1, −3); 8 units
4   x = 2   5   (a) −9;   (b) x = −1 or +5
6   Maximum; +8   7   (a) Max. +13; Min. −14;   (b) y = +6
8   (2, 2) and (−2, −2)   9   (a) (8, 3);   (b) (8, 9)

Exercise 8.2 (*p.* 123)

1   (a) 5;   (b) 10;   (c) 13;   (d) $\sqrt{101}$;   (e) $6\sqrt{5}$;   (f) $a(p - q)\sqrt{(p + q)^2 - 4}$
2   k = 9 or −7

4   (a) (4, 9);   (b) (1, −4);   (c) $(a + 2, b + 3)$;   (d) $\left(\dfrac{ap^2 + aq^2}{2}, ap + aq\right)$

5   k = 4; l = 0   6   P is (5, $4\frac{1}{3}$)

7   (a) 2;   (b) −$\frac{4}{3}$;   (c) 5;   (d) −$\frac{1}{7}$;   (e) 2;   (f) $\frac{1}{2}$;   (g) −$\dfrac{1}{pq}$;   (h) $\dfrac{2}{t_2 + t_1}$

9   (a) Collinear – gradient of +2;   (b) Not collinear;
    (c) Collinear – gradient of −$\frac{4}{9}$
10) A is (−2, 8) or (6, 2)

Exercise 8.3 (*p.* 126)

1   (a) 10 unit^2;   (b) 9 unit^2;   (c) 11·5 unit^2;   (d) $2pq$ unit^2;   (e) $\frac{1}{2}k$ unit^2
2   Area = 0. The three points are collinear
3   Area = $\pm\frac{1}{2}(t^2 - 5t + 6)$. t = 2 or 3 for collinear points
4   Area = $4b - 8$, so independent of a; (10, 5)
5   $2a - b + 2 = 0$ and $2a - b - 2 = 0$
6   (a) $14\frac{1}{2}$ unit^2;   (b) 24 unit^2;   (c) 19·5 unit^2 7   9 unit^2
8   (a) AB = 13; AC = 17;   (b) 70 unit^2;   (c) $\frac{140}{221}$

Exercise 8.4 (*p.* 131)

1 (a) $y = 2x + 3$;   (b) $y = 4x - 5$;   (c) $y = -2x + 6$;   (d) $y = -\frac{3}{2}x - \frac{3}{4}$

2 (a) $y = 3x - 5$;   (b) $y = -2x + 12$;   (c) $y = \frac{3}{4}x + 9$;   (d) $y = -\frac{1}{2}x - 7$

3 (a) $y = \frac{1}{4}x + 3\frac{1}{2}$;   (b) $y = -\frac{4}{3}x - \frac{7}{3}$;   (c) $y = x - 3$;   (d) $y = -3$

4 Mid-point of **AB** is $(\frac{7}{2}, \frac{5}{2})$; mid-point of **BC** is $(4, 0)$ **5** $y = 2x + 11$

6 (a) $y = mx - 2m + 7$;   (b) $y = 2x - 2t + \dfrac{1}{t}$;   (c) $y = -3x + c$

7 $y = -\frac{5}{2}x + \frac{1}{2}$ **8** Intercept is -9

9 (i) Intercept on x-axis is $(3 - \dfrac{2}{m})$; intercept on y-axis is $(-3m + 2)$;

   (ii) area of triangle is $\left(-\dfrac{9}{2}m + 6 - \dfrac{2}{m}\right)$

10 $m = \dfrac{k}{h}$; $c = k$

Exercise 8.5 (*p.* 134)

1 (a) $2x - y + 8 = 0$;   (b) $3x + 5y - 5 = 0$;   (c) $6x - 8y - 1 = 0$;
   (d) $5x + 2y - 10 = 0$;   (e) $x - 3y + 3 = 0$;   (f) $4x + 6y + 24 = 0$

2 (a) $\dfrac{x}{2} + \dfrac{y}{6} = 1$;   (b) $\dfrac{x}{-6} + \dfrac{y}{8} = 1$;   (c) $\dfrac{x}{-5} + \dfrac{y}{-3} = 1$;

   (d) $\dfrac{x}{-6} + \dfrac{y}{6} = 1$;   (e) $\dfrac{x}{4} + \dfrac{y}{8} = 1$;   (f) $\dfrac{x}{\frac{2}{3}} + \dfrac{y}{\frac{1}{2}} = 1$

3 (a) $y = \frac{1}{2}x + \frac{2}{3}$;   (b) $y = -\frac{1}{2}x + 2$;   (c) $y = -\frac{1}{3}x + \frac{7}{3}$
   (d) $y = -4x - 8$;   (e) $y = \frac{3}{4}x + 5$;   (f) $y = -\frac{3}{7}x - \frac{2}{7}$

4 $\dfrac{x}{3} + \dfrac{y}{-5} = 1$; gradient is $+\dfrac{5}{3}$ **5** $p = \dfrac{3q}{2 + q}$

6 (a) $(2, 1)$;   (b) $(3, -4)$;   (c) $(-2, -3)$ **7** $(4, 5)$

8 x-intercept is $(1 + \dfrac{2}{m})$; y-intercept is $-(m + 2)$; $x + y + 1 = 0$

9 $9x - 2y - 6 = 0$ or $x - 2y + 2 = 0$

10 P is $\left(\dfrac{3a - 12}{a}, 4\right)$; area is 6 unit^2, i.e. independent of a as the area is

   constant

Exercise 8.6 (*p.* 140)

1 (a) 4;   (b) 1;   (c) 2;   (d) $\dfrac{3\sqrt{5}}{5}$;   (e) $\frac{12}{5}$;   (f) $\frac{24}{13}$

2 (a) $\frac{66}{13}$;   (b) $-\frac{33}{13}$;   (c) $2\sqrt{5}$;   (d) $\dfrac{7\sqrt{10}}{5}$ **3** $\frac{29}{5}$

4 2   **5** $y = \frac{5}{12}x + \frac{10}{3}$ **6** $\frac{19}{20}$

7 On opposite sides of the line; $-\dfrac{\sqrt{5}}{5}$ and $+\dfrac{6\sqrt{5}}{5}$

8 $\dfrac{11}{\sqrt{13}} = 3\cdot2$

Exercise 8.7 (*p.* 148)

1 $5x + 12y - 28 = 0$ and $5x + 12y + 24 = 0$

2 $x + 2y + 2 = 0$; $\dfrac{6\sqrt{5}}{5}$

3 $x - y - 1 = 0$; Both points are not on the line.

4 $3x + 2y + 7 = 0$; $\dfrac{2\sqrt{13}}{13}$

5 $13x - y - 10 = 0$ and $11x + 143y - 100 = 0$

6 $y = 2x$　　　　**7** $21x - 7y + 59 = 0$　　　　**10** Area $= 2$ unit^2

Exercise 8.8 (*p.* 149)

1 15 unit^2　　　**2** AP is $5x + y = 16$; BQ is $5x + 3y = 20$; CR is $y = 2$

3 $\left(\dfrac{4h}{h + 2}, \dfrac{8}{h + 2} \right)$　　　**4** (a) $13, 17$;　(b) 70;　(c) $39°$ $18'$ or $19'$

5 (ii) $\frac{1}{2}(bx + ay - ab)$

Exercise 8.9 (*p.* 153)

1 $m = 30$; $n = 2 \cdot 4$; $x = 10$　　　　**2** $y = 50x^2$

3 $p = 35$; $n = 0 \cdot 5$; $y = 111$　　　　**4** $m = 40$; $n = 6 \cdot 6$; $x = 2 \cdot 5$

5 $b = 2$; $S = 10$; $\lambda = 1 \cdot 125$　　　　**6** $n = 2 \cdot 7$; $a = 5$

7 $a = $ approx. 4　　　　　　　　　　**8** $a = 2$; $b = 4$

Exercise 9.1 (*p.* 160)

1 $4x$　**2** $4x$　**3** $4x$　**4** $2x + 1$　**5** $3x^2$　**6** $3x^2 + 2x - 2$　**7** $-\dfrac{2}{x^2}$

8 $6x^2$　**9** 12　**10** -8　**11** 12　　　**12** 5　　**13** 12　　　　　　**14** $-\frac{1}{8}$

Exercise 9.2 (*p.* 163)

1 $12x$　　　　**2** x　　　　**3** $6x^2$　　　**4** $2x - 1$　　　**5** $4x + 3$

6　　　$x = -1$　　$x = 0$　　$x = 2$
　　(1)　　-12　　　0　　　24
　　(2)　　-1　　　　0　　　2
　　(3)　　6　　　　　0　　　24
　　(4)　　-3　　　　-1　　　3
　　(5)　　-1　　　　3　　　11

Exercise 9.3 (*p.* 165)

1 $10x$　　**2** $9x^2$　　　　　**3** $40x^3$　　**4** 8　　　　　**5** 0

6 $2x + 5$　**7** $1 + \dfrac{1}{x^2}$　　　**8** $9 - \dfrac{5}{x^2}$　**9** $12x^2 - 2$　**10** $\frac{1}{3}x^{-\frac{2}{3}}$

11 $-\dfrac{3}{x^4}$　**12** $3x^2 + \dfrac{1}{x^2} - \dfrac{6}{x^3}$　**13** $-\dfrac{1}{2x^{\frac{3}{2}}}$　**14** $\dfrac{2}{\sqrt{x}}$　　**15** $12x^2 + \dfrac{12}{x^5}$

16 $4t + \dfrac{1}{t^2}$　**17** $9t^2 + 4t - 7$　　**18** 5　　　**19** $\frac{1}{2}$　　　　**20** 2

Exercise 10.1 (*p.* 167)

1 $20x^3 - 6x + 7$ **2** $2x + 2$ **3** $8x - 12$ **4** $4x - 1$

5 $-\dfrac{1}{x^2}$ **6** $1 - \dfrac{5}{x^2}$ **7** $3x - \dfrac{2}{x^2}$ **8** $3x^2 - 12x + 12$

9 $\frac{1}{2} - \dfrac{9}{2x^2}$ **10** $9x^2 + 8x - 1$

Exercise 10.2 (*p.* 168)

1 $6(2x + 5)^2$ **2** $10(2x + 5)^4$ **3** $9(3x - 1)^2$ **4** $21(3x - 1)^6$

5 $27(3x - 1)^8$ **6** $16(4x - 3)^3$ **7** $28(4x - 3)^6$ **8** $2(4x - 3)^{-\frac{1}{2}}$

9 $21(7x + 1)^2$ **10** $4a(ax + b)^3$ **11** $8x(x^2 + 1)^3$ **12** $10x(x^2 - 1)^4$

13 $12x(3x^2 - 1)$ **14** $18x^2(2x^3 - 1)^2$ **15** $24x^2(2x^3 - 1)^3$

16 $3(2x + 1)(x^2 + x + 1)^2$ **17** $2(2x + 3)(x^2 + 3x - 1)$

18 $4(2x - 2)(x^2 - 2x + 5)^3$ **19** $3(8x - 1)(4x^2 - x + 5)^2$

20 $4(6x^2 - 2x)(2x^3 - x^2 + 1)^3$ **21** $2(3x^2 + 2x + 1)(x^3 + x^2 + x + 1)$

22 $\dfrac{3}{2\sqrt{3x - 1}}$ **23** $\dfrac{2}{\sqrt{4x + 3}}$ **24** $\dfrac{2x - 1}{2\sqrt{x^2 - x + 1}}$ **25** $\dfrac{3x - 1}{\sqrt{3x^2 - 2x + 1}}$

26 $\dfrac{-4}{(4x - 3)^2}$ **27** $\dfrac{-3}{(3x + 5)^2}$ **28** $120x^2(5x^3 - 4)^3$ **29** $(3x + 4)^{-\frac{2}{3}}$

30 $\frac{1}{2}(1 + \dfrac{1}{x^2})(x - \dfrac{1}{x})^{-\frac{1}{2}}$ **31** $2x(2x^3 - 5)^{-\frac{2}{3}}$ **32** $\dfrac{-(2x + 1)}{(x^2 + x - 3)^2}$

33 $3(16x^3 - 2x)(4x^4 - x^2 + 2)^2$ **34** $-\frac{3}{2}(3x - 5)^{-\frac{9}{2}}$

35 $\dfrac{-3x^2}{(x^3 - 6)^2}$ **36** $-(x + 1)(x^2 + 2x - 4)^{-\frac{3}{2}}$ **37** $-4(2x - 3)^{-3}$

38 $-2(4x - 1)(2x^2 - x - 6)^{-3}$

Exercise 10.3 (*p.* 171)

1 $6(x + 4)(x + 1)$ **2** $(x^2 - 1)^4(11x^2 - 1)$

3 $2x(x^3 + 1)(4x^3 - 3x + 1)$ **4** $2(3x - 1)(6x^2 - x - 15)$

5 $3x^2(x + 5)^3(7x + 15)$ **6** $30(3x - 2)(2x - 3)^2(x - 1)$

7 $4(x^2 - 2x - 8)(5x^2 - 9x - 5)$ **8** $2(2x^3 - 3x + 4)$

9 $\dfrac{(x + 3)(5x + 3)}{2\sqrt{x}}$ **10** $3x^2 - 4x - 3$

11 $x(x + 2)(5x^2 + 4x - 4)$ **12** $2(x - 1)(2x^2 + 5x + 1)$

Exercise 10.4 (*p.* 173)

1 $\dfrac{1}{(x + 2)^2}$ **2** $\dfrac{3}{(x + 1)^2}$ **3** $\dfrac{-10}{(x - 5)^2}$ **4** $\dfrac{-4}{(x - 4)^2}$

5 $\dfrac{x^2 + 4x + 1}{(x + 2)^2}$ **6** $\dfrac{2x(x - 3)}{(2x - 3)^2}$ **7** $\dfrac{x(x - 2)}{(x - 1)^2}$ **8** $\dfrac{-2(x^2 + 1)}{(x^2 - x - 1)^2}$

9 $\dfrac{-2x(x^3 - 3x - 8)}{(x^3 + 4)^2}$ **10** $\dfrac{x - 4}{2(x - 2)^{\frac{3}{2}}}$ **11** $\dfrac{3x + 4}{(2x + 1)^{\frac{3}{2}}}$ **12** $\dfrac{6x^2}{(x^3 + 1)^2}$

Exercise 10.5 (*p.* 174)

1 $9x^2$, $18x$ **2** $4, 0$ **3** $4x^3 + 6x^2 - 6x$, $12x^2 + 12x - 6$

4 $-1/x^2$, $2/x^3$ **5** $1/2\sqrt{x}$, $-1/4x^{\frac{3}{2}}$ **6** $-1/(x+1)^2$, $2/(x+1)^3$

7 $-2/(x-1)^2$, $4/(x-1)^3$ **8** $12t^2 - 6t + 1$, $24t - 6$

9 $6t + 1/t^2$, $6 - 2/t^3$ **10** nax^{n-1}, $n(n-1)ax^{n-2}$

11 $10x^4 + 16x^3 - 9x^2 + 4x$, $40x^3 + 48x^2 - 18x + 4$, $120x^2 + 96x - 18$

12 $-1/x^2$, $2/x^3$, $-6/x^4$, $24/x^5$, $-120/x^6$; $-(1 \times 2 \times 3 \times 4 \times 5 \times 6 \times 7)/x^8$;
   $-(1 \times 2 \times 3 \times 4 \times 5 \times 6 \times 7 \times 8 \times 9)/x^{10}$

13 $1/(1+x)^2$; $-2/(1+x)^3$

Exercise 10.6 (*p.* 175)

1 $15x^4 - 4x - \dfrac{4}{x^2}$ **2** $6x^2 - 10x + 4$ **3** $\dfrac{3x+4}{2\sqrt{x+2}}$

4 $-\dfrac{1}{(2x-3)^2}$ **5** $18x(3x^2-1)^2$ **6** $\dfrac{x^2(3+x^2)}{(1+x^2)^2}$

7 $\dfrac{4x+1}{2\sqrt{2x^2+x-1}}$ **8** $\dfrac{-2}{(4x-3)^{\frac{3}{2}}}$ **9** $1 + \dfrac{1}{2x^{\frac{3}{2}}}$

10 $1 + \dfrac{2}{x^2} - \dfrac{2}{x^3}$ **11** $2\left(x - \dfrac{2}{x}\right)\left(1 + \dfrac{2}{x^2}\right)$ **12** $\dfrac{1}{2\sqrt{x}}\left(1 - \dfrac{1}{x}\right)$

13 $\dfrac{1}{x^{\frac{2}{3}}}\left(1 + \dfrac{1}{3x^{\frac{2}{3}}}\right)$ **14** $\dfrac{3x^2}{2\sqrt{x^3+1}}$ **15** $\dfrac{x+1}{\sqrt{x^2+2x}}$

16 $-\dfrac{3}{2x^{\frac{5}{2}}}$ **17** $\dfrac{3(1+\sqrt{x})^2}{2\sqrt{x}}$ **18** $\dfrac{-4x^3}{3(3-x^4)^{\frac{2}{3}}}$

19 $\dfrac{1}{3} - \dfrac{5}{3x^2} - \dfrac{3}{x^4}$ **20** $3(x-1)^2(3x-2)(5x-4)$

21 $na(1+ax)^{n-1}$ **22** $\dfrac{12x^2}{5} + \dfrac{3}{2x^4}$

23 $4(6x-5)(3x^2-5x+7)^3$ **24** $\dfrac{-(15x+7)}{(x-3)^3}$

25 $\frac{7}{4}$ **26** $\frac{3}{4}$ **27** $12\frac{1}{8}$ **29** $-\frac{8}{3}$

32 $20, -20$

Exercise 11.1 (*p.* 177)

1 $y = 36x - 58$ **2** $y + 2x + 1 = 0$ **3** $y = 6x - 4$

4 $y + 3x = 2$ **5** $25y = x + 3$ **6** $y = 12x + 28$

7 $54y + x = 27$ **8** $y = 2x - 1$ **9** $y = x - 6$; $x = 6$

10 $(\frac{1}{3}, \frac{121}{27})$, $(1, 3)$; $27y + 54x = 139$; $y + 2x = 5$

11 $y + 4x + 5 = 0$ **12** $y = x - 1$

13 $y = 0$, $y = 4x + 4$; $(3, 16)$ **14** $y = 8x - 14$

15 $-1/t^2$; **(a)** small and $-$ **(b)** large and $-$ **(c)** small and $-$
   (d) large and $-$; $y - 1/t = 2 - (x - t)/t^2$; L$(2t, 0)$; M$(0, 2/t)$; 2

Exercise 11.2 (*p.* 182)

1 Min. -1 where $x = 1$ **2** Max. 9, $x = 2$
3 No stat. points   **4** Max. 0, $x = -1$; min. -4, $x = 1$
5 Max. $\frac{9}{4}$, $x = \frac{3}{2}$ **6** Max. 86, $x = -3$; min. -39, $x = 2$
7 Point of inflexion, $x = 1$ **8** Max. 4, $x = -1$; min. 6, $x = 1$
9 Point of inflexion, $x = 0$ **10** Max. 3, $x = 1$; min. -1, $x = 3$
11 Point of inflexion, $x = 0$; min. 1, $x = 1$
12 $1 + 4/x^2$; $4x^2 + 2/x^3$; $-1 < x < 1$; no
13 $2 \pm \sqrt{5}$ **14** Max. 1, $x = 2$; min. -4, $x = -\frac{1}{2}$

Exercise 11.3 (*p.* 185)
14 Max. $-815/243$, $x = -5/9$; min. -9, $x = 1$

Exercise 11.4 (*p.* 187)

1 9   **2** $\frac{9}{2}$ **3** Square side 15 m, area 225 m^2
4 8   **5** 384 cm^3 **6** 144 units
7 6 cm $\times$ 12 cm $\times$ 4 cm   **8** $A = 2\pi r^2 + 108\pi/r$; 3
9 $V = x(16 - 2x)(10 - 2x)$; $x = 2$ **10** $(x^2 - 12x + 81)/9$; 5 cm

11 $\dfrac{150 - x}{100} + \dfrac{\sqrt{3600 + x^2}}{60}$; 45; 2·3 min   **12** 24 cm^2

13 $8x - 2x^2$; 8 cm^2; 3·2   **14** $\dfrac{1}{\sqrt[3]{\pi}}$ **15** $32/3\sqrt{3}$

Exercise 11.5 (*p.* 191)

1 0·0075   **2** -0.016 **3** 0·04π **4** -0.006
5 $\delta V \approx 4\pi r^2 \, \delta r$; 3%   **6** -19.5% **7** -0.05 **8** 10%
9 4%   **10** 0·06π **11** 1·2π **12** 0·03   **13** 2%

14 $\sqrt{\dfrac{250}{\pi}} = 8.92$ **15** 3%

Exercise 11.6 (*p.* 195)
7 32 m, 0 m; -12, -12 m s^{-2}

Exercise 11.7 (*p.* 198)

1 1·6π cm s^{-1} **2** 1250 mm^2 s^{-1} **3** 0·125 mm s^{-1} **4** 32π cm^3 s^{-1}
5 1/10π m s^{-1} **6** 1/π mm s^{-1} **7** 1·2 units/s
8 400/81π = 1·57 cm s^{-1} **9** 0·003 75 m^3 s^{-1} **10** 41π/5 = 25·8 cm^2 s^{-1}
11 0·012 cm^2 s^{-1}; 5 km   **12** 144 cm^2; $\frac{4}{3}$ cm s^{-1}
13 $\sqrt{2 \times 10^6 - 5 \times 10^4 t + 325 t^2}$; (i) 17·95 ms^{-1}; (ii) $\dfrac{1000}{3}$ s
14 20 cm^2 s^{-1}

Exercise 12.1 (*p.* 202)

1 5! = 120   **2** 7! = 5040   **3** 7!   **4** 504   **5** 6840
6 120   **7** 12   **8** 80   **9** 300   **10** 12!
11 360   **12** 360   **13** (a) 6 (b) 27   **14** 6!; 5!; 5!; 4!
15 2$\times$5!; 144; 96   **16** 6; 24; 120; 720   **18** 1320   **19** 66; 56; 3003
20 $n(n - 1)$

Exercise 12.2 (p. 205)

1 192 **2** 1260 **3** (a) 24 (b) 36   **4** 291
5 102 **6** 14 400 **7** 144 **8** 5040   **9** 60
10 12 **11** 75 **12** 7! × 8 = 8!
13 21 × 5 × 10 × 10 × 10 = 105 000   **14** (i) 60; (ii) 60; 24
15 300 **16** 144; 36 **17** 4464   **18** (a) 240; (b) 36

Exercise 12.3 (p. 208)

1 10 **2** 15 **3** 120 **4** 78 **5** 10
6 150; 90 **7** 200 **8** 1008 **9** 25 **10** (a) 9; (b) 10
11 (a) 70; (b) 40 (If no distinction is made between cars, 35, 20)
12 1680 **13** 448 **14** 6; 70; 84; 35

Exercise 12.4 (p. 212)

1 $\frac{5}{18}$ **2** $\frac{1}{3}$ **3** (a) $\frac{1}{13}$; (b) $\frac{1}{4}$; (c) $\frac{3}{13}$ **4** $\frac{4}{9}$
5 $\frac{1}{12}$; $\frac{11}{12}$ **6** $\frac{1}{216}$ **7** $\frac{1}{6}$ **8** $\frac{1}{2}$ **9** $\frac{5}{42}$
10 $\frac{5}{9}$ **11** (a) $\frac{1}{6}$; (b) $\frac{7}{12}$ **12** $\frac{1}{7}$
13 (a) $\frac{27}{100}$; (b) $\frac{33}{100}$; (c) $\frac{11}{50}$ **14** (a) $\frac{1}{6}$; (b) $\frac{1}{9}$; $\frac{1}{6}$; $\frac{1}{12}$; 7 **15** $\frac{8}{125}$

Exercise 13.1 (p. 215)

1 7·3 km **2** 9·7 km; 12·1 km   **3** 7·5 km; 044°
4 4·0 km **5** 41 km **6** 431 m   **7** 8·0 km

Exercise 13.2 (p. 223)

1 (a) 13·34 cm; (b) 75° 58′; 67° 23′; (c) 240 cm³
2 (a) 30·31 m; (b) 28° 18′   **3** (a) 349·7 m; (b) 110·3 m
4 (a) 10·99 cm; (b) 62° 46′   **5** 454 m
6 Height = 66·77 m; distance = 221·4 m
7 (a) 54° 44′; (b) 70° 32′; (c) 117·8 cm³
8 (a) 58° 0′; (b) 48° 31′; (c) 19° 16′

9 (a) 11·08 cm; (b) 47° 17′   **10** $\frac{\pi a}{3}$; 141`

11 7·1 m; (a) 64° 54′; 55° 33′; 45°; (b) 25° 07′
12 157½°   **13** 62·35 m³
14 (a) 67° 53′; (b) 78·75 cm²; 4·57 cm
15 PQ = 147 m; 241°

Exercise 14.1 (p. 229)

1 $\frac{56}{65}$
2 (a) sin (A − B) = −0·28; (b) cos (A − B) = 0·96; (c) A + B = 90°
3 (a) sin (A + B) = $-\frac{16}{65}$; (b) cos (A + B) = $-\frac{63}{65}$; (c) cos (A − B) = $-\frac{33}{65}$
4 (a) sin 65°; (b) cos 36°; (c) sin 66°; (d) cos 81°
5 $\frac{\sqrt{6} - \sqrt{2}}{4}$ **6** $\frac{\sqrt{6} + \sqrt{2}}{4}$

8 (a) sin (θ + 30°) or cos (θ − 60°); (b) sin (θ − 45°) or cos (θ + 45°)
10 (a) $\frac{\sqrt{2} - \sqrt{6}}{4}$; (b) $\frac{\sqrt{6} - \sqrt{2}}{4}$; (c) $-\left(\frac{\sqrt{6} - \sqrt{2}}{4}\right)$

Exercise 14.2 (*p.* 231)

1 (a) $2 + \sqrt{3}$; (b) $2 - \sqrt{3}$; (c) $-(2 + \sqrt{3})$; (d) $-(2 + \sqrt{3})$

2 $-\frac{1}{7}$ **3** $\frac{2}{9}$ **4** (i) $\tan(\theta - 60°)$; (ii) $\tan(\theta + 45°)$

5 (i) 1; (ii) $\sqrt{3}$; (iii) -1 **6** $\tan B = \frac{7}{6}$ **7** $\cot A = \frac{2}{9}$

8 $\tan(A + B) = -\frac{56}{33}$; $\tan(A - B) = -\frac{16}{33}$

Exercise 14.3 (*p.* 236)

1 (a) $\sin 80°$; (b) $\cos 64°$; (c) $\cos 130°$; (d) $\tan 110°$; (e) $\cos 4\theta$

2 (a) $\frac{1}{4}$; (b) $\sqrt{3}$; (c) $\sqrt{3}/2$; (d) $\frac{1}{2}$; (e) $\frac{1}{2}$

3 (a) $\frac{120}{169}$; (b) $\sqrt{3}/2$ **4** (a) $\sin\theta = \frac{5}{13}$; (b) $\cos\theta = \frac{12}{13}$

5 (a) $\tan\theta = \frac{5}{12}$; (b) $\cos\theta = \frac{12}{13}$; (c) $\sin\theta = \frac{5}{13}$

6 (a) $\frac{4}{3}$; (b) $-\frac{24}{7}$; (c) $-\frac{17}{31}$

7 (a) $\tan A = 4$; (b) $\sin A = 4/\sqrt{17}$; (c) $\cos 2A = -\frac{15}{17}$

8 (a) $\dfrac{-b}{1 - c}$; (b) $\dfrac{\pm(1 - c)}{\sqrt{1 - 2c + c^2 - b^2}}$

9 (a) $\tan\theta = 2$; (b) $\tan 4\theta = \frac{24}{7}$

10 $\cos 3A = 4\cos^3 A - 3\cos A$

Exercise 14.4 (*p.* 239)

1 (a) $2\sin 4\theta \cos\theta$;   (b) $2\cos 4\theta \cos\theta$;   (c) $2\cos 3\theta/2 \sin\theta/2$;
  (d) $-2\sin 4\theta \sin\theta$;   (e) $-2\cos 5\theta \sin 2\theta$;   (f) $2\sin 5\theta \sin 3\theta$

2 (a) $\sin 4\theta + \sin 2\theta$;   (b) $\sin 4A - \sin 2A$;   (c) $\cos 8x + \cos 2x$;
  (d) $\cos 8\theta - \cos 2\theta$ (e) $\sin 4A - \sin 6A$;   (f) $\cos 6\theta - \cos 8\theta$

3 (a) $2\cos(\theta + 30°)\cos(\theta - 30°)$;   (b) $2\cos(2\theta + 60°)\sin(2\theta - 60°)$;
  (c) $-2\sin(\theta + 30°)\sin(\theta - 30°)$

4 (a) $\sin 4\theta + \sin 60°$; (b) $\sin 2\theta - \sin 80°$;   (c) $\cos 2A - \cos 124°$

5 (a) $2\sin(\theta + 10°)\cos 10°$;   (b) $2\cos(45° - \dfrac{\theta}{2})\cos(\dfrac{5\theta}{2} - 45°)$;

  (c) $-2\sin(\theta + 30°)\sin 30°$;   (d) $2\cos 2\theta \cos 10°$

Exercise 14.6 (*p.* 245)

1 $-\frac{24}{65}$ **2** (i) $\frac{5}{12}; \frac{120}{119}; \frac{-1}{239}$; (ii) 1

3 (i) $\pm\frac{4}{5}$; $\pm\frac{3}{4}$; (ii) 0.7002; -0.7431; -0.8660;
  (iii) (a) 0.8; (b) 0.936; (c) 1.333

4 (a) $\frac{4}{5}$; (b) $\sqrt{0.1}$; (c) $\frac{1}{3}$; (d) $\frac{13}{9}$

5 $\sin 75° = \dfrac{\sqrt{2} + \sqrt{6}}{4}$; $\cos 105° = \dfrac{\sqrt{2} - \sqrt{6}}{4}$; $\tan 285° = \dfrac{1 + \sqrt{3}}{1 - \sqrt{3}}$; K $= 4$

7 (a) $\dfrac{p}{1 - q}$; (b) $\dfrac{p^2}{p^2 + (1 - q)^2}$; (c) $\dfrac{(1 - q)^2 - p^2}{(1 - q)^2 + p^2}$

8 $\tan A = \frac{1}{2}$; $\tan B = \frac{1}{3}$ or $\tan A = -2$; $\tan B = -3$

Exercise 15 (*p.* 249)

1 $45°$ **2** $26° 34'$ **3** $45°$ **4** $9° 44'$

5 (a) $x + 6y + 5 = 0$; (b) $4x + y - 18 = 0$ and $4x - y - 18 = 0$;
  (c) $8x - 2y + 15 = 0$

Exercise 15 (*continued*)

6 $x + y + 1 = 0$

7 Tangent $3x + y - 13 = 0$; normal $x - 3y + 9 = 0$

8 Normal $x + y - 3 = 0$; intersection $(-6, 9)$

9 $5y - 2x + 11 = 0$ **10** $x/6 + y/2 = 1$ **11** $3y - 2x + 7 = 0$

12 AB $= 10$ units; $4x - 3y - 2 = 0$; $3x + 4y - 14 = 0$;
$4x - 3y - 27 = 0$; 5 units

13 (a) $x + y = 10$; (b) $x - y = 2$; (c) $3\sqrt{2}$ **14** $(5, 9)$; $\sqrt{26}$

15 $P\left(\dfrac{4m - 7}{m}, 0\right)$; Q$(3m + 8, 0)$; R$(0, 7 - 4m)$;

$S\left(0, \dfrac{3m + 8}{m}\right)$ OP.OQ $= 12m + 11 - \dfrac{56}{m}$;

OR.OS $= \dfrac{56}{m} - 11 - 12m$

Exercise 16 (*p.* 257)

1 $\sin x/\cos^2 x$ **2** $-\cos x/\sin^2 x$ **3** $-\operatorname{cosec}^2 x$ **4** $4\cos 4x$

5 $-6\sin 6x$ **6** $2\sec^2 2x$ **7** $-3\operatorname{cosec}^2 3x$ **8** $-2\sin(2x - 8)$

9 $3\cos(3x + \pi/3)$ **10** $x\sec^2 x + \tan x$ **11** $\cos 2x - 2x\sin 2x$

12 $\sin 3x + 3x\cos 3x$ **13** $-2\sin 2x + \cos x$ **14** $\cos x + \sin x$

15 $-2\cos x\sin x = -\sin 2x$ **16** $10\sin 5x\cos 5x = 5\sin 10x$

17 $2\tan x\sec^2 x$ **18** $2\sin(x - \pi/4)\cos(x - \pi/4) = \sin(2x - \pi/2)$

19 $8x + 4\cos 4x$ **20** $1/(1 + \cos x)$ **21** $2/(\cos x - \sin x)^2$

22 $4\sin 2x/(1 + \cos 2x)^2$ **23** $\cos x\tan x + \sin x\sec^2 x = \sin x(1 + \sec^2 x)$

24 $2\sin x\sec^3 x$ **25** $\frac{1}{2}\sec^2 x/2$ **26** $3x^2\tan x + x^3\sec^2 x$

27 $2x\tan x + (1 + x^2)\sec^2 x$ **28** $\dfrac{\sin x - (1 + x)\cos x}{\sin^2 x}$

29 $-6\cos x\sin x = -3\sin 2x$ **30** $(\sin x + \cos x)(1 + \sin x - \cos x)$

31 $2\cos 2x$, $-4\sin 2x$, $-8\cos 2x$

32 Max. $\sqrt{5}$, $x = 26° 34'$; min. $-\sqrt{5}$, $x = 206° 34'$

35 $\frac{1}{3}$; $\frac{1}{2} + \pi/8$; 1 **38** $1, -1$ **40** Max. $\frac{73}{24}$, min. -4

Exercise 17.1 (*p.* 264)

1 $x^4 + 8x^3 + 24x^2 + 32x + 16$

2 $x^5 - 15x^4 + 90x^3 - 270x^2 + 405x - 243$

3 $8x^3 - 12x^2 + 6x - 1$ **4** $81x^4 - 216x^3 + 216x^2 - 96x + 16$

5 $1 + 6x + 15x^2 + 20x^3 + 15x^4 + 6x^5 + x^6$

6 $1 + 2x + 3x^2/2 + x^3/2 + x^4/16$

7 $x^5 - 5x^3 + 10x - 10/x + 5/x^3 - 1/x^5$

8 $x^4 + 2x^2 + \frac{3}{2} + 1/2x^2 + 1/16x^4$

9 $1 - 20x + 180x^2 - 960x^3 + 3360x^4$

10 $2\cdot488\,32$; $2\cdot489$ **11** $1 - 6x + 21x^2 - 50x^3$

12 $1 + 5x - 30x^3$ **13** $0\cdot9044$ **14** $1178\sqrt{2}$

15 $-1792x^2$; 1120 **16** $-\frac{280}{27}$

17 $x^8 + 8x^7y + 28x^6y^2 + 56x^5y^3 + 70x^4y^4$; $1\cdot032\,45$

18 (i) $160x + 80x^3 + 2x^5$; (ii) $9, \frac{4}{3}$

19 $1 - 20x + 180x^2 - 960x^3 + 3360x^4$; $0.980\ 179\ 0$

20 (i) $1 - 8x + 24x^2 - 32x^3 + 16x^4$;

$\quad\quad 1 - 8x + 28x^2 - 56x^3 + 70x^4 - 56x^5 + 28x^6 - 8x^7 + x^8$;

$\quad\quad 0, -\frac{1}{8}$

$\quad$ (ii) $0.132\ 651 \times 2^{15}$

21 $3, 4, 5$

Exercise 17.2 (*p.* 269)

1 $1 + 2x + 3x^2 + 4x^3$; $-1 < x < 1$

2 $1 - 2x + 3x^2 - 4x^3$; $-1 < x < 1$

3 $1 - 4x + 10x^2 - 20x^3$; $-1 < x < 1$

4 $1 + x/2 - x^2/8 + x^3/16$; $-1 < x < 1$

5 $1 + x - x^2/2 + x^3/2$; $-\frac{1}{2} < x < \frac{1}{2}$

6 $1 + x + x^2 + x^3$; $-1 < x < 1$

7 $1 - 4x + 12x^2 - 32x^3$; $-\frac{1}{2} < x < \frac{1}{2}$

8 $1 - x - 7x^2/2 - 35x^3/2$; $-\frac{1}{8} < x < \frac{1}{8}$

9 $1 + x - x^2 + 5x^3/3$; $-\frac{1}{3} < x < \frac{1}{3}$

10 $1 - x/2 + 3x^2/8 - 5x^3/16$; $-1 < x < 1$

11 $1 + 4x + 12x^2 + 32x^3$; $-\frac{1}{2} < x < \frac{1}{2}$

12 $\frac{1}{3}(1 + x/3 + x^2/9 + x^3/27)$; $-3 < x < 3$

13 $\sqrt{2}(1 + x/4 - x^2/32 + x^3/128)$; $-2 < x < 2$

14 $1 - x/4 - x^2/32 - x^3/128$; $-2 < x < 2$

15 $\frac{1}{4}(1 + 3x/4 + 9x^2/16 + 27x^3/64)$; $-\frac{4}{3} < x < \frac{4}{3}$

16 $1 - x + x^2 - x^3$; $-1 < x < 1$ **17** 1.0050

18 1.0291 **19** 0.9975 **20** $1 + 3x/2 + 7x^2/8$

21 $-x/2 + 9x^2/8 - 15x^3/16$

22 $1, 0.04, -8 \times 10^{-4}, 3.2 \times 10^{-5}$; $1.732\ 05$ **23** $2 - 2x + \frac{5}{2}x^2 - \frac{7}{2}x^3$

Exercise 18.1 (*p.* 272)   (The constant is omitted from each answer)

1 $x^3/3$ **2** $x^4/4$ **3** $2x^5/5$ **4** $3x^2/2$ **5** $2x^6/3$

6 $8x$ **7** $x^8/8$ **8** $-\frac{1}{2}x^{-2}$ **9** $4x^{\frac{3}{2}}/3$ **10** $8x^{\frac{3}{2}}/3$

11 $3x^{\frac{5}{3}}/5$ **12** $2x^{\frac{1}{2}}$ **13** x **14** $4x^{\frac{7}{4}}/7$ **15** $x^7/14$

16 $3x^{\frac{2}{3}}/2$ **17** $x^4/4 + x^3/3 + x^2/2 + x$ **18** $3x^5/5 - x^2/2 + 2x$

19 $x^2/2 - 2x^{\frac{2}{3}}/3$ **20** $x^2/2 - 1/x - 1/x^2$ **21** $3x/2 - 1/2x + 1/x^2$

22 $x^3/3 + 3x^2 + 9x$ **23** $x^3/3 - x^2 + x$ **24** $x^3/3 - 2x - 1/x$

25 $x^5/5 + 2x^3/3 + x$ **26** $x^4/4 - x^3 + 3x^2/2 - x$

27 $-3/2x^3 - 3/2x^2 - 1/2x$

Exercise 18.2 (*p.* 275)   (Constant omitted)

1 $x^2/2$ **2** $3x$ **3** $x^3/3 + x$ **4** $-\frac{1}{2}\cos 2x$

5 $x^3/3 + 3x^2 + 9x$ **6** $-1/x$ **7** $\sin 5x/5$

8 $2x^{\frac{3}{2}}$ **9** $x^3/3 - 2x - 1/x$ **10** $2\tan x/2$

11 $2x^{\frac{3}{2}}/3 - 2x^{\frac{1}{2}}$ **12** $x - 1/x$ **13** $\frac{1}{2}\sin 2x + \frac{1}{4}\cos 4x$

14 $x^2/2 - x - 1/x$ **15** $x^3/3 - 3x^2/2$ **16** $x^3/3 - x^2/2 - 2x$

17 $3t^2/2 + 4t^3/3$ **18** $x^2/2 + 3\sin x/3$ **19** $t^4/4 - t^2/2$

20 $\frac{3}{2}\tan 2\theta/3$ **21** $x^2/2 - 4x^{\frac{1}{2}} - 1/x$

Exercise 18.2 (*continued*)

22 $\sin x + \frac{1}{2}\sin 2x + \frac{1}{3}\sin 3x$ **23** $-2/x - 3/x^2$

24 $y = x^2/2 + 2x - 6$ **25** $y = x^2 + x$ **26** $A = t^3/3 - t + 2$

27 $s = 2t^3/3 - 3t^2/2$ **28** (a) 32 m   (b) 12 m s^{-1}

29 $y = -1/0 + \frac{1}{4}\sin 2\theta + 2/\pi$ **30** 3; 189 m   **31** (ii) 16 m

32 $y = x^3 - 2x^2 + 3x$ **33** (a) $x/2 - (\sin 2x)/4$ (b) $x/2 + (\sin 4x)/8$

Exercise 18.3 (*p.* 280)

1 1	**2** $\frac{9}{2}$	**3** 4	**4** 2	**5** 0
6 1	**7** 0	**8** $\frac{1}{6}$	**9** $8\frac{2}{3}$	**10** $-\frac{1}{6}$
11 1	**12** $\frac{5}{6}$	**13** $t^3/3 - 3t$	**14** $-\frac{1}{6}$	**15** 0

Exercise 18.4 (*p.* 284)

1 9	**2** $8\frac{1}{2}$	**3** $4\frac{1}{2}$	**4** $\frac{2}{3}$	**5** 20
6 $\frac{2}{3}$	**7** $1\frac{1}{3}$	**8** $\frac{11}{6}$	**9** $4\frac{1}{2}$	**10** 2
11 2	**12** $\frac{2}{3}$	**13** $\frac{1}{6}$	**14** $3, -2, 6; \frac{8}{3}$	**16** $\frac{1}{3}$
17 $\frac{128}{3}$	**18** $k, k/h^2$	**19** $(1, 0), (3, 2), \frac{4}{3}$	**20** $y = 12x - 28; \frac{4}{3}$	

Exercise 18.5 (*p.* 287)

1 $19\pi/3$	**2** $8\pi/5$	**3** $\pi/30$	**4** $15\pi/2$	**5** $\pi m^2 h^3/3$
6 $64\pi/15$	**7** (a) $13\cdot5; 24\cdot3\pi$	**8** $\frac{1}{5}; 5\pi/28$	**29** $80\pi/3$	
10 $2\pi r^3/3$	**11** $83\pi/15$	**12** $412\pi/15$		

Exercise 18.6 *Miscellaneous* (*p.* 288)

1 $4\frac{1}{2}$	**2** $20\frac{5}{6}$	**3** $\dfrac{3\pi}{2}$	**4** $11\frac{5}{24}$
5 $\sqrt{2} - 1$	**6** $y + x = 1; \frac{1}{2}$		**7** $4\sqrt{2} - \frac{1}{3} = 5\cdot32$
8 54, 50; $-6, +6$	**9** $\sqrt{2} + \frac{1}{2} = 1\cdot91$	**10** $\frac{8}{3}$	

Exercise 19.1 (*p.* 290)

1 (i) $x^2 + y^2 = 16$;   (ii) $x^2 + y^2 = 36$;   (iii) $x^2 + y^2 = 3$;

  (iv) $x^2 + y^2 = m^2 + n^2$;   (v) $x^2 + y^2 = 13$;   (vi) $x^2 + y^2 = 25$;

  (vii) $x^2 + y^2 = 169$;   (viii) $x^2 + y^2 = 7$

2 (i) 11 units; (ii) 7 units; (iii) $2\sqrt{3}$ units; (iv) $(a + b)$ units; (v) $\sqrt{pq}$ units

4 Inside the circle

5 (i) $x^2 + y^2 - 6x - 10y - 2 = 0$; (ii) $x^2 + y^2 + 8x - 2y + 13 = 0$;

  (iii) $4x^2 + 4y^2 + 16x + 12y + 5 = 0$

6 $x^2 + y^2 + 2x - 12y + 19 = 0$; radius $= 3\sqrt{2}$ units

7 $x^2 + y^2 + 6x + 8y = 0$; radius $= 5$ units

8 $x^2 + y^2 - 2px - 2qy - 3p^2 = 0$; radius $= \sqrt{4p^2 + q^2}$

Exercise 19.2 (*p.* 291)

1 (i) Centre (1, 3); radius 5;   (ii) Centre (3, −7); radius 3;

  (iii) Centre $(-\frac{1}{2}, -2)$; radius $\frac{5}{2}$;   (iv) Centre $(\frac{1}{2}, -2)$; radius $\frac{5}{2}$;

  (v) Centre (3, 0); radius 3

2 (a) Circle; centre $(\frac{3}{2}, -2)$; radius $\sqrt{97}/2$;
   (b) Not a circle; has a term in xy;
   (c) Circle; centre $(-\frac{1}{3}, 0)$; radius $\frac{1}{3}$;
   (d) Circle; centre $(2, -\frac{5}{2})$; radius $\sqrt{41}/2$;
   (e) Not a circle; signs of x^2 and y^2 are not the same;
   (f) Circle; centre $(0, 0)$; radius 4;
   (g) Not a circle; coefficients of x^2 and y^2 are not the same;
   (h) Not a circle; $(\text{radius})^2 = -5$, no real values
3 (i) Centre $(0, -b)$; radius 2;   (ii) Centre $(c, -c)$; radius $\sqrt{2}c$;
   (iii) Centre $(at, a/t)$; radius a

Exercise 19.3 (*p.* 296)
1 (a) $x^2 + y^2 + 2y - 12 = 0$;   (b) $3x^2 + 3y^2 - 9x + 11y + 6 = 0$;
   (c) $x^2 + y^2 - x - y - 2 = 0$
3 (a) $x^2 + y^2 - 5x + 3 = 0$;   (b) $x^2 + y^2 - 4x + 5y + 6 = 0$;
   (c) $x^2 + y^2 + x - y - 14 = 0$;   (d) $x^2 + y^2 - x + 5y - 6 = 0$
6 A – outside; B – inside the circle; C – inside the circle
7 $y = x - 3$
8 $x^2 + y^2 - 2x - y - 10 = 0$; centre $(1, \frac{1}{2})$; outside the circle

Exercise 19.4 (*p.* 299)
1 (a) $3x - 4y + 25 = 0$;   (b) $x - 2y + 10 = 0$;
   (c) $6x - \sqrt{3}y - 39 = 0$;   (d) $4x - 3y - 25 = 0$;
3 (a) $5x - 4y - 11 = 0$;   (b) $3x + 14 = 0$;   (c) $2x + 3y - 10 = 0$;
   (d) $x + 2y + 4 = 0$;   (e) $5x + 22 = 0$
5 (a) Yes;   (b) No;   (c) Yes
6 $x^2 + y^2 - 4x + 8y + 4 = 0$ or $x^2 + y^2 - 4x - 8y + 4 = 0$
7 $x^2 + y^2 - 10x - 6y + 9 = 0$
8 $x^2 + y^2 - 6x + 2y - 15 = 0$; centre $(3, -1)$
9 $x^2 + y^2 - 8x - 2y + 13 = 0$

Exercise 19.5 (*p.* 303)
1 $(-3\frac{1}{2}, 4\frac{1}{2})$ 2 (i) 2;   (ii) -3; $2x - 2y = 5$
3 (i) $x^2 + 4x - 6y + 13 = 0$;   (ii) $4x = 3y + 1$
4 (i) $x^2 - 6x - 14y + 16 = 0$;   (ii) $x^2 + y^2 - 6x - 6y + 13 = 0$
5 $(6, 9)$, 10; $(x - 3)^2 + (y - 5)^2 = 25$; $3x + 4y - 4 = 0$
7 $(-1, 4)$; $x^2 + y^2 + 2x - 8y + 4 = 0$

Exercise 19.6 (*p.* 310)
1 (i) $x^2 + y^2 = 9$;   (ii) $(x-1)^2 + (y-3)^2 = 16$;   (iii) $9x = 2y^2$;
   (iv) $y = x - 2$;   (v) $x^{\frac{2}{3}} + y^{\frac{2}{3}} = 1$;   (vi) $x = y^2 - 4y + 5$;
   (vii) $y + 1 = (x-2) \tan 30°$;   (viii) $xy + 2x = y^2 + 4y + 5$.
2 $x = -2 + r \cos \alpha$, $y = 3 + r \sin \alpha$
3 $x = 1 + 3 \cos t$, $y = 5 + 3 \sin t$
4 (i) $- \cot \theta$;   (ii) $- \cot t$;   (iii) $\dfrac{3t}{4}$;   (iv) 1;   (v) $- \tan \theta$;   (vi) $\dfrac{1}{2t}$;

Exercise 19.6 (*continued*)

(vii) $- \cot 30° = -\sqrt{3}$; (viii) $\dfrac{t^2}{t^2 - 1}$.

5 Tangent: $ty = x + at^2$; normal: $y + tx = 2at + at^3$

6 $y = 3x - 1$ **7** $2y = x + 7$ **8** $y^2 = 4(x - 1)$

12 $(\dfrac{t}{2}, t)$; $y = 2x$ **13** $4xy = 1$

14 $2t + 1$; (i) -1, (ii) $-\frac{1}{2}$; $y = x^2 + 3x + 2$

15 $\dfrac{2t - 1}{2}$; $4y + 4x = 13$; $2y = x - 1$

16 $4y = 2x - 1$ **17** (b) $x = 4\cos^2 \alpha$, $y = 4\cos \alpha \sin \alpha$

18 $\dfrac{t^2 + 1}{t^2 - 1}$; $3y = 5x - 8$; $x^2 - y^2 = 4$ **19** $y = x - 1$; $x + y = 5$

20 $(\dfrac{5t}{2}, 0)$, $(0, 5t)$; $y = 2x$ **21** $t = \pm 1$; $(2, 1)$, $(0, -3)$

22 $3y = 2x - 9$, $2y + 3x = 33$ **23** $(-t^2, 0)$; $y^2 = -2x$

24 $(\dfrac{6t}{5}, -\dfrac{3t}{5})$; $x + 2y = 0$ **25** $(-m^2, -m)$; $y^2 = -x$

26 $ty = x + 3t^2$; $(-6, 3)$ **27** $2\sqrt{r - 1}$; $y^2 = 4(x + 1)$

28 $P = \left(\dfrac{t}{2}, \dfrac{-t(t - 3)}{2}\right)$: Locus $y + 2x^2 = 3x$

29 $27y^2 = 4x^3$ **30** $y(x^2 + 4) = 8$

Exercise 20.1 (*p.* 320)

1 209 m   **2** (i) $\frac{1}{24}$ m s^{-2}; (ii) $\frac{1}{24}$ m s^{-2}; (iii) $7\cdot5$ km

3 (i) 275 m; (ii) $\frac{55}{7}$ m s^{-2}; (iii) $18\cdot75$ s   **4** $\frac{55}{3}$ s, $\frac{400}{3}$ m

5 1 km/min $= \frac{50}{3}$ m s^{-1} **6** (i) $4\cdot2$ m s^{-2}; (ii) 179 m approx.

7 (a) 6300 m; (b) $0\cdot07$ m s^{-2}; (c) 666 m

8 45 s, 900 m from start   **9** $2\cdot5$ m s^{-2}, $\frac{5}{6}$ m s^{-2}, 5 m s^{-1}

10 18 s, 150 s

Exercise 20.2 (*p.* 324)

1 (i) $4\cdot5$; (ii) 8; (iii) $3 + 0\cdot5t$ (m s^{-1});

(i) $11\cdot25$; (ii) 55; (iii) $3t + 0\cdot25t^2$ (m)

2 22 m s^{-1}, 260 m, $302\cdot5$ m   **3** 20 s, 20 m s^{-1} **4** 16 m, 4 s

5 50 m s^{-1}, 125 m   **6** (a) $\frac{5}{3}$ m s^{-2} (b) $\frac{71}{6}$ m s^{-1} (c) $20\cdot1$ m

7 $\frac{8}{3}$ m s^{-1}, $\frac{320}{3}$ m   **8** 10 m s^{-1}, 15 m

9 45 s, 1600 km/h$^2 = \frac{10}{81}$ m s^{-2}, $\frac{1}{32}$ km   **10** $\frac{12}{7}$

Exercise 20.3 (*p.* 328)

1 $7\cdot2$ m, $1\cdot2$ s   **2** 10 m s^{-1} **3** 2 s, 20 m s^{-1} **4** 5 s, 35 m s^{-1}

5 20 m   **6** $(5 \pm \sqrt{21})/2 = 4\cdot79, 0\cdot21$ s

7 (a) 6 s (b) 45 m (c) 2 s   **8** $3\cdot2$ m   **9** $19\frac{11}{16}$ m

10 $(3 + 2\sqrt{3}) = 6\cdot464$ s

Exercise 20.4 (*p.* 331)
4 They form a parallelogram   **5** $b = a$

Exercise 20.5 (*p.* 336)(Answers by drawing will be approximately equal to these)
1 (a) 5·75 km/h, 031°   (b) 9·43 m/s, 013°   (c) 13·61 m/s, 068° 27′
   (d) 69·28 km/h, 080°
2 036° 48′   3 10·77 m/s at 21° 48′ to line of ship
4 155·2 km/h at 014° 56′   5 (a) 2·24 km/h at 63° 26′ to bank (b) 0·2 km
6 229·1 km/h at 340° 53′   7 53° 8′ to bank, 6 min
8 (a) 081° 52′   (b) 170 km/h   (c) 1·76 h   (d) 278° 8′   (e) 1·34 h
9 21·36 km/h, 172° 24′; 186° 43′   10 From 236° or 304°
11 A to B 153·1 km/h; B to A 238 km/h   12 056° 25′; 48 min

Exercise 20.6 (*p.* 344)
1 250 km/h, 126° 53′
2 72·1 km/h, 90° ± 56° 18′ or 270° ± 56° 18′
   (depending on directions of the cars)
3 32·8 km/h, 077° 36′   4 4° 28′ or 4° 34′ (both to the horizontal)
5 108·2 km/h, 236° 19′, 50·6 m   6 33·3 km/h in direction 064° 52′
7 5·66 km/h from the SE   8 009°, 44·3 min
9 19·21 km/h, 231° 23′   10 9·22 km/h, 77° 27′ to bank
11 14·26 km, 18·4 min   12 2·94 km, 40·8 min
13 (a) 100° 33′;   (b) 13 h 26 min   14 315°

Exercise 21.1 (*p.* 351)
1 (a) $P \cos 30°$, $P \sin 30°$; $-Q \cos 60°$, $Q \sin 60°$; $-R \sin 10°$, $-R \cos 10°$;
   $S \cos 20°$, $-S \sin 20°$;
   (b) $A \cos 50°$, $A \sin 50°$; $B \cos 20°$, $-B \sin 20°$; $-W \cos 70°$, $-W \sin 70°$;
   (c) (i) 25·98, 15 N; (ii) -25, $-43·3$ N; (iii) 0, 80 N; (iii) $-57·36$, 81·92 N
2 (a) 15·6 N at 39° 48′ to 12 N;   (b) 50 N at 36° 52′ to 40 N;
   (c) 9·67 N at 11° 56′ to 6 N;   (d) 13·91 N at 27° 32′ to 20 N;
   (e) 17·32 N, bisecting angle
3 (a) 35·98, 12·32 N;   (b) 5, 27·32 N;   (c) 0, $-24·64$ N
4 5 N, 3 N   5 5 N, 120°   6 112° 37′, 12 N

Exercise 21.2 (*p.* 361)
1 10 m s^{-2} 2 $6\frac{2}{3} \text{ m s}^{-2}$ 3 $1\frac{9}{11} \text{ m s}^{-2}$ 4 0·4 N
5 1·2 N   6 8 : 49   7 8 m s^{-1} 8 $3·4 \text{ m s}^{-2}$
9 $44·72 \text{ m s}^{-2}$ at 26° 34′ to 20 N   10 $4·9 \text{ m s}^{-2}$ 11 $2·5 \text{ m s}^{-2}$
12 3 s   13 8 N   14 $6·16 \text{ m s}^{-2}$ 15 $\frac{13}{70}(0·186) \text{ m s}^{-2}$
16 (i) $2·5 \text{ m s}^{-2}$; (ii) 37·5 N; (iii) 75 N
17 (i) 784 N; (ii) 824 N; (iii) 752 N   18 4 m s^{-2}, 12 N
19 $5·29 \text{ m s}^{-2}$ 20 46·8 N   21 2070 N; 6670 N
22 3·5 N   23 1230 N   24 1044 N
25 $4·2 \text{ m s}^{-2}$, 14 N, 22·4 N   26 $1·05 \text{ m s}^{-2}$ 27 $0·245 \text{ m s}^{-2}$

Exercise 22 (*p.* 369)
1 15·68 N   2 0·51   3 $1/\sqrt{3}$, 30°

Exercise 22 (*continued*)

4 (i) 1.505 m s^{-2}; (ii) 15.05 N; (iii) 83.1 N   **5** 1.31 m s^{-2}

6 6.53 N   **7** 98 N   **8** 9.9 m s^{-2}

9 1.51 m s^{-2}, 3.48 m s^{-1} **10** 0.32 **11** 0.403 m s^{-2}

12 96.9 N   **13** 0.063 m

14 (a) 11.76 N; (b) 4.9 m s^{-2}; (c) 8.82 N   **15** 0.53

Exercise 23.1 (*p.* 377)

1 10 N 20 N 15 N   **2** $6\frac{2}{3}$ N   **3** 14 N

    A 0 $+56.6$ $+150$ N m   **4** 257 N

    B -40 0 $+150$ N m   **5** 102.5 N at A, 147.5 N at D

    C -40 -141.4 0 N m   **6** Weight 480 N, C.G. 1.7 m from A

7 100 N, 0.8 m from A   **8** 11 kg; $\frac{140}{11}$ cm

9 80 N at B, 140 N at C   **10** (i) $\dfrac{7\sqrt{3}}{3}$ N m;   (ii) $\dfrac{29\sqrt{3}}{2}$ N m

11 16 N, 18 N

12 (i) 64;   (ii) -38.4;   (iii) 62.2 N m

Exercise 23.2 (*p.* 387)

1 (a) 80 N, 15 cm from 50 N force;   (b) 20 N, 60 cm beyond 50 N force

2 30 N, between B and C, $3\frac{1}{3}$ cm from C.

3 $25\sqrt{3}$ (43.3) N to wall, 156 N at angle $73°$ $54'$ to floor

4 62.5 N; 62.5 N at $53°$ $7'$ to rod.   **5** 101.4 N, 16.9 N

6   (i) $P = 25\sqrt{3}$ N, $Q = 25$ N;   (ii) $R = 67.13$ N, $S = 104.4$ N;

    (iii) $P = 50$ N, $T = 50\sqrt{5}$ N at $116°$ $34'$ to P;

    (iv) $T_1 = 48.8$ N, $T_2 = 33.3$ N, angle between T_1 and $T_2 = 87°$ $58'$

8 10.9 N, 65.5 N   **9** 0.25 m, 120 N, $60\sqrt{3}$ N

10 14.7 N, 7.4 N, $43°$ $24'$ **11** 120 N; 108 N, $48°$ $36'$ to wall

12 (i) 13.25 cm   (ii) 21.9 N, 11.6 N

13 77 N in AC, 54 N in CD, 152 N in DB; $W = 123$ N (approx.)

14 36 N; 54 N in CD, 42 N in DE, 128 N in EF (approx.)

Exercise 23.3 (*p.* 391)

1 41.85 N, $23°$ $21'$ **2** 60.7 N, $19°$ $14'$ **3** $R = 47.95$ N, $S = 53.75$ N

4 37.71 N, $201°$ $50'$ **5** 125 N, $16°$ $16'$ **6** 50.04 N, $147°$ $45'$

Exercise 23.4 (*p.* 395)

1 23.1 N, 46.2 N   **2** 37.51 N, 46.12 N   **3** 51.96 N, 30 N

4 51.96 N, 79.38 N   **5** $166\frac{2}{3}$ N, 51.7 N   **6** 5.8 N, 9.2 N, 3.02 m

7 4.8 N, 0.49 **8** 158.2 N

9 (a) 37.5 N;   (b) $18°$ $18'$ **10** (a) $30°$ $58'$;   (b) $36°$ $52'$

11 0.14 **12** 409.8 N

13 $18\frac{1}{2}°$, $0.669g$ $(=6.56)$ N   **14** 160 cm; 12 N

15 (a) 393.8 N;   (b) 693.8 N   **16** 35.75 N, 35.75 N, 100 N

Exercise 23.5 (*p.* 400)
1 (i) $4\sqrt{2}$ N at 45° to Ox; (ii) -7 m, -7 m from O
2 (i) 17·46 N at 23° 38′ to Ox; (ii) $-\frac{12}{7}$ m, $-\frac{3}{4}$ m from O
3 (i) 14·18 N at 42° 13′ to Ox; (ii) $-\frac{1}{11}$ m, 0·08 m from O
4 $8\sqrt{2}$ N   **5** (a) 165 N (b) 9 m
6 $4\sqrt{5}$ N at 63° 26′ to BC; 18 m, 9 m from B
7 $5\sqrt{5}$ N at 26° 34′ to AD; 3 m, 6 m from A
8 $\sqrt{34}$ (5·83) N at 30° 58′ to AD; $\frac{26}{3}$ m from A along AD
9 15·72 N at 82° 41′ to AB; $\frac{8}{9}$ m from B on AB produced.
10 11·6 N at 87° 42′ to BA; 19 cm from A, 475 cm from A
11 9·54 N through midpoint of AB at 65° 12′ to BA

Exercise 24 (*p.* 407)
   From AY   From AX
1 $\frac{113}{19}$ (5·95) $\frac{83}{38}$ (2·18)   cm
2 6·5   3   cm
3 1·5 $\frac{61}{16}$ (3·81)   cm
4 $\frac{307}{140}$ (2·85) $\frac{177}{70}$ (2·53)   cm
5 (i) 37° 16′; (ii) 50° 32′; (iii) 51° 38′; (iv) 41° 35′

6 $3g$ (29·4) N; $\dfrac{6g}{\sqrt{13}}$ (16·3) N perpendicular to AC

7 (fig 24.9) 2·1 cm from AY, 5·4 cm from AX; (fig 24.10) 4·26 cm from AY, 3 cm from AX
8 $19\frac{4}{9}$ cm   **9** $\frac{100}{21}$ (4·76) cm   **10** $\frac{77}{38}$ (2·02), $\frac{63}{38}$ (1·66) cm
11 5 cm from AB and BC; 18° 26′   **12** $5\frac{1}{3}$ cm   **13** 6·7 cm
14 5·7 cm   **15** 1·5 cm   **16** $\frac{30}{29}$ (1·03) cm; 35° 24′
17 9·83 cm from both   **18** $\frac{5}{8}$ cm; 86° 24′ approx
19 1 cm   **20** $23\frac{1}{3}$ cm

Exercise 25.1 (*p.* 412)
1 (a) 10 N s; (b) 0·5 N s; (c) 0·12 N s; (d) $\frac{5}{3} \times 10^3$ N s
2 5 N s; 7·5 m s^{-1}; 3·125 m   **3** 4 N   **4** 2·25 N s
5 20 N s; 40 N   **6** 16·8 N   **7** $9\frac{3}{8}$ N   **8** 4·17 N; 18 s
9 37·5 N   **10** 50·4 N   **11** 250 N s; 250 N
12 3·77 kg; 45·2 N

Exercise 25.2 (*p.* 419)
1 2·8 m s^{-1} **2** $1\frac{1}{11}$ m s^{-1}; $\frac{3}{11}$ N s   **3** $1\frac{1}{3}$ m s^{-1}; 60 N
4 0·1 m s^{-1} **5** $\frac{10}{3}$ m s^{-1}
6 1 m s^{-1} in opposite direction; 10·4 N s   **7** $13\frac{1}{4}$ m s^{-1}
8 New speed = $2933\frac{1}{3}$ m s^{-1} **9** 6·13 m s^{-2}
10 5·6 m s^{-1}; 4·2 m s^{-2} **11** 4·2 m s^{-2}; 8·4 m s^{-1}, 3·6 m; 2·4 m s^{-1}
12 (i) $\frac{5}{3}$ m s^{-1} north; (ii) 20 N s   **13** $\dfrac{Mu}{M+m}$; $\dfrac{Mmu}{M+m}$
14 3·25, 3·75 m s^{-1} **15** 2 : 5

Exercise 26.1 (*p.* 424)

1 500 J   **2** 40 J   **3** 156·8 J   **4** $13\frac{1}{3}$ J
5 $1\frac{9}{16}$ J; 0·16   **6** $1\frac{1}{8}$ N   **7** 89·3 m   **8** $6\frac{2}{3}$ N
9 332·5 N   **10** 81 N   **11** 4·17 × 10^4 N   **12** 1·67 × 10^3 N
13 (i) 0·36 m;   (ii) 0·08 m   **14** 0·36 m
15 25 m s^{-1}; $8\frac{1}{3}$ m s^{-1} **16** 1·67 × 10^4 N   **17** 3·5 × 10^4 N
18 4·287 × 10^3 N **19** 100 m s^{-1} in opposite direction; 10^5 J, 2·5 × 10^5 J
20 2 × 10^4 N   **21** 1·48 m   **22** 476·4 N

Exercise 26.2 (*p.* 429)

1 96 J, 16·5%   **2** 77·1%   **3** 2 m s^{-1}; 15 J   **4** $2\frac{2}{3}$ m s^{-1}, $1\frac{1}{3}$ J
5 2·43 m s^{-1} **6** 1 m s^{-1}; 84%   **7** 19° 30′   **8** 5 kg, 52·5 J
9 2·6 m s^{-1} **10** 80%   **11** 1 m s^{-1}; 135 J
12 0·154 m s^{-1}; 48%   **13** 0·44 m
14 (i) 8·4 m s^{-1};   (ii) 494 J   **15** 32° 48′

Exercise 26.3 (*p.* 435)

1 3 kW   **2** 345 W   **3** 1 kW   **4** 28 m s^{-1}
5 7350 W   **6** 600 kW   **7** 5·7 kW   **8** 400 N
9 19·7 kW   **10** 50 m s^{-1}; 40·54 m s^{-1}; 0·26 m s^{-2}
11 344 N; 34·9 m s^{-1} **12** 9 × 10^4 N; 0·04 m s^{-2}
13 378 N, 15·12 kW   **14** 30·9 m s^{-1}
15 112 N; 252 N; 10·92 kW   **16** 980 kW

Exercise 27.1 (*p.* 442)

1 1250 kg, 25   **2** 500 N   **3** 66·7 cm
4 (i) 8   (ii) $62\frac{1}{2}$%   **5** $168\frac{8}{19}$ N   **6** 2600 N

Exercise 27.2 (*p.* 449)

1 45 kg   **2** (i) 70%;   (ii) 10%   **3** 8; 400 N; 4800 J
4 (i) 62·5%;   (ii) 600 J   **5** $142\frac{1}{22}$ N   **6** $44\frac{9}{14}$ N; 375 J
7 (i) 19;   (ii) 42·9;   (iii) 3·2 t   **8** 19; 152 kg; 55·7%

Exercise 27.3 (*p.* 453)

1 $a = 0·1$, $b = 60$; $82\frac{1}{7}$%   **2** (i) $5\frac{5}{11}$;   (ii) $68\frac{2}{11}$%;   (iii) 5280 J
3 2963 N   **4** 4; $P = (W + 1·5g)/4$; 97·5%

5 250; 15·9% ($\frac{100}{2\pi}$)   **6** (a) 65·625%;   (b) $\frac{1}{8}$

7 (a) $416\frac{2}{3}$ N;   (b) 10 kJ   **8** $33\frac{1}{3}$ kW

Revision Papers (p. 455)

I

1 0·2528

2 $\dfrac{a+b}{a+1}$, $(\dfrac{a+b}{a+1})^2$

3 26

4 (a) -18; (b) $\dfrac{2}{3}(1 - \dfrac{1}{\sqrt{2}})$

5 $-3, -11$; $(x - 3)$

6 $y = 4x - 3$; $(0, -3)$; 5

7 $2(1 - \dfrac{x}{8} - \dfrac{x^2}{64})$; $1·9975$

8 29·5

9 7·55 km at 14·29 h; A is 29·74 km N of original position; B is 22·3 km N and 1·37 km E

10 93·4 N, N 2° 38′ E

II

1 (i) 77; (ii) $\dfrac{x^2 + 4x - 4}{(x + 2)^2}$

2 (i) $(2, 3), 3$; (iii) $5y + 12x = 0$; (iv) $112°\ 38′$

3 (i) $0°, 60°, 180°, 300°, 360°$; (ii) $120°, 180°, 240°$;
(iii) $7\frac{1}{2}°, 37\frac{1}{2}°, 97\frac{1}{2}°, 127\frac{1}{2}°, 187\frac{1}{2}°, 217\frac{1}{2}°, 277\frac{1}{2}°, 307\frac{1}{2}°$

4 (i) $1\frac{11}{12}$ (ii) $-3 < x < 2$; $6\frac{1}{4}, -\frac{1}{2}$ **5** $1 + x + \dfrac{x^2}{2} + \dfrac{x^3}{2}$; $3·317$

6 (i) $y = -3x$; (ii) $3x^2 + 3y^2 + 18x - 4y + 15 = 0$;
(iii) $x^2 + y^2 + 8x - 4y + 2xy + 10 = 0$

7 (i) $14·4$ (ii) $\dfrac{17\pi}{6}$ (iii) $\sin x - x \cos x + c$

8 0·49

9 $-2\frac{1}{2}a, 2\frac{1}{2}a$

10 8·5 s

III

1 $(1 - \dfrac{1}{m}, 1 - m)$; $x + y = xy$

2 (i) $x \to 2x + 2, x \to 2x + 5$

3 $a = 3·5, n = 1·2$

4 (i) $12, -12$; (ii) -220×2^9

5 (i) $(-5, -18)$; (ii) $y + 7x = -53$

6 (i) $e^{\frac{2u-v}{5}}$; (ii) $\pm 1·496$

7 (ii) $0°, 116°\ 34′, 180°, 296°\ 34′$

8 (i) 588; (ii) 660; (iii) 480 N

9 $(\frac{17}{7}, \frac{19}{14})$; 56°

10 4350 W

IV

1 (i) $\frac{11}{2}, \frac{5}{4}$ (ii) 5

2 (i) $x^2 - y^2 = 4$; (ii) 192

3 $\dfrac{2}{y}$; $(t^2 + 1, t)$; $x = y^2 + 1$

4 (i) (a) $\dfrac{4}{(2x + 1)^2}$; (b) $2x(\tan 2x + x \sec^2 2x)$; (ii) $y = \dfrac{x^3}{3} - \dfrac{x^2}{2} + \dfrac{7}{6}$;

(iii) $\frac{9}{16}$

5 (i) $4x^2 + x - 2 = 0$; (ii) $\frac{1}{2} \leqslant x \leqslant 3$

6 (i) $11·7$; $72°\ 5′$

7 1·36 m

8 $\frac{5}{7}, \frac{45}{14}$ m/s

9 848 W

10 009° 36′

V

1 (ii) 6; (iii) $x^2 - y^2 = 1$; $\dfrac{1}{\sin t}$ **2** (i) 1, 4, 7, ...; (ii) 2, 1, $(x + 2)$

3 $62°\,35'$ **4** $x^2 + y^2 - 8x - 6y + 9 = 0$

5 420   **6** (ii) $22°\,31'$, $157°\,29'$

7 (i) 3 m/s; (ii) 17 m/s; (iii) 0·4 m/s^2

8 VR = 5, MA = 4·62; efficiency = 92·3 %

9 N 12° 15′ E; 153 km/h; 238 km/h   **10** 3·82 m/s

VI

1 $204\frac{4}{5}\pi$ **2** (i) $y = 3x + 1$; (ii) $(\frac{1}{3}, 2)$

3 (i) (a) $2y + x$; (b) $y - \dfrac{x}{2}$; (c) $2y - \dfrac{3x}{2} + 2$; (ii) $x = \sqrt{\dfrac{y+1}{2}}$

4 (ii) $\pi/3$, π, $5\pi/3$

5 (i) $3x^2 - 12x + 5 = 0$ (ii) 0·017 cm^2/s

6 (ii) 0

7 (i) 35; (ii) 35; (iii) 30; (iv) 65

8 (i) 9·9 m/s (ii) 7·67 m/s (iii) 3·51 N s (iv) 4·92 J

9 (i) 7·5; (ii) 2·25 m   **10** (i) 510 N; (ii) 0·2 m/s^2

VII

1 (i) 3 or 4; (ii) $x^2 - 20x + 79 = 0$

2 $25°\,51'$ **3** (i) 15·3; (ii) 1

4 $216x + 24x^2$; 21·624   **5** (ii) (1, 6)

6 Q is $(0, \dfrac{-t}{t-1})$; M is $(\dfrac{t}{2}, \dfrac{t^2 - 2t}{2(t-1)})$; $y = \dfrac{2x(x-1)}{2x-1}$

7 (i) (a) $x \to 2(x^2 - 1)$; (b) $\dfrac{x+1}{2}$; hgf   (ii) 2, -1

8 41 cm   **9** $\sin \theta = \frac{1}{49}$ **10** 1·1 km

VIII

1 (i) $p = 2$, $q = 4$; (ii) $(x+4)(x-1)(x+4)$

2 $a = 0·6 \pm 0·05$, $b = 2 \pm 0·5$ **3** 8 cm^2/min

4 (i) 56; (ii) 28; (iii) $\frac{1}{2}$; (iv) $\frac{1}{7}$

5 $7x + 2y = 26$ and $7x + 2y + 14 = 0$

6 1·194 05   **7** (i) 2, 3 and 6, 2; (ii) $9 + 16x - 4x^2$

8 0·5 kg; 3·6 m   **9** (i) 0·5 m/s^2 (ii) 2400 N

10 4·5, 6 m/s; 225 N

IX

1 (i) $\dfrac{1}{b-2}$; (ii) $\dfrac{b^3}{8(b-2)}$ **2** (a) $\dfrac{t}{2(t^2+1)}$; (b) $y^2 = x - 2$

3 (ii) $y = 9x^{-1} + 9x - 18$

4 (i) $y^2 + 6y - 4x + 13 = 0$; (ii) $y = \sqrt{3x}$; (iii) (0, 2)

5 (i) $m = 3$, $n = -4$ or $m = -3$, $n = 4$; (ii) $(x-3)(2x+1)(x+1)$

   (iii) $\dfrac{10!}{(2!)^3} = 453\,600$

6 (i) $y + 7x = 13$ (ii) 3   **7** (i) $x = 2$;   (ii) $c = 2, N = 81$
8 (i) $8, -2$ m/s;   (ii) 120 J;   (iii) 4000 N
9 1·13 h   **10** 4·2 to 0·77 kg; 1·39 m/s^2

X

1 (i) 0·09;   (ii) 5

2 (i) (a) $-2x(1 - 3x + 2x^2)$;   (b) $\dfrac{1 - x^2}{(1 - x + x^2)^2}$;   (c) $\cos 5x - 5x \sin 5x$

  (ii) $\dfrac{5}{12}$; $\dfrac{25\pi}{126}$

3 P is $(t, 2 - t)$; M is $(\dfrac{t}{2}, 2 - \dfrac{3t}{2})$; $3x + y = 2$

4 (i) $y = \dfrac{-1}{8(4x + 1)^2}$;   (ii) $y + 2x = 3$; $2y = x + 1$

5 (i) $2 < x < 3$;   (ii) $1 - 10x + 45x^2 - 120x^3 + 210x^4 - 252x^5$; 0·904 382 1
6 (i) 13·34;   (ii) 22°   **7** (i) 1·04g;   (ii) 0·56g;   (iii) 15·4 m/s
8 0·9   **9** 51·5 N; 64·7, 172 N
10 0·32, 0·02 m; 60° 38′

XI

1 (i) $-2, 0, 4, 10$; 15th;   (ii) $\pm(4 - 2\sqrt{3})$ **2** 6340 m^2
4 $-2, 3$; $(0, 0)$, $(\sqrt{3}/2, 9/8)$, $(-\sqrt{3}/2, 9/8)$ **5** (ii) $-2 \leqslant x \leqslant 3$; $6\frac{1}{4}$

6 (i) (a) $3x - \dfrac{1}{x^3}$;   (b) $\dfrac{3x}{\sqrt{3x^2 - 1}}$;   (c) $6 \sin^2 2x \cos 2x$;   (ii) (a) $7\frac{1}{2}$;   (b) $\frac{254}{7}$

8 12 s; 16 m/s   **9** (ii) 11·7, $-17·5$ cm
10 1·75 m/s; 1312·5 J

XII

1 (i) $\dfrac{19 - 5\sqrt{3}}{26}$;   (ii) $x = -2$, $y = 1$ or $x = \frac{13}{4}$, $y = -\frac{3}{4}$;   (iii) 1·961

2 (i) $4x^2 - 3x - 2 = 0$;   (ii) $2 < x < 5$
3 A $= 27°$ 32′, B $= 110°$ 48′, C $= 41°$ 50′, area $= 16·25$ cm^2
4 $1 - 18x + 144x^2 - 672x^3 + 2016x^4$; 0·833 748
5 4 or 2 cm; 57° or 229°   **6** (i) 0;   (iii) $-1, \frac{13}{3}$
7 0·68 m/s   **8** 4·9 m; 1 s; 3·3 m/s
9 17 min   **10** 0·124

XIII

1 (i) $\frac{5}{9}\%$ increase;   (ii) 18·75 cm^3/s

2 (i) (a) $\dfrac{2x - 1}{2\sqrt{x^2 - x - 1}}$ (b) $2 \cos 2x \cos x - \sin 2x \sin x$; (ii) $\frac{10}{3}$;   (iii) 8·1π

3 (i) (a) $(6, 9)$;   (b) 5;   (c) $\frac{18}{5}$;   (d) 18;   (ii) $17y = 15x + 24$; $15y + 17x = 127$
4 (i) $\sqrt{3}$;   (ii) 24° 17′, 65° 42′, 204° 17′, 245° 42′;   (iii) 120°
5 80 m; $\frac{2}{3}$ s; $\frac{2}{27}$ m   **6** (i) $10r - r^2$; 25 cm^2;   (ii) 2, -1
7 15 m/s; 1000 N; 0·1 m/s^2; 1500 N
8 13 kg; $\frac{42}{13}$ m   **9** 9·8 N; $\sqrt{7}\,g\,(=25·9)$ N
10 3·76 m/s; 20° 24′

XIV

1 (i) 5040; $\frac{1}{4}$; (ii) $\frac{4}{7}$ **2** (i) 26; (ii) $2, 6, 18, \ldots$

3 (i) $-1, 2$; $-3, \frac{10}{3}$; (ii) Min $(3, -23)$, max $(-1, 9)$

4 $(\frac{14}{3}, \frac{7}{3})$, $(\frac{4}{3}, \frac{8}{3})$, $10y + x = 28$ **5** 40.4 m

6 (i) (a) $\frac{4}{5}$; (b) $\dfrac{1}{\sqrt{10}}$; (c) $\dfrac{13}{5\sqrt{10}}$; (ii) $116 - 90\sqrt{2}$; (iii) $(x-2)(x-2)(x+3)$

7 (i) (a) $\dfrac{5}{(1-2x)^2}$; (b) $\dfrac{3 \sin 3x}{\cos^2 3x}$; (c) $2(2x-1)(x^2-x+5)$; (ii) 1.528 cm/s

8 $15g \ (=147)$ N **9** 16.3 J; $1\frac{1}{9}$ m **10** 6.6 s

XV

1 (i) 1; (ii) $-2, 0$ **2** $\dfrac{ax(h - x)}{h}$; $\dfrac{ah}{4}$

3 (i) (a) 2; (b) $111° \, 28'$, $248° \, 32'$; (ii) $(\dfrac{2m + 3}{2}, \dfrac{2 - 3m}{2})$; $6x + 4y = 13$

4 (i) (a) $3x^2 - 2x + 1$; (b) $\dfrac{2x^2 + 1}{\sqrt{x^2 + 1}}$; (ii) (a) $\frac{7}{8}$; (b) $-\frac{27}{2}$; (iii) $-\frac{50}{3}, -8$

5 (i) $2x^2 - 3x - 4 = 0$; (ii) 6; (iii) $-\frac{2}{3}, 2$

6 (i) $\frac{1}{10}$; (ii) $\frac{1}{2}(1 + \dfrac{x}{2} + \dfrac{x^2}{4} + \dfrac{x^3}{8})$; $-2 < x < 2$; $0.502\,512\,6$

7 (i) $057° \, 2'$; (ii) 48 min **8** 6.88 N; $0.80, -0.52$

9 108.4 kW **10** $\frac{185}{9}, \frac{275}{18}$ cm; $27° \, 26'$